# Diabetes, Insulin and Alzheimer's Disease



For further volumes:
http://www.springer.com/series/1175

RESEARCH AND PERSPECTIVES IN ALZHEIMER'S DISEASE

Suzanne Craft • Yves Christen
Editors

# Diabetes, Insulin and Alzheimer's Disease

Springer

*Editors*
Dr. Suzanne Craft
University of Washington
School of Medicine
Department of Neurology
1660 S. Columbian Way
Seattle WA 98108
Mailstop 127
USA
scraft@u.washington.edu

Dr. Yves Christen
Fondation IPSEN pour la
Recherche Therapeutique
65 quai George Gorse
92650 Boulogne Billancourt
Cedex
France
yves.christen@beaufour-ipsen.com

ISSN 0945-6066
ISBN 978-3-642-26281-4 e-ISBN 978-3-642-04300-0
DOI 10.1007/978-3-642-04300-0
Springer Heidelberg Dordrecht London New York



*Cover design:* WMXDesign GmbH, Heidelberg, Germany

Printed on acid-free paper

Springer is part of Springer Science+Business Media (www.springer.com)

# Forword

The importance of insulin in the regulation of corporal aging has been established by the dramatic increases in longevity experienced by animals in which the adipose insulin receptor or the insulin-related daf genes have been genetically modified. However, a long-held belief, described as recently as ten years ago in endocrinology textbooks, declared that the brain was an insulin-insensitive organ. This pervasive belief was challenged by leaders like Jesse Roth, Daniel Porte, and others, who established the existence of insulin receptors in the central nervous system and a clear role for insulin in CNS control of feeding. New research demonstrates that, analogous to its influence on corporal aging, insulin also makes important contributions to brain aging and the expression of late-life neurodegenerative disease. Insulin plays a key role in cognition and other aspects of normal brain function. Insulin resistance induces chronic peripheral insulin elevations and is associated with reduced insulin activity both in periphery and brain. The insulin resistance syndrome underlies conditions such as Type 2 diabetes mellitus and hypertension, which are associated with age-related cognitive impairment and Alzheimer's disease.

This volume contains the proceedings of the 24$^{th}$ *Colloque Médecine et Recherche dedicated to Alzheimer's disease* organized by *the Fondation IPSEN* entitled "Diabetes, Insulin and Alzheimer's Disease" which brought together experts from basic and clinical science to provide a broad survey of the role of insulin in the brain, and to discuss the mechanisms through which insulin dysregulation contributes to the development of cognitive impairment and late-life neurodegenerative disease. Each author has greatly furthered our understanding of the relationships among insulin, diabetes, and Alzheimer's disease, moving us far beyond the belief that the brain is an insulin-insensitive organ. Given the recent pandemic of conditions associated with insulin resistance, it is imperative that we achieve a comprehensive knowledge of the mechanisms through which insulin resistance affects brain function in order to develop therapeutic strategies to address these effects.

Suzanne Craft
Yves Christen

# Acknowledgments

The editors wish to thank Jacqueline Mervaillie, Béatrice André and Sonia Le Cornec for the organization of the meeting and Mary Lynn Gage for the editing of the book.

# Contents

# List of contributors

**Alkon Daniel L.** Blanchette Rockefeller Neurosciences Institute, at West Virginia University, Medical Center Drive, PO Box 9301, Morgantown, WV 26505, USA, dalkon@engram.brni-jhu.org

**Biessels Geert Jan** Department of Neurology, G03.228, University Medical Center, PO Box 85500, 3508 GA Utrecht, The Netherlands, g.j.biessels@umcutrecht.nl

**Cole Greg M.** Greater Los Angeles Veteran's Administration Healthcare, Geriatric Research Education and Clinical Core, Department of Medicine, University of California, Los Angeles, Department of Neurology, UCLA, 16111 Plummer Street, Bldg. 7, Room A102, North Hills, CA 91343, USA, gmcole@ucla.edu

**Craft Suzanne** GRECC S-182, VAPSHCS, 1660 South Columbian Way, Seattle, WA 98108, USA, scraft@u.washington.edu

**Deshpande Atul** Greater Los Angeles Veteran's Administration Healthcare, Geriatric Research Education and Clinical Core,Department of Medicine, University of California, 16111 Plummer Street, Bldg. 7, Room A102, North Hills, CA 91343, USA

**Frautschy Sally A.** Greater Los Angeles Veteran's Administration Healthcare, Geriatric Research Education and Clinical Core, Department of Medicine, University of California, Department of Neurology, UCLA, 16111 Plummer Street, Bldg. 7, Room A102, North Hills, CA 91343, USA

**Gerozissis Kyriaki** NMPA, University Paris Sud-11, 15 rue Georges Clémenceau - Bat. 447, 91405 - Orsay Cedex, France, gerozissis@yahoo.co.uk

**Kahn C. Ronald** Joslin Diabetes Center, One Joslin Place, Boston, MA 02215, USA, c.ronald.kahn@joslin.harvard.edu

**Killick Richard** Institute of Psychiatry, King's College London, De Crespigny Park, London, SE5 8AF, UK

**Launer Lenore J.** Laboratory of Epidemiology, Demography and Biometry, National Institute on Aging, Gateway Building 3C309, 7201 Wisconsin Avenue, Bethesda, MD 20892, USA, launerl@nia.nih.gov

**Lovestone Simon** Institute of Psychiatry, King's College London, De Crespigny Park, London, SE5 8AF, UK, s.lovestone@iop.kcl.ac.uk

**Luchsinger José A.** 630 West 168th St., PH19, New York, NY 10032, USA, jal94@columbia.edu

**Ma Qiu-Lan** Greater Los Angeles Veteran's Administration Healthcare, Geriatric Research Education and Clinical Core, Department of Medicine, University of California, 16111 Plummer Street, Bldg. 7, Room A102, North Hills, CA 91343, USA

**Mattson Mark P.** National Institutes of Health, National Institute on Aging, Biomedical Research Center, Laboratory of Neurosciences, 251 Bayview Blvd, BALTIMORE, MD 21224-6825, USA

**Nelson Thomas J.** Blanchette Rockefeller Neurosciences Institute, at West Virginia University, Medical Center Drive, PO Box 9301, Morgantown, WV 26505, USA

**Nishijima Takeshi** Cajal Institute, CSIC, and Ciberned, Avenida Doctor Arce 37, Madrid 28002, Spain

**Piriz Joaquin** Cajal Institute, CSIC, and Ciberned, Avenida Doctor Arce 37, Madrid 28002, Spain

**Reagan Lawrence P.** Department of Pharmacology, Physiology and Neuroscience, University of South Carolina School of Medicine, 6439 Garner's Ferry Road, D40, Columbia, SC 29208 USA, lpreagan@uscmed.sc.edu

**Stranahan Alexis M.** National Institutes of Health, National Institute on Aging, Biomedical Research Center, Laboratory of Neurosciences, 251 Bayview Blvd, Baltimore MD 21224-6825, USA, alexis.stranahan@jhu.edu

**Sun Miao-Kun** Blanchette Rockefeller Neurosciences Institute, at West Virginia University, Medical Center Drive, PO Box 9301, Morgantown, WV 26505, USA

**Suzuki Ryo** Joslin Diabetes Center, One Joslin Place, Boston, MA 02215, USA, c.ronald.kahn@joslin.harvard.edu

**Torres Aleman Ignacio** Cajal Institute, CSIC, and Ciberned, Avenida Doctor Arce 37, Madrid 28002, Spain, torres@cajal.csic.es

**Trejo Jose Luis** Cajal Institute, CSIC, and Ciberned, Avenida Doctor Arce 37, Madrid 28002, Spain

**Ubeda Oliver** Greater Los Angeles Veteran's Administration Healthcare, Geriatric Research Education and Clinical Core, Department of Medicine, University of California, 16111 Plummer Street, Bldg. 7, Room A102, North Hills, CA 91343, USA

**Yang Fusheng** Greater Los Angeles Veteran's Administration Healthcare, Geriatric Research Education and Clinical Core, Department of Medicine, University of California, 16111 Plummer Street, Bldg. 7, Room A102, North Hills, CA 91343, USA

# Insulin Action in the Brain and the Pathogenesis of Alzheimer's Disease

**C. Ronald Kahn and Ryo Suzuki**

**Abstract** Over 24 million people in the U.S. have diabetes mellitus, and about 90% of these have the type 2 form of the disease. In addition, an estimated 40–60 million people have pre-type 2 diabetes, impaired glucose tolerance or the cluster of abnormalities referred to variably as the metabolic syndrome or syndrome X (Reaven 1988). In all of these disorders, a central component of the pathophysiology is insulin resistance. Insulin resistance is also closely linked to other common health problems, including obesity, polycystic ovarian disease, hyperlipidemia, hypertension and atherosclerosis (Biddinger and Kahn 2006). Recent data also indicate a link between insulin resistance, type 2 diabetes and Alzheimer's disease (Craft 2007). Cross-sectional studies have suggested an association between type 2 diabetes and cognitive decline, especially in aspects of verbal memory (Strachan et al. 1997). Longitudinal studies have revealed that patients with type 2 diabetes have a 1.5-fold greater change over time in measures of cognitive function than those without diabetes (Cukierman et al. 2005). While some of this change may certainly be due to the increased prevalence of atherosclerosis in diabetic patients, there is increasing evidence that insulin resistance itself may affect CNS function and risk of Alzheimer's disease. In this review we will explore this relationship, focusing on experiments we have performed in mice.

## 1 The Insulin Signaling System

The insulin/IGF-1 signaling system is evolutionarily very ancient. Homologues of these receptors have been identified in *Drosophila*, *C. elegans*, *Porifera* and many other species (Petruzzelli et al. 1986; Skorokhod et al. 1999; Dorman et al. 1995;

C.R. Kahn (✉)
Joslin Diabetes Center and Harvard Medical School, One Joslin Place, Boston, MA 02215, USA
e-mail: c.ronald.kahn@joslin.harvard.edu

S. Craft and Y. Christen (eds.), *Diabetes, Insulin and Alzheimer's Disease*,
Research and Perspectives in Alzheimer's Disease
DOI 10.1007/978-3-642-04300-0_1, 

Renteria et al. 2008).The insulin receptor (IR) was initially identified using $^{125}$I-insulin binding (Freychet et al. 1972). Biosynthetic and affinity labeling revealed the IR to be glycoproteins, consisting of two 135 kDa α-subunits and two ~95 kDa β-subunits linked by disulfide bonds to form an $\alpha_2\beta_2$ heterotetramer (Massague et al. 1980; Kasuga et al. 1982a). On pulse-chase labeling, both subunits were derived from a single chain precursor or proreceptor (Hedo et al. 1983). In 1982, we demonstrated that the IR possessed tyrosine kinase activity, placing it biochemically in the family of receptor tyrosine kinases and opening up new avenues in the study of insulin signaling (White et al. 1987; Kasuga et al. 1982b). In 1985, two groups succeeded in cloning the cDNA of the human IR, confirming these structural features (Ullrich et al. 1985; Ebina et al. 1985).

The IR gene is present on chromosome 19p13.3 in humans and chromosome 8 in the mouse. In both, the gene is >120 kb in length and is composed of 22 exons, which to some extent encode functional domains of the receptor (Fig. 1). The IR cDNA in both humans and rodents possesses an open reading frame of 4,146 nucleotides that encodes the 1,382-amino acid precursor of the receptor (Ullrich et al. 1985; Ebina et al. 1985), including a 27-amino acid signal peptide, a 721-amino acid α-subunit, a four-amino acid processing site, and a β-subunit of 619 amino acids. During the biosynthesis of the proreceptor, both subunits undergo glycosylation, disulfide bond formation and proteolytic cleavage by a furin-related protease to give the mature receptor (Hedo et al. 1983).

Functionally, the IR behaves as a classic allosteric enzyme. The α-subunit of the IR serves as both the insulin binding subunit and the regulatory subunit. Insulin binding to the α-subunit induces conformational changes in the receptor and activates the kinase activity in the β-subunit. The β-subunit is a transmembrane protein linked by disulfide bonds to the α-subunit and contains the tyrosine kinase activity critical for insulin action (Kasuga et al. 1982b). Following stimulation, the β-subunit undergoes autophosphorylation on seven Tyr residues in an ordered cascade; three of these at Tyr 1158, 1162, 1163 result in activation of the receptor kinase toward other substrates (Feener et al. 1993; Hubbard 1997). The IR occurs as two splice variants based on inclusion (IR-B) or exclusion (IR-A) of a 12-residue segment encoded by exon 11 and inserted between residues 716 and 717 (IR-A numbering) near the C-terminus of the α-subunit. In the brain, the major isoform of the insulin receptor is the A isoform (Kenner et al. 1995). The molecular weights of the denatured α- and β-subunits from brain insulin receptors are 5-10 kDa smaller than their counterparts in other tissues, which appears to be due to differences in N-linked glycosylation (Heidenreich et al. 1983). Whether they are differences in IR isoform splicing or glycosylation in Alzheimer's brain versus normal brain has not been studied.

cDNA cloning and functional studies have revealed two other members of the IR family: the highly homologous IGF-1 receptor (Ullrich et al. 1986; Abbott et al. 1992) and the IR-related receptor (IRR; Shier and Watt 1989; Zhang and Roth 1992). Insulin, IGF-1 and IGF-2 can bind to both the IR and IGF-1R, albeit with differing affinities. No ligand has thus far been identified for the IRR, and thus its physiological function is unknown. All three receptors are normally disulfide-linked

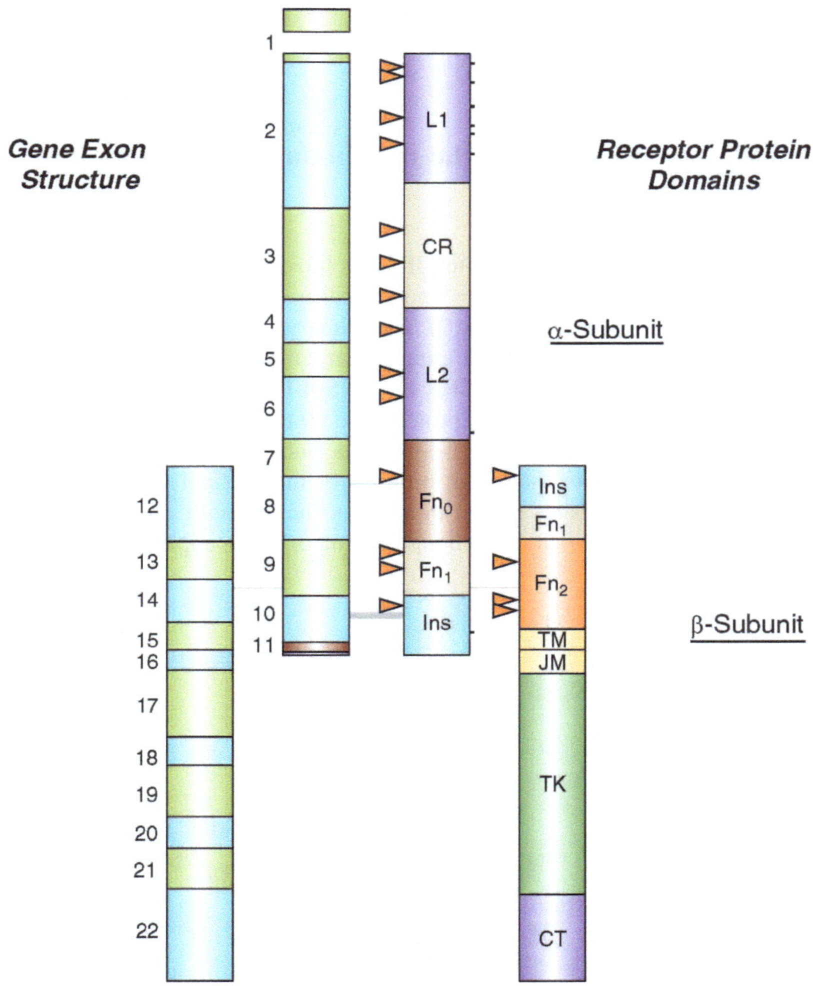


**Fig. 1** Modular structure of insulin receptor (IR) gene and protein. Schematic of the $\alpha_2\beta_2$ structure of the IR. On the left, the half-receptor heterodimer is depicted by its genomic structure, which is encoded by the 22-exon sequences. On the right, the half-receptor heterodimer is depicted by predicted protein modules. L1: large domain 1; CR: cystein-rich domain; L2: large domain 2; Fn: fibronectin type III domains; Ins: Insert; TM: transmembrane domain; JM: juxtamembrane domain; TK: tyrosine kinase domain; CT: C-terminal domain. The orange arrowheads indicate the N-glycosylation sites. Adapted from De Meyts and Whittaker (2002)

homodimers but may also function as heterodimer hybrids, like IR/IGF-1R hybrids, in tissues that express both receptors (Slaaby et al. 2006; Benyoucef et al. 2007). IGF-1 receptors are abundant in brain and widely distributed therein and they have a somewhat different distribution from IRs (Dore et al. 1997; Baron Van Evercooren et al. 1991). IRR mRNA is also found in brain, but its distristribution is highly restricted to the forebrain, primarily cholinergic neurons and neurons that coexpress trkA, a high-affinity receptor for nerve growth factor (Tsuji et al. 1996).

## 2 Linking the Receptor to the Insulin and IGF-1 Signaling Pathways

Following ligand binding, the activated IRs and IGF-1 receptors initiate signaling networks that share many similarities and critical nodes of signal divergence and regulation (Taniguchi et al. 2006; Fig. 2). The primary action of these receptors is to phosphorylate a family of at least 12 intracellular substrate proteins. The first four of these identified were designated IRS-1 to IRS-4 (IR substrates 1-4; Taniguchi et al. 2006; White 1998). These IRS proteins are 60 to 180 kDa and are characterized by a pleckstrin homology (PH) domain, a phosphotyrosine binding (PTB) domain, which account for their high affinity for the IR, and up to 20 potential tyrosine phosphorylation sites spread throughout the molecule. All four IRS proteins have been identified in brain, although the major forms appear to be IRS-1 and

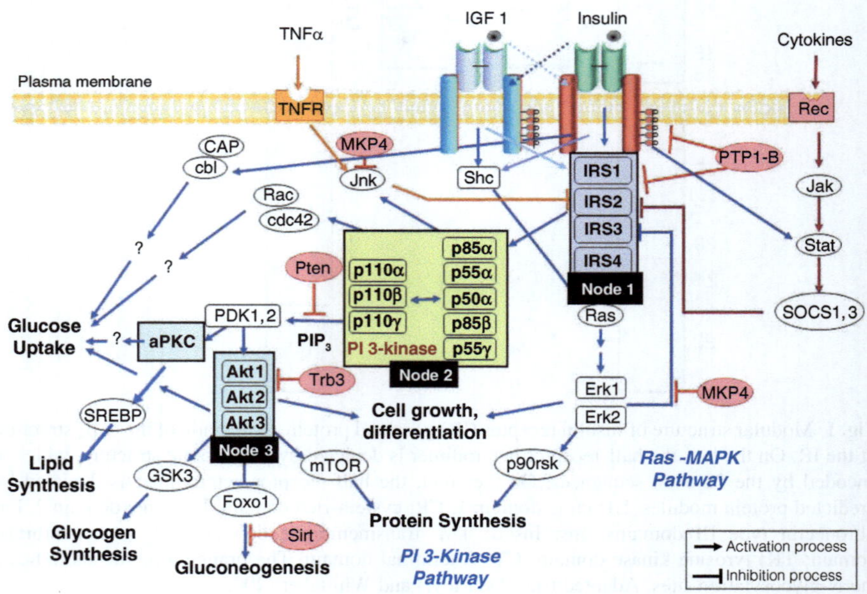


**Fig. 2** Signaling networks of insulin and IGF-I receptors. The insulin and IGF-1 signaling networks are complex and contain at least three critical nodes. The three major nodes in this pathway are the IR coupled to IRS proteins (purple box), PI 3-Kinase (green box), and Akt (blue box). Plain arrows represent an activation process, blocked arrows represent an inhibition process, and dashed arrows represent an activation process with less intensity. Cytokine (TNFα, IL6, leptin) signaling pathways have been shown to interfere with insulin signaling and are also represented on this figure (orange and red arrows). Adapted from Taniguchi et al. (2006)

IRS-2, similar to most insulin-sensitive tissues. Other direct substrates of the insulin/IGF-1 receptor kinases include the various isoforms of Shc, DOK-4 and DOK-5 (also referred to as IRS-5 and IRS-6), Gab-1, p62dok, Cbl, FAK, Sam68, DAPP1, and CEACAM1 (Ribon et al. 1998; Poy et al. 2002; Najib and Sanchez-Margalet 2002; Okamura-Oho et al. 2001; Wick et al. 2001; Cai et al. 2003). Following phosphorylation, these substrates function as key intermediates in signal transduction by interacting with other intracellular molecules. The best-characterized of these are SH2 domain proteins that bind to phosphotyrosines in specific sequence motifs on the IRS proteins. These SH2 proteins fall into two major categories: adaptor molecules, such as the regulatory subunit of PI 3-kinase, Grb2, which associates with SOS to activate the Ras-MAP kinase pathway (Baltensperger et al. 1993; Valverde et al. 2001), CrkII (Karas et al. 2001) and Nck2 (Tu et al. 2001), and enzymes, such as the phosphotyrosine phosphatase SHP2 (Rocchi et al. 1996) and the tyrosine kinase Fyn (Sun et al. 1996). The IRS proteins also interact with proteins that do not contain SH2 domains, including the calcium ATPases SERCA 1 and 2 (Algenstaedt et al. 1997), SV40 large T antigen (Prisco et al. 2002), Rho-kinases (Begum et al. 2002), PH domain-interacting protein (PHIP; Farhang-Fallah et al. 2000), IRAS (Sano et al. 2002) and others (Kruger et al. 2008; Hanke and Mann 2009). Through extensive studies, each of these has been shown to play important roles in the downstream actions of insulin and IGF-1, with the enzyme PI 3-kinase forming the most important link in insulin signaling to its metabolic effects (reviewed in Taniguchi et al. 2006).

In addition to these primary pathways of insulin/signal transduction, there are a number of other pathways activated, including pathways involving Cbl, CAP and the GTPase TC10 (Ribon et al. 1998; Chang et al. 2007), activation of GTPase of the Rac and Rho family (Usui et al. 2003), and interactions with the adaptor protein APS (Barres et al. 2006). Indeed, in collaboration with Mathias Mann using phosphoproteomics, we have identified as many as 40 proteins involved in insulin/IGF-1 action via tyrosine phosphorylation (Kruger et al. 2008).

## 3 Regulation of the IRS and IGF-1 Receptors in Physiology and Pathophysiology

The insulin/IGF-1 signaling pathway is subject to regulation at multiple levels in normal physiology and disease states. Over 100 patients with syndromes of severe insulin resistance have been reported with mutations in the IR gene (Taylor et al. 1994; Rouard et al. 1999). In addition to mutation, there are a number of mechanisms that play a role in acquired alterations of IR signaling in disease. The most common mechanism is down-regulation of the IR, which occurs to variable degrees in all hyperinsulinemic states (Gavin et al. 1974; Haft et al. 1994). This down-regulation occurs through internalization and subsequent degradation of the receptor. It is not clear if the brain shows similar down-regulation of the IR in obesity and type 2 diabetes in humans; in rodents, studies on this point have provided conflicting results (Figlewicz et al. 1986; Havrankova et al. 1979)

In addition to changes in receptor concentration, inhibition of receptor kinase activity can occur in diabetes and obesity secondary to phosphorylation of the IR or its substrates by serine kinases activated by increased levels of cytokines, such as TNFα and IL-6 (Hotamisligil et al. 1993; Fernandez-Real et al. 2000; Takayama et al. 1988). IR and IGF-1 receptor function may also be modified by protein-protein interaction. Interacting proteins include the suppressors of cytokine signaling (SOCS) proteins (Emanuelli et al. 2000; Ueki et al. 2004), the growth factor receptor-bound proteins Grb10 and Grb14 (He et al. 1998; Kasus-Jacobi et al. 1998) and PC-1, also termed ectonucleotide pyrophosphatase phosphodiesterase 1 (ENPP1; Goldfine et al. 2008).

## 4 Creation and Characterization of the Brain IR Knockout Mouse

IRs are present on virtually all tissues in mammals, including the classic insulin-responsive tissues (muscle, fat and liver) and non-classical tissues, such as brain, β-cells, endothelial cells, etc. A major initiative of the past 10 years was based on the use of the Cre-*lox* system to create tissue-specific IR knockout (KO) mice and to use these to define more precisely the role of insulin action in each tissue of the body. Since IRs are widely distributed throughout the central nervous system (CNS; Havrankova et al. 1978) and have been suggested to play a role in feeding behavior (Schwartz et al. 2000), we decided to study the physiological role of insulin in the brain. We created mice with a neuron-specific disruption of the IR gene (NIRKO mice) using the nestin promoter (Bruning et al. 2000). Inactivation of the brain IR had no impact on brain development but, as expected, resulted in a loss of insulin-stimulated $PIP_3$ in the hypothalamus, while response to leptin remained normal (Fig. 3, top; Schubert et al. 2004). As a result of CNS insulin resistance, NIRKO

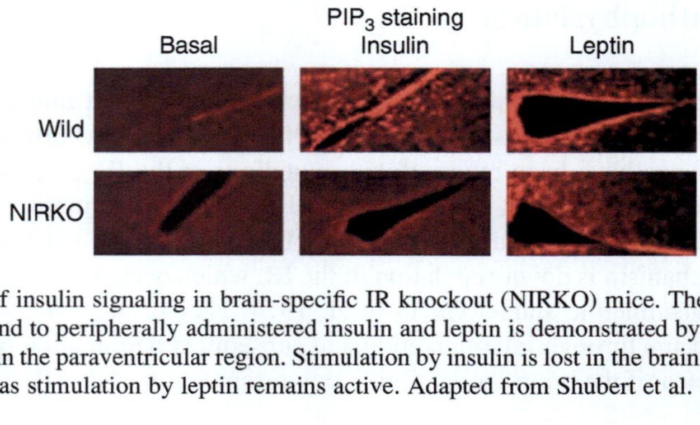


**Fig. 3** Loss of insulin signaling in brain-specific IR knockout (NIRKO) mice. The ability of the brain to respond to peripherally administered insulin and leptin is demonstrated by an increase in $PIP_3$ staining in the paraventricular region. Stimulation by insulin is lost in the brain of the NIRKO mouse, whereas stimulation by leptin remains active. Adapted from Shubert et al. (2004)

mice showed increased food intake, and both male and female mice developed diet-sensitive obesity, with increases in body fat and plasma leptin levels, mild insulin resistance, elevated insulin levels, and hypertriglyceridemia (Fig. 3, middle). In addition, loss of insulin action in the CNS had an effect on liver metabolism. Thus, while peripheral insulin suppressed hepatic glucose production by 74% in control mice, insulin action on hepatic glucose production (HGP) was markedly blunted in NIRKO mice (Fig. 4, bottom; Fisher et al. 2005). This finding is complementary to those of Rossetti and Accili that insulin action on the brain can regulate hepatic glucose output (Obici et al. 2002; Okamoto et al. 2004). In NIRKO mice, insulin-stimulated brain glucose uptake was reduced ~46%,

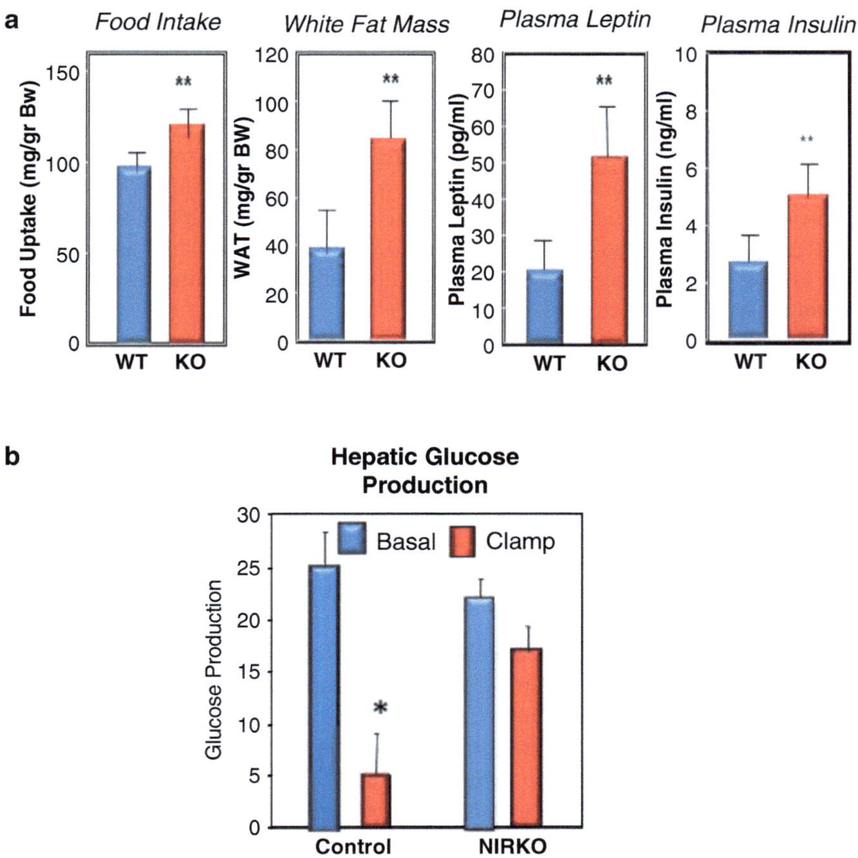


**Fig. 4** Metabolic phenotypes of brain-specific IR knockout (NIRKO) mice. Knockout of the IR in the brain results in mild hyperphagia and obesity with increased leptin levels and increased plasma insulin levels (Panel a). WT: wild type; KO: knockout. The hyperglycemia is due to a defect in insulin's ability to suppress hepatic glucose output (Panel b). Adapted from Bruning et al. (2000)

whereas glucose transport in muscle or fat was not altered. Finally, NIRKO mice exhibited defects in counter-regulatory response to hypoglycemia, especially increases in epinephrine and nor-epinephrine (Fisher et al. 2005), and impaired testicular and ovarian function due to hypothalamic hypogonadism (Bruning et al. 2000). Thus, IR signaling in the CNS plays an important role in regulation of appetite, energy disposal, hepatic metabolism, hypoglycemic counter-regulation and reproduction.

To define the specific cells in the brain involved in control of metabolism, in collaboration with Jens Bruning, we generated mice with selective inactivation of the IR in pro-opiomelanocortin (POMC) or agouti-related peptide (AgRP)-expressing neurons (Konner et al. 2007). While neither POMC- nor AgRP- IR KO mice exhibited obesity or altered energy homeostasis, IR KO in AgRP neurons resulted in a loss of insulin's ability to normally suppress HGP. AgRP-IRKO mice also exhibited reduced hepatic IL-6 expression and increased hepatic expression of glucose-6-phosphatase. In addition, we created two mouse models with inducible IR inactivation, one in the whole body including brain (IRΔwb) and a second restricted to peripheral tissues (IRΔper) (Koch et al. 2008). While both strains developed severe hyperinsulinemia, hyperglycemia was more pronounced in IRΔwb mice, consistent with the additional role of insulin action in brain control of glucose metabolism also observed by Accili (Okamoto et al. 2004). Interestingly, the IRΔwb mice also had a more pronounced reduction in the white adipose tissue (WAT) mass than IRΔper, suggesting an additional role of central insulin action in control of fat mass (Koch et al. 2008).

## 5 Impairment of insulin Signaling in Brain is Linked to Neurodegenerative Disease

There is a growing body of evidence linking insulin resistance and insulin action in the brain to neurodegenerative disease, especially Alzheimer's disease (Craft 2007). Low concentrations of insulin and reduced receptor numbers and signaling events in the CNS with Altzheimer's disease have been reported (Frolich et al. 1998; Hoyer 2002). Insulin administration while maintaining euglycemia improves memory in both healthy adults and Alzhermer's disease patients (Craft et al. 1999). In addition to Alzheimer's disease, Parkinson's disease is reported to accompany insulin resistance with a high prevalence (Pressley et al. 2003). Likewise, some studies have found that patients with Huntington's disease have a higher prevalence of diabetes and insulin resistance (Farrer 1985). Since insulin has neuroprotective effects in vivo (Hui et al. 2005; Rizk et al. 2006; Collino et al. 2009), impaired insulin action in the brain may have a critical role for pathogenesis of those neurodegenerative diseases.

One specific potential molecular link between insulin and neurodegeneration is the enzyme glycogen synthase kinase 3 (GSK3; Hooper et al. 2008). GSK3 activity is negatively modulated by insulin via an activation of Akt. GSK3 induces the hyperphosphorylation of Tau in vitro, and its overexpression in the adult brain of conditional transgenic mice causes Tau-hyperphosphorylation and neurodegeneration (Lucas et al. 2001).

To directly determine whether the brain IR is an important regulator of GSK3 in vivo, we performed additional studies in NIRKO mice. These studies revealed a markedly reduced phosphorylation of Akt and GSK3ß in the brains of NIRKO mice leading to a parallel and substantial increase in Tau-phosphorylation (Fig 5, bottom right); Schubert et al. 2004). In vitro neurons of NIRKO mice exhibit a complete loss of insulin-mediated activation of PI 3-kinase and inhibition of neuronal apoptosis. Thus, lack of insulin signaling in neurons can induce some markers of neurodegeneration and increased susceptibility to cell death. Nevertheless, NIRKO mice exhibit no alteration in neuronal survival or memory function measured by water maze test (Schubert et al. 2004), suggesting that, for development of Allzheimer's disease, some other mechanisms might be crucial besides insulin signal deficiency in the brain. Surprisingly, one model of Alzheimer's disease, the Tg2576 Swedish amyloid precursor protein mutant-overexpressing transgenic, shows improvement in premature mortality or Aß deposition when the mice lack IGF-1 receptor or IRS-2 in the hippocampus (Freude et al. 2009; Killick et al. 2009); in this model, IR deficiency did not affect mortality (Freude et al. 2009). These data suggest distinct roles for IRs and IGF-1 receptors in the hippocampus and in the pathogenesis of Alzheimer's disease.

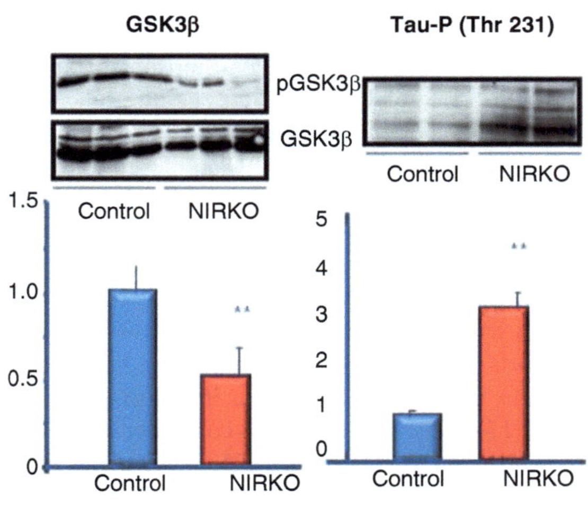


**Fig. 5** Altered GSK3 and Tau phosphorlation in brain-specific IR knockout (NIRKO) mice. The left panel shows reduced GSK3β phosphorylation in the NIRKO mouse brain as determined by immunoblotting. Since phosphorylation decreases GSK3β activity, this decrease would correspond to increased kinase activity. The right panel shows increased Tau threonine-231 phosphorylation, which is presumably the result of the increased GSK3β activity. Increased Tau phosphorylation is a marker of abnormalities in Alzheimer's disease. Adapted from Shubert et al. (2004)

## 6 Insulin-degrading Enzyme in Pathogenesis of Alzheimer's Disease and Metabolic Diseases

Another potential candidate that links insulin resistance/diabetes and Alzheimer's disease is insulin-degrading enzyme (IDE). More than 50 years ago, Mirsky and Broh-Kahn described "insulinase," a 110 kDa zinc metalloendopeptidase present in liver exract (Mirsky and Broh-Kahn 1948). This enzyme, currently named IDE or insulysin, is highly expressed in the brain, testis and muscle, as well as in the liver (Kuo et al. 1993). IDE is predominantly cytosolic, with smaller amounts in peroxisomes, endoplasmic reticulum and plasma membranes (Miners et al. 2008). Interestingly, up to 10% fraction of the total IDE is trafficked to the extracellular space, despite its lack of a classical signal peptide, presumably via an unconventional protein secretion pathway (Zhao et al. 2009). Several peptides with molecular weights of 3-10 kDa have been shown to serve as the substrates of IDE, including insulin, IGF-I, IGF-II, amylin, and Aß. The peptide substrates share little to no homology of primary amino acid sequence but have a similar secondary structure with "amyloidogenic" character (Qiu and Folstein 2006), as demonstrated by recent crystallographic data (Shen et al. 2006).

Levels of IDE protein and transcripts are reduced in the hippocampi from Alzheimer's disease patients with an apolipoprotein E (apoE)-ε4 allele compared to either patients without this allele or normal subjects (Cook et al. 2003). A recent report exhibited that Aß degradation extracellularly by IDE is facilitated by apoE (Jiang et al. 2008). The IDE region of chromosome 10q has been shown to have genetic linkage to late-onset Alzheimer's disease (Bertram et al. 2000). A lot of evidence indicates that the same region of chromosome 10q is also genetically linked to type 2 diabetes (Saxena et al. 2007; Zeggini et al. 2007). In addition, a well-characterized rat model of type 2 diabetes, Goto-Kakizaki (GK), has been found to harbor two missense mutations in IDE gene that decrease its ability to degrade both insulin and Aß (Fakhrai-Rad et al. 2000; Farris et al. 2004). Furthermore, genetic discuption of IDE gene in mice causes increased levels of cerebral Aß and glucose intolerance with hyperinsulinemia (Farris et al. 2003).

Because of the strong Aß-degrading ability of IDE, defects in IDE activity in the brain can be a direct trigger of Aß deposition to develop Alzheimer's disease (Fig. 6). Several reports suggest that insulin signaling regulates IDE expression. Incubation with insulin increases IDE protein in primary hippocampal neurons, whereas reduction of PI 3-kinase p85 subunit is correlated with a decrease of IDE in human Alzheimer's disease brains and in Tg2576 transgenic mice fed a high-fat diet (Zhao et al. 2004). Insulin-deficient diabetes induced by streptozotocin (STZ) administration also reduces IDE protein in the brain (Jolivalt et al. 2008). Thus, secondary reduction of IDE caused by insufficient insulin action in the brain might accelerate onset of Alzheimer's disease.

In contrast, insulin-resistant/glucose-intolerant phenotypes of IDE-deficient rodents and a genetic linkage of human IDE chromosomal region with type 2 diabetes susceptibility strongly suggest that IDE has a role in maintaining

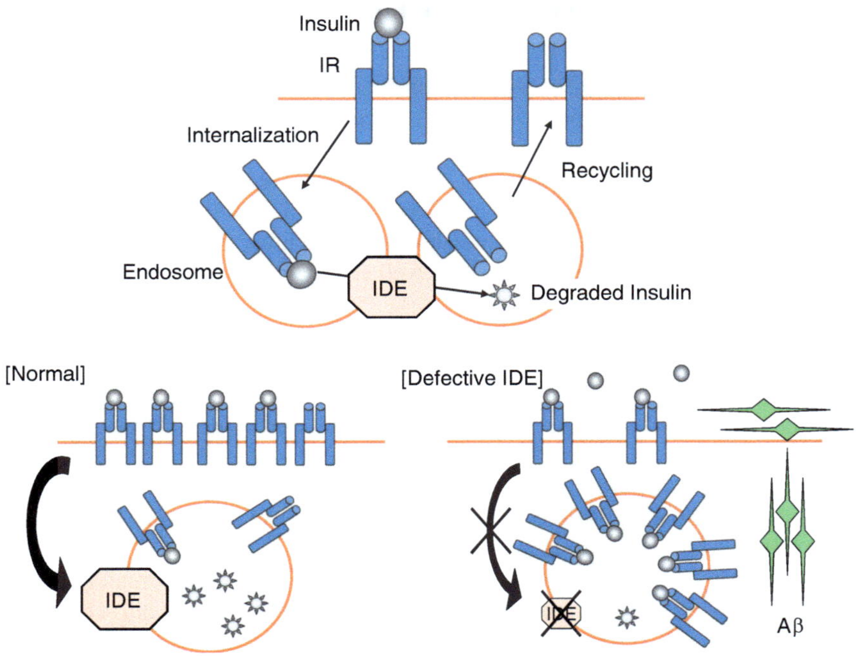


**Fig. 6** Hypothetical mechanism of insulin resistance caused by IDE insufficiency. IR bound with insulin receives internalization. IDE degrades the ligand insulin at endosome. Free receptor is transferred to membrane and recycled (top). When IDE is functional, sufficient numbers of receptors are recycled to the cell surface, and the downstream signal maintains expression of IDE (bottom left). In case IDE has insufficient function, ligand-bound IRs are trapped and unable to be transferred/recycled. Reduction of available IR causes insulin resistance, and consequent impairment of insulin action causes IDE downregulation, which aggravates IDE insufficiency as a "vicious cycle"

insulin sensitivity in the body. However, the mechanism remains unclear. IDE knockout mice have about a 3-fold increase in fasting insulin levels in plasma (Farris et al. 2003), possibly as a consequence of reduced insulin degradation, but hyperinsulinemia itself does not always cause systemic insulin resistance or impaired glucose tolerance (Hennige et al. 2003). Fakhrai-Rad proposed a hypothesis that a decreased intracellular degradation of insulin bound to its receptor would inhibit receptor-mediated signal transduction by lowering the number of available receptors on the cell membrane and/or compromising the downstream signaling from the receptor (Fakhrai-Rad et al. 2000). When either IDE or IR has a defective activity a priori or posteriori, mutual regulation between IDE and insulin action in the CNS may behave as a "vicious cycle" that may trigger development of cognitive dysfunction and onset of Alzheimer's disease in diabetes patients (Fig. 6).

## 7 Insulin, Diabetes, and Brain Cholesterol Metabolism

Insulin plays a crucial role for glucose homeostasis, cell survival, and lipid metabolism. Both type 1 and type 2 diabetes are frequently accompanied by dyslipidemia, which can occur as a consequence of alterations in lipogenesis, lipoprotein secretion, and lipolysis in the body. A number of studies have been reported concerning the effects of insulin on circulating lipid or lipid contents in the peripheral tissues. However, lipid metabolism, especially cholesterol in the CNS with insulin resistance/diabetes, is not yet well characterized in spite of its potential importance.

In preliminary studies, we have observed a possible connection between insulin action in the brain and cholesterol metabolism. In studies using Affymetrix microarrays to identify genes differentially expressed in the hypothalami from STZ-diabetic mice (a model of type 1 diabetes), ob/ob mice (a model of type 2 diabetes) and NIRKO mice, a model of CNS insulin resistance we found that the cholesterol biosynthesis pathway was one of the most highly regulated gene sets in the hypothalamus of the STZ-diabetic mouse, with a decrease in expression of cholesterol synthesis-related genes (Fig. 7).

The brain is the most cholesterol-rich organ, containing approximately 25% of the cholesterol present in the body. Disturbances of intracellular cholesterol

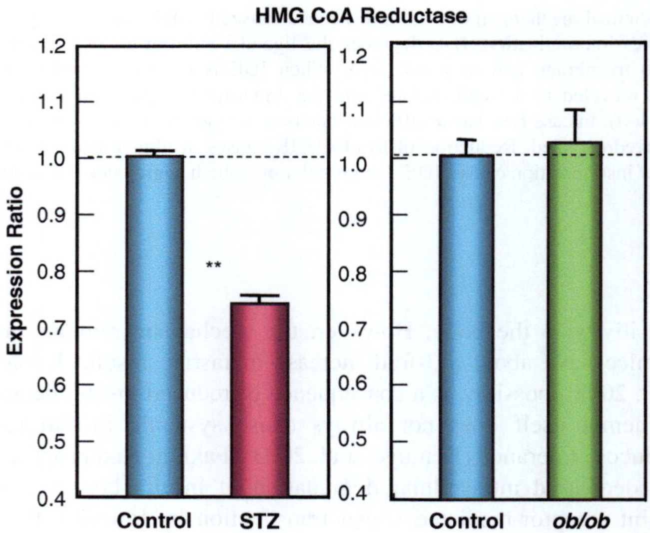


**Fig. 7** Suppression of a cholesterol synthetic gene in streptozotocin-induced diabetes mouse brain. Expression of HMG-CoA reductase (HMGCR), a rate-limiting enzyme for cholesterol biosynthesis, was measured using RNA from hypothalami of brains from control mice, mice with streptozotocin (STZ)-induced diabetes (a model of type 1 diabetes) and *ob/ob* mice (a model of obesity and type 2 diabetes) using quantitative real-time PCR. Data are expressed relative to control levels of 1.0

metabolism by gene mutations cause some congenital diseases, i.e., Smith-Lemli-Opitz syndrome (DHCR7 gene) and Niemann-Pick disease type C (NPC1 and NPC2 genes), which exhibit CNS manifestations including mental retardation. The majority of cholesterol present in the CNS resides in two different pools, the myelin sheaths of oligodendroglia and the plasma membranes of astrocytes and neurons. It has been estimated that up to 70% of the brain cholesterol is associated with myelin (Bjorkhem and Meaney 2004). Although neurons are capable of synthesizing cholesterol, it has been suggested that, in the adult state, neurons rely on delivery of cholesterol from nearby cells such as astrocytes (Mauch et al. 2001; Pfrieger 2003). Many studies indicate that cholesterol content in neurons is crucial for biogenesis of the synaptic vesicles (Thiele et al. 2000; Hering et al. 2003) and their exocytosis (Chamberlain et al. 2001; Lang et al. 2001; Mitter et al. 2003). Cholesterol delivered to neurons is thought to be important in synaptogenesis and may be incorporated into synaptic vesicles. For the delivery of cholesterol, apoE-containing lipoproteins serve as cholesterol carriers. Transcripts for apoE are distributed throughout all regions of the brain and are localized to astrocytes and microglia (Beffert et al. 1998).

A growing body of evidence now implicates a possible link between cholesterol and neurodegenerative disorders, including Alzheimer's disease (Canevari and Clark 2007). The earliest observation of the link was the recognition of apo E-ε4 allele as an important risk factor for late-onset Alzheimer's disease (Corder et al. 1993). The association between diabetes and Alzheimer's disease is especially strong among carriers of the apo E-ε4 allele (Peila et al. 2002). Furthermore, possible associations have been reported between Alzheimer's disease and polymorphisms within the lipoprotein-related protein 1 (LRP1) gene (Kang et al. 1997; Lendon et al. 1997), or HMG-CoA reductase (HMGCR) gene (Park et al. 2003; Recuero et al. 2009), strengthening the putative relationship between cholesterol biology and Alzheimer's disease.

The role of cholesterol in Alzheimer's disease has been controversial for a long time (Shobab et al. 2005; Ledesma and Dotti 2006; Jaeger and Pietrzik 2008). One controversy is about the possible beneficial effect of cholesterol-lowering statins in reducing the risk of Alzheimer's disease. Early case-control studies suggested that statin use was associated with a significant decrease in prevalence of Alzheimer's disease or dementia. However, several recent prospective cohort studies did not find any significant decrease in incidence of Alzheimer's disease or dementia among statin users (Shobab et al. 2005). Multiple in vitro studies have identified a role for cholesterol in promoting Aß production, but not all results are consistent (Abad-Rodriguez et al. 2004; Park et al. 2003; Liu et al. 2007). A significant reduction of brain cholesterol in Alzheimer's disease patients has been observed, particularly in areas loaded with amyloid plaques (Ledesma et al. 2003). It is possible that the heterogeneity of Alzheimer's disease pathology makes the questions very complicated. Further studies examing the potential reduction of cholesterol synthesis in brains with diabetes may provide insights into one mechanism of reduced synapse plasticity in diabetes (Biessels et al. 1996; Kamal et al. 2006) and pathogenesis of cognitive dysfunction in diabetes patients.

# References


Abad-Rodriguez J, Ledesma MD, Craessaerts K, Perga S, Medina M, Delacourte A, Dingwall C, De Strooper B, Dotti CG (2004) Neuronal membrane cholesterol loss enhances amyloid peptide generation. J Cell Biol 167:953–960

Abbott AM, Bueno R, Pedrin MT, Murray JM, Smith RJ (1992) Insulin-like growth factor I receptor gene structure. J Biol Chem 267:10759–10763

Algenstaedt P, Antonetti DA, Yaffe MB, Kahn CR (1997) Insulin receptor substrate proteins create a link between the tyrosine phosphorylation cascade and the Ca2+-ATPases in muscle and heart. J Biol Chem 272:23696–23702

Baltensperger K, Kozma LM, Cherniack AD, Klarlund JK, Chawla A, Banerjee U, Czech MP (1993) Binding of the Ras activator son of sevenless to insulin receptor substrate-1 signaling complexes. Science 260:1950–1952

Baron Van Evercooren A, Olichon-Berthe C, Kowalski A, Visciano G, Van Obberghen E (1991) Expression of IGF-I and insulin receptor genes in the rat central nervous system: a developmental, regional, and cellular analysis. J Neurosci Res 28:244–253

Barres R, Gremeaux T, Gual P, Gonzalez T, Gugenheim J, Tran A, Marchand-Brustel Y, Tanti JF (2006) Enigma interacts with adaptor protein with PH and SH2 domains to control insulin-induced actin cytoskeleton remodeling and glucose transporter 4 translocation. Mol Endocrinol 20:2864–2875

Beffert U, Danik M, Krzywkowski P, Ramassamy C, Berrada F, Poirier J (1998) The neurobiology of apolipoproteins and their receptors in the CNS and Alzheimer's disease. Brain Res Brain Res Rev 27:119–142

Begum N, Sandu OA, Ito M, Lohmann SM, Smolenski A (2002) Active Rho kinase (ROK-alpha ) associates with insulin receptor substrate-1 and inhibits insulin signaling in vascular smooth muscle cells. J Biol Chem 277:6214–6222

Benyoucef S, Surinya KH, Hadaschik D, Siddle K (2007) Characterization of insulin/IGF hybrid receptors: contributions of the insulin receptor L2 and Fn1 domains and the alternatively spliced exon 11 sequence to ligand binding and receptor activation. Biochem J 403:603–613

Bertram L, Blacker D, Mullin K, Keeney D, Jones J, Basu S, Yhu S, McInnis MG, Go RC, Vekrellis K, Selkoe DJ, Saunders AJ, Tanzi RE (2000) Evidence for genetic linkage of Alzheimer's disease to chromosome 10q. Science 290:2302–2303

Biddinger SB, Kahn CR (2006) From mice to men: insights into the insulin resistance syndromes. Annu Rev Physiol 68:123–158

Biessels GJ, Kamal A, Ramakers GM, Urban IJ, Spruijt BM, Erkelens DW, Gispen WH (1996) Place learning and hippocampal synaptic plasticity in streptozotocin-induced diabetic rats. Diabetes 45:1259–1266

Bjorkhem I, Meaney S (2004) Brain cholesterol: long secret life behind a barrier. Arterioscler Thromb Vasc Biol 24:806–815

Bruning JC, Gautam D, Burks DJ, Gillette J, Schubert M, Orban PC, Klein R, Krone W, Muller-Wieland D, Kahn CR (2000) Role of brain insulin receptor in control of body weight and reproduction. Science 289:2122–2125

Cai D, Dhe-Paganon S, Melendez PA, Lee J, Shoelson SE (2003) Two new substrates in insulin signaling, IRS5/DOK4 and IRS6/DOK5. J Biol Chem 278:25323–25330

Canevari L, Clark JB (2007) Alzheimer's disease and cholesterol: the fat connection. Neurochem Res 32:739–750

Chamberlain LH, Burgoyne RD, Gould GW (2001) SNARE proteins are highly enriched in lipid rafts in PC12 cells: implications for the spatial control of exocytosis. Proc Natl Acad Sci USA 98:5619–5624

Chang L, Chiang SH, Saltiel AR (2007) TC10alpha is required for insulin-stimulated glucose uptake in adipocytes. Endocrinology 148:27–33

Collino M, Aragno M, Castiglia S, Tomasinelli C, Thiemermann C, Boccuzzi G, Fantozzi R (2009) Insulin reduces cerebral ischemia/reperfusion injury in the hippocampus of diabetic rats: a role for glycogen synthase kinase-3beta. Diabetes 58:235–242

Cook DG, Leverenz JB, McMillan PJ, Kulstad JJ, Ericksen S, Roth RA, Schellenberg GD, Jin LW, Kovacina KS, Craft S (2003) Reduced hippocampal insulin-degrading enzyme in late-onset Alzheimer's disease is associated with the apolipoprotein E-epsilon4 allele. Am J Pathol 162:313–319

Corder EH, Saunders AM, Strittmatter WJ, Schmechel DE, Gaskell PC, Small GW, Roses AD, Haines JL, Pericak-Vance MA (1993) Gene dose of apolipoprotein E type 4 allele and the risk of Alzheimer's disease in late onset families. Science 261:921–923

Craft S (2007) Insulin resistance and Alzheimer's disease pathogenesis: potential mechanisms and implications for treatment. Curr Alzheimer Res 4:147–152

Craft S, Asthana S, Newcomer JW, Wilkinson CW, Matos IT, Baker LD, Cherrier M, Lofgreen C, Latendresse S, Petrova A, Plymate S, Raskind M, Grimwood K, Veith RC (1999) Enhancement of memory in Alzheimer disease with insulin and somatostatin, but not glucose. Arch Gen Psychiat 56:1135–1140

Cukierman T, Gerstein HC, Williamson JD (2005) Cognitive decline and dementia in diabetes–systematic overview of prospective observational studies. Diabetologia 48:2460–2469

De Meyts P, Whittaker (2002) Structural biology of insulin and IGF-I receptors: implications from drug design. Nature Rev Struct Biol 1:769–783

Dore S, Kar S, Rowe W, Quirion R (1997) Distribution and levels of [125I]IGF-I, [125I]IGF-II and [125I]insulin receptor binding sites in the hippocampus of aged memory-unimpaired and -impaired rats. Neuroscience 80:1033–1040

Dorman JB, Albinder B, Shroyer T, Kenyon C (1995) The age-1 and daf-2 genes function in a common pathway to control the lifespan of Caenorhabditis elegans. Genetics 141:1399–1406

Ebina Y, Ellis L, Jarnagin K, Edery M, Graf L, Clauser E, Ou J-H, Masiar F, Kan YW, Goldfine ID, Roth RA, Rutter WJ (1985) The human insulin receptor cDNA: The structural basis for hormone activated transmembrane signalling. Cell 40:747–758

Emanuelli B, Peraldi P, Filloux C, Sawka-Verhelle D, Hilton D, Van Obberghen E (2000) SOCS-3 is an insulin-induced negative regulator of insulin signaling. J Biol Chem 275:15985–15991

Fakhrai-Rad H, Nikoshkov A, Kamel A, Fernstrom M, Zierath JR, Norgren S, Luthman H, Galli J (2000) Insulin-degrading enzyme identified as a candidate diabetes susceptibility gene in GK rats. Human Mol Genet 9:2149–2158

Farhang-Fallah J, Yin X, Trentin G, Cheng AM, Rozakis-Adcock M (2000) Cloning and characterization of PHIP, a novel insulin receptor substrate-1 pleckstrin homology domain interacting protein. J Biol Chem 275:40492–40497

Farrer LA (1985) Diabetes mellitus in Huntington disease. Clin Genet 27:62–67

Farris W, Mansourian S, Chang Y, Lindsley L, Eckman EA, Frosch MP, Eckman CB, Tanzi RE, Selkoe DJ, Guenette S (2003) Insulin-degrading enzyme regulates the levels of insulin, amyloid beta -protein, and the beta -amyloid precursor protein intracellular domain in vivo. Proc Natl Acad Sci USA 100:4162–4167

Farris W, Mansourian S, Leissring MA, Eckman EA, Bertram L, Eckman CB, Tanzi RE, Selkoe DJ (2004) Partial loss-of-function mutations in insulin-degrading enzyme that induce diabetes also impair degradation of amyloid beta-protein. Am J Pathol 164:1425–1434

Feener EP, Backer JM, King GL, Wilden PA, Sun XJ, Kahn CR, White MF (1993) Insulin stimulates serine and tyrosine phosphorylation in the juxtamembrane region of the insulin receptor. J Biol Chem 268:11256–11264

Fernandez-Real JM, Broch M, Vendrell J, Gutierrez C, Casamitjana R, Pugeat M, Richart C, Ricart W (2000) Interleukin-6 gene polymorphism and insulin sensitivity. Diabetes 49:517–520

Figlewicz DP, Ikeda H, Hunt TR, Stein LJ, Dorsa DM, Woods SC, Porte D, Jr. (1986) Brain insulin binding is decreased in Wistar Kyoto rats carrying the 'fa' gene. Peptides 7:61–65

Fisher SJ, Bruning JC, Lannon S, Kahn CR (2005) Insulin signaling in the central nervous system is critical for the normal sympathoadrenal response to hypoglycemia. Diabetes 54:1447–1451

Freude S, Hettich MM, Schumann C, Stohr O, Koch L, Kohler C, Udelhoven M, Leeser U, Muller M, Kubota N, Kadowaki T, Krone W, Schroder H, Bruning JC, Schubert M (2009) Neuronal IGF-1 resistance reduces A{beta} accumulation and protects against premature death in a model of Alzheimer's disease. FASEB J

Freychet P, Roth J, Kahn CR, Neville DM, Jr. (1972) Insulin-receptor interaction in the liver: Specific binding of 125I-insulin to the plasma membrane. In: INSERM (ed) Colloque sur les Hormones Pancreatiques. Paris, France, pp. 335–357

Frolich L, Blum-Degen D, Bernstein HG, Engelsberger S, Humrich J, Laufer S, Muschner D, Thalheimer A, Turk A, Hoyer S, Zochling R, Boissl KW, Jellinger K, Riederer P (1998) Brain insulin and insulin receptors in aging and sporadic Alzheimer's disease. J Neural Transm 105:423–438

Gavin JR, III, Roth J, Neville DM, Jr., De Meyts P, Buell DN (1974) Insulin-dependent regulation of insulin receptor concentrations: a direct demonstration in cell culture. Proc Natl Acad Sci USA 71:84–88

Goldfine ID, Maddux BA, Youngren JF, Reaven G, Accili D, Trischitta V, Vigneri R, Frittitta L (2008) The role of membrane glycoprotein plasma cell antigen 1/ectonucleotide pyrophosphatase phosphodiesterase 1 in the pathogenesis of insulin resistance and related abnormalities. Endocr Rev 29:62–75

Haft CR, Klausner RD, Taylor SI (1994) Involvement of dileucine motifs in the internalization and degradation of the insulin receptor. J Biol Chem 269:26286–26294

Hanke S, Mann M (2009) The phosphotyrosine interactome of the insulin receptor family and its substrates IRS-1 and IRS-2. Mol Cell Proteomics 8:519–534

Havrankova J, Roth J, Brownstein M (1978) Insulin receptors are widely distributed in the central nervous system of the rat. Nature 272:827–829

Havrankova J, Roth J, Brownstein MJ (1979) Concentrations of insulin and insulin receptors in the brain are independent of peripheral insulin levels. Studies of obese and streptozotocin-treated rodents. J Clin Invest 64:636–642

He W, Rose DW, Olefsky JM, Gustafson TA (1998) Grb10 interacts differentially with the insulin receptor, insulin-like growth factor I receptor, and epidermal growth factor receptor via the Grb10 Src homology 2 (SH2) domain and a second novel domain located between the pleckstrin homology and SH2 domains. J Biol Chem 273:6860–6867

Hedo JA, Kahn CR, Hayoshi M, Yamada KM, Kasuga M (1983) Biosynthesis and glycosylation of the insulin receptor. Evidence for a single polypeptide precursor of the two major subunits. J Biol Chem 258:10020–10026

Heidenreich KA, Zahniser NR, Berhanu P, Brandenburg D, Olefsky JM (1983) Structural differences between insulin receptors in the brain and peripheral target tissues. J Biol Chem 258:8527–8530

Hennige AM, Burks DJ, Ozcan U, Kulkarni RN, Ye J, Park S, Schubert M, Fisher TL, Dow MA, Leshan R, Zakaria M, Mossa-Basha M, White MF (2003) Upregulation of insulin receptor substrate-2 in pancreatic beta cells prevents diabetes. J Clin Invest 112:1521–1532

Hering H, Lin CC, Sheng M (2003) Lipid rafts in the maintenance of synapses, dendritic spines, and surface AMPA receptor stability. J Neurosci 23:3262–3271

Hooper C, Killick R, Lovestone S (2008) The GSK3 hypothesis of Alzheimer's disease. J Neurochem 104:1433–1439

Hotamisligil GS, Shargill NS, Spiegelman BM (1993) Adipose expression of tumor necrosis factor-α: Direct role in obesity-linked insulin resistance. Science 259:87–91

Hoyer S (2002) The aging brain. Changes in the neuronal insulin/insulin receptor signal transduction cascade trigger late-onset sporadic Alzheimer disease (SAD). A mini-review. J Neural Transm 109:991–1002

Hubbard SR (1997) Crystal structure of the activated insulin receptor tyrosine kinase in complex with peptide substrate and ATP analog. EMBO J 16:5572–5581

Hui L, Pei DS, Zhang QG, Guan QH, Zhang GY (2005) The neuroprotection of insulin on ischemic brain injury in rat hippocampus through negative regulation of JNK signaling pathway by PI3K/Akt activation. Brain Res 1052:1–9

Jaeger S, Pietrzik CU (2008) Functional role of lipoprotein receptors in Alzheimer's disease. Curr Alzheimer Res 5:15–25

Jiang Q, Lee CY, Mandrekar S, Wilkinson B, Cramer P, Zelcer N, Mann K, Lamb B, Willson TM, Collins JL, Richardson JC, Smith JD, Comery TA, Riddell D, Holtzman DM, Tontonoz P, Landreth GE (2008) ApoE promotes the proteolytic degradation of Abeta. Neuron 58:681–693

Jolivalt CG, Lee CA, Beiswenger KK, Smith JL, Orlov M, Torrance MA, Masliah E (2008) Defective insulin signaling pathway and increased glycogen synthase kinase-3 activity in the brain of diabetic mice: parallels with Alzheimer's disease and correction by insulin. J Neurosci Res 86:3265–3274

Kamal A, Biessels GJ, Gispen WH, Ramakers GM (2006) Synaptic transmission changes in the pyramidal cells of the hippocampus in streptozotocin-induced diabetes mellitus in rats. Brain Res 1073–1074:276–280

Kang DE, Saitoh T, Chen X, Xia Y, Masliah E, Hansen LA, Thomas RG, Thal LJ, Katzman R (1997) Genetic association of the low-density lipoprotein receptor-related protein gene (LRP), an apolipoprotein E receptor, with late-onset Alzheimer's disease. Neurology 49:56–61

Karas M, Koval AP, Zick Y, LeRoith D (2001) The insulin-like growth factor I receptor-induced interaction of insulin receptor substrate-4 and Crk-II. Endocrinology 142:1835–1840

Kasuga M, Hedo JA, Yamada KM, Kahn CR (1982a) The structure of the insulin receptor and its subunits: Evidence for multiple non-reduced forms and a 210K possible proreceptor. J Biol Chem 257:10392–10399

Kasuga M, Karlsson FA, Kahn CR (1982b) Insulin stimulates the phosphorylation of the 95,000-dalton subunit of its own receptor. Science 215:185–187

Kasus-Jacobi A, Perdereau D, Auzan C, Clauser E, Van Obberghen E, Mauvais-Jarvis F, Girard J, Burnol AF (1998) Identification of the rat adapter Grb14 as an inhibitor of insulin actions. J Biol Chem 273:26026–26035

Kenner KA, Kusari J, Heidenreich KA (1995) cDNA sequence analysis of the human brain insulin receptor. Biochem Biophys Res Commun 217:304–312

Killick R, Scales G, Leroy K, Causevic M, Hooper C, Irvine EE, Choudhury AI, Drinkwater L, Kerr F, Al Qassab H, Stephenson J, Yilmaz Z, Giese KP, Brion JP, Withers DJ, Lovestone S (2009) Deletion of Irs2 reduces amyloid deposition and rescues behavioural deficits in APP transgenic mice. Biochem Biophys Res Commun 386:257–262

Koch L, Wunderlich FT, Seibler J, Konner AC, Hampel B, Irlenbusch S, Brabant G, Kahn CR, Schwenk F, Bruning JC (2008) Central insulin action regulates peripheral glucose and fat metabolism in mice. J Clin Invest 118:2132–2147

Konner AC, Janoschek R, Plum L, Jordan SD, Rother E, Ma X, Xu C, Enriori P, Hampel B, Barsh GS, Kahn CR, Cowley MA, Ashcroft FM, Bruning JC (2007) Insulin action in AgRP-expressing neurons is required for suppression of hepatic glucose production. Cell Metab 5:438–449

Kruger M, Kratchmarova I, Blagoev B, Tseng YH, Kahn CR, Mann M (2008) Dissection of the insulin signaling pathway via quantitative phosphoproteomics. Proc Natl Acad Sci USA 105:2451–2456

Kuo WL, Montag AG, Rosner MR (1993) Insulin-degrading enzyme is differentially expressed and developmentally regulated in various rat tissues. Endocrinology 132:604–611

Lang T, Bruns D, Wenzel D, Riedel D, Holroyd P, Thiele C, Jahn R (2001) SNAREs are concentrated in cholesterol-dependent clusters that define docking and fusion sites for exocytosis. EMBO J 20:2202–2213

Ledesma MD, Dotti CG (2006) Amyloid excess in Alzheimer's disease: what is cholesterol to be blamed for? FEBS Lett 580:5525–5532

Ledesma MD, Abad-Rodriguez J, Galvan C, Biondi E, Navarro P, Delacourte A, Dingwall C, Dotti CG (2003) Raft disorganization leads to reduced plasmin activity in Alzheimer's disease brains. EMBO Rep 4:1190–1196

Lendon CL, Talbot CJ, Craddock NJ, Han SW, Wragg M, Morris JC, Goate AM (1997) Genetic association studies between dementia of the Alzheimer's type and three receptors for apolipoprotein E in a Caucasian population. Neurosci Lett 222:187–190

Liu Q, Zerbinatti CV, Zhang J, Hoe HS, Wang B, Cole SL, Herz J, Muglia L, Bu G (2007) Amyloid precursor protein regulates brain apolipoprotein E and cholesterol metabolism through lipoprotein receptor LRP1. Neuron 56:66–78

Lucas JJ, Hernandez F, Gomez-Ramos P, Moran MA, Hen R, Avila J (2001) Decreased nuclear beta-catenin, tau hyperphosphorylation and neurodegeneration in GSK-3beta conditional transgenic mice. EMBO J 20:27–39

Massague J, Pilch PF, Czech MP (1980) Electrophoretic resolution of three major insulin receptor structures with unique subunits stoichiometries. Proc Natl Acad Sci USA 77:7137–7141

Mauch DH, Nagler K, Schumacher S, Goritz C, Muller EC, Otto A, Pfrieger FW (2001) CNS synaptogenesis promoted by glia-derived cholesterol. Science 294:1354–1357

Miners JS, Baig S, Palmer J, Palmer LE, Kehoe PG, Love S (2008) Abeta-degrading enzymes in Alzheimer's disease. Brain Pathol 18:240–252

Mirsky IA, Broh-Kahn RH (1948) The role of insulinase in the regulation of carbohydrate metabolism. J Clin Invest 27:549

Mitter D, Reisinger C, Hinz B, Hollmann S, Yelamanchili SV, Treiber-Held S, Ohm TG, Herrmann A, Ahnert-Hilger G (2003) The synaptophysin/synaptobrevin interaction critically depends on the cholesterol content. J Neurochem 84:35–42

Najib S, Sanchez-Margalet V (2002) Sam68 associates with the SH3 domains of Grb2 recruiting GAP to the Grb2-SOS complex in insulin receptor signaling. J Cell Biochem 86:99–106

Obici S, Zhang BB, Karkanias G, Rossetti L (2002) Hypothalamic insulin signaling is required for inhibition of glucose production. Nature Med 8:1376–1382

Okamoto H, Nakae J, Kitamura T, Park BC, Dragatsis I, Accili D (2004) Transgenic rescue of insulin receptor-deficient mice. J Clin Invest 114:214–223

Okamura-Oho Y, Miyashita T, Yamada M (2001) Distinctive tissue distribution and phosphorylation of IRSp53 isoforms. Biochem Biophys Res Commun 289:957–960

Park IH, Hwang EM, Hong HS, Boo JH, Oh SS, Lee J, Jung MW, Bang OY, Kim SU, Mook-Jung I (2003) Lovastatin enhances Abeta production and senile plaque deposition in female Tg2576 mice. Neurobiol Aging 24:637–643

Peila R, Rodriguez BL, Launer LJ (2002) Type 2 diabetes, APOE gene, and the risk for dementia and related pathologies: The Honolulu-Asia Aging Study. Diabetes 51:1256–1262

Petruzzelli L, Herrera R, Arenas Garcia R, Fernandez R, Birnbaum MJ, Rosen OM (1986) Isolation of a Drosophila genomic sequence homologous to the kinase domain of the human insulin receptor and detection of the phosphorylated Drosophila receptor with an anti-peptide antibody. Proc Natl Acad Sci USA 83:4710–4714

Pfrieger FW (2003) Role of cholesterol in synapse formation and function. Biochim Biophys Acta 1610:271–280

Poy MN, Yang Y, Rezaei K, Fernstrom MA, Lee AD, Kido Y, Erickson SK, Najjar SM (2002) CEACAM1 regulates insulin clearance in liver. Nat Genet 30:270–276

Pressley JC, Louis ED, Tang MX, Cote L, Cohen PD, Glied S, Mayeux R (2003) The impact of comorbid disease and injuries on resource use and expenditures in parkinsonism. Neurology 60:87–93

Prisco M, Santini F, Baffa R, Liu M, Drakas R, Wu A, Baserga R (2002) Nuclear translocation of insulin receptor substrate-1 by the simian virus 40 T antigen and the activated type 1 insulin-like growth factor receptor. J Biol Chem 277:32078–32085

Qiu WQ, Folstein MF (2006) Insulin, insulin-degrading enzyme and amyloid-beta peptide in Alzheimer's disease: review and hypothesis. Neurobiol Aging 27:190–198

Reaven GM (1988) Banting lecture: Role of insulin resistance in human disease. Diabetes 37:1595–1607

Recuero M, Vicente MC, Martinez-Garcia A, Ramos MC, Carmona-Saez P, Sastre I, Aldudo J, Vilella E, Frank A, Bullido MJ, Valdivieso F (2009) A free radical-generating system induces the cholesterol biosynthesis pathway: a role in Alzheimer's disease. Aging Cell 8:128–139

Renteria ME, Gandhi NS, Vinuesa P, Helmerhorst E, Mancera RL (2008) A comparative structural bioinformatics analysis of the insulin receptor family ectodomain based on phylogenetic information. PLoS ONE 3:e3667

Ribon V, Printen JA, Hoffman NG, Kay BK, Saltiel AR (1998) A novel, multifunctional c-Cbl binding protein in insulin receptor signaling in 3T3-L1 adipocytes. Mol Cell Biol 18:872–879

Rizk NN, Rafols JA, Dunbar JC (2006) Cerebral ischemia-induced apoptosis and necrosis in normal and diabetic rats: effects of insulin and C-peptide. Brain Res 1096:204–212

Rocchi S, Tartare-Deckert S, Sawka-Verhelle D, Gamha A, Van Obberghen E (1996) Interaction of SH2-containing protein tyrosine phosphatase 2 with the insulin receptor and the insulin-like growth factor-I receptor: Studies of the domains involved using the yeast two-hybrid system. Endocrinology 137:4944–4952

Rouard M, Bass J, Grigorescu F, Garrett TP, Ward CW, Lipkind G, Jaffiole C, Steiner DF, Bell GI (1999) Congenital insulin resistance associated with a conformational alteration in a conserved beta-sheet in the insulin receptor L1 domain. J Biol Chem 274:18487–18491

Sano H, Liu SC, Lane WS, Piletz JE, Lienhard GE (2002) Insulin receptor substrate 4 associates with the protein IRAS. J Biol Chem 277:19439–19447

Saxena R, Voight BF, Lyssenko V, Burtt NP, de Bakker PI, Chen H, Roix JJ, Kathiresan S, Hirschhorn JN, Daly MJ, Hughes TE, Groop L, Altshuler D, Almgren P, Florez JC, Meyer J, Ardlie K, Bengtsson BK, Isomaa B, Lettre G, Lindblad U, Lyon HN, Melander O, Newton-Cheh C, Nilsson P, Orho-Melander M, Rastam L, Speliotes EK, Taskinen MR, Tuomi T, Guiducci C, Berglund A, Carlson J, Gianniny L, Hackett R, Hall L, Holmkvist J, Laurila E, Sjogren M, Sterner M, Surti A, Svensson M, Tewhey R, Blumenstiel B, Parkin M, DeFelice M, Barry R, Brodeur W, Camarata J, Chia N, Fava M, Gibbons J, Handsaker B, Healy C, Nguyen K, Gates C, Sougnez C, Gage D, Nizzari M, Gabriel SB, Chirn GW, Ma Q, Parikh H, Richardson D, Ricke D, Purcell S (2007) Genome-wide association analysis identifies loci for type 2 diabetes and triglyceride levels. Science 316:1331–1336

Schubert M, Gautam D, Surjo D, Ueki K, Baudler S, Schubert D, Kondo T, Alber J, Galldiks N, Kustermann E, Arndt S, Jacobs AH, Krone W, Kahn CR, Bruning JC (2004) Role for neuronal insulin resistance in neurodegenerative diseases. Proc Natl Acad Sci USA 101:3100–3105

Schwartz MW, Woods SC, Porte D, Jr., Seeley RJ, Baskin DG (2000) Central nervous system control of food intake. Nature 404:661–671

Shen Y, Joachimiak A, Rosner MR, Tang WJ (2006) Structures of human insulin-degrading enzyme reveal a new substrate recognition mechanism. Nature 443:870–874

Shier P, Watt VM (1989) Primary structure of a putative receptor for a ligand of the insulin family. J Biol Chem 264:14605–14608

Shobab LA, Hsiung GY, Feldman HH (2005) Cholesterol in Alzheimer's disease. Lancet Neurol 4:841–852

Skorokhod A, Gamulin V, Gundacker D, Kavsan V, Muller IM, Muller WE (1999) Origin of insulin receptor-like tyrosine kinases in marine sponges. Biol Bull 197:198–206

Slaaby R, Schaffer L, Lautrup-Larsen I, Andersen AS, Shaw AC, Mathiasen IS, Brandt J (2006) Hybrid receptors formed by insulin receptor (IR) and insulin-like growth factor I receptor (IGF-IR) have low insulin and high IGF-1 affinity irrespective of the IR splice variant. J Biol Chem 281:25869–25874

Strachan MW, Deary IJ, Ewing FM, Frier BM (1997) Is type II diabetes associated with an increased risk of cognitive dysfunction? A critical review of published studies. Diabetes Care 20:438–445

Sun XJ, Pons S, Asano T, Myers MG, Jr., Glasheen EM, White MF (1996) The fyn tyrosine kinase binds IRS-1 and forms a distinct signaling complex during insulin stimulation. J Biol Chem 271:10583–10587

Takayama S, White MF, Kahn CR (1988) Phorbol ester-induced serine phosphorylation of the insulin receptor decreases its tyrosine kinase activity. J Biol Chem 263:3440–3447

Taniguchi CM, Emanuelli B, Kahn CR (2006) Critical nodes in signalling pathways: insights into insulin action. Nat Rev Mol Cell Biol 7:85–96

Taylor SI, Wertheimer E, Accili D, Cama A, Hone J, Roach P, Quon MJ, Suzuki Y, Levy-Toledano R, Taouis M, Luz Sierra M, Barbetti F, Gorden P (1994) Mutations in the insulin receptor gene: update. Endocr Rev 2:58–65

Thiele C, Hannah MJ, Fahrenholz F, Huttner WB (2000) Cholesterol binds to synaptophysin and is required for biogenesis of synaptic vesicles. Nature Cell Biol 2:42–49

Tsuji N, Tsujimoto K, Takada N, Ozaki K, Ohta M, Itoh N (1996) Expression of insulin receptor-related receptor in the rat brain examined by in situ hybridization and immunohistochemistry. Brain Res Mol Brain Res 41:250–258

Tu Y, Liang L, Frank SJ, Wu C (2001) Src homology 3 domain-dependent interaction of Nck-2 with insulin receptor substrate-1. Biochem J 354:315–322

Ueki K, Kondo T, Kahn CR (2004) Suppressor of cytokine signaling 1 (SOCS-1) and SOCS-3 cause insulin resistance through inhibition of tyrosine phosphorylation of insulin receptor substrate proteins by discrete mechanisms. Mol Cell Biol 24:5434–5446

Ullrich A, Bell JR, Chen EY, Herrera R, Petruzzelli LM, Dull TJ, Gray A, Coussens L, Liao YC, Tsubokawa M, Mason A, Seeburg PH, Grunfeld C, Rosen OM, Ramachandran J (1985) Human insulin receptor and its relationship to the tyrosine kinase family of oncogenes. Nature 313:756–761

Ullrich A, Gray A, Tam AW, Yang-Feng T, Tsubokawa M, Collins C, Henzel W, Le Bon T, Kathuria S, Chen E (1986) Insulin-like growth factor I receptor primary structure: comparison with insulin receptor suggests structural determinants that define functional specificity. EMBO J 5:2503–2512

Usui I, Imamura T, Huang J, Satoh H, Olefsky JM (2003) Cdc42 is a Rho GTPase family member that can mediate insulin signaling to glucose transport in 3T3-L1 adipocytes. J Biol Chem 278:13765–13774

Valverde AM, Mur C, Pons S, Alvarez AM, White MF, Kahn CR, Benito M (2001) Association of insulin receptor substrate 1 (IRS-1) Y895 with Grb-2 mediates the insulin signaling involved in IRS-1-deficient brown adipocyte mitogenesis. Mol Cell Biol 21:2269–2280

White MF (1998) The IRS-signalling system: a network of docking proteins that mediate insulin action. Mol Cell Biochem 182:3–11

White MF, Stegmann EW, Dull TJ, Ullrich A, Kahn CR (1987) Characterization of an endogenous substrate of the insulin receptor in cultured cells. J Biol Chem 262:9769–9777

Wick MJ, Dong LQ, Hu D, Langlais P, Liu F (2001) Insulin receptor-mediated p62dok tyrosine phosphorylation at residues 362 and 398 plays distinct roles for binding GTPase-activating protein and Nck and is essential for inhibiting insulin-stimulated activation of Ras and Akt. J Biol Chem 276:42843–42850

Zeggini E, Weedon MN, Lindgren CM, Frayling TM, Elliott KS, Lango H, Timpson NJ, Perry JR, Rayner NW, Freathy RM, Barrett JC, Shields B, Morris AP, Ellard S, Groves CJ, Harries LW, Marchini JL, Owen KR, Knight B, Cardon LR, Walker M, Hitman GA, Morris AD, Doney AS, McCarthy MI, Hattersley AT (2007) Replication of genome-wide association signals in UK samples reveals risk loci for type 2 diabetes. Science 316:1336–1341

Zhang B, Roth RA (1992) The insulin receptor-related receptor: tissue expression, ligand binding specificity, and signaling capabilities. J Biol Chem 267:18320–18328

Zhao J, Li L, Leissring MA (2009) Insulin-degrading enzyme is exported via an unconventional protein secretion pathway. Mol Neurodegener 4:4

Zhao L, Teter B, Morihara T, Lim GP, Ambegaokar SS, Ubeda OJ, Frautschy SA, Cole GM (2004) Insulin-degrading enzyme as a downstream target of insulin receptor signaling cascade: implications for Alzheimer's disease intervention. J Neurosci 24:11120–11126

# The Brain-insulin Connection, Metabolic Diseases and Related Pathologies

**Kyriaki Gerozissis**

**Abstract** By its action in the brain, insulin controls neuronal survival, energy balance, glucose and lipid metabolism, cognition and additional vital functions. Metabolic, endocrine and neural signals interact with the hormone in the central nervous system, in particular in the hypothalamus and the hippocampus, and regulate its efficiency. Insulin, leptin and serotonin share common signaling routes involved in food intake, energy and glucose homeostasis, such as phosphatidylinositol-3-kinase (PI3K), STAT-3 and MAP kinase pathways. Alterations of brain levels and brain signaling of either insulin or its partners, associated with deficient beta-cell secretion and/or peripheral insulin resistance, contribute to the initiation and progress of metabolic and related pathologies, Alzheimer's disease and depressive syndromes.

Despite the availability of numerous therapeutic options for diabetes, current approaches are not adequately effective. Most of them do not take into account either the complex interactions among the various sites of insulin action or the importance of central insulin resistance or its interplay with neurotransmitters and peptides. Most antidiabetic therapies induce many adverse effects, in particular obesity, and thus may initiate a vicious cycle of problems. Furthermore, inefficient diabetic therapy is a high risk for the development of mood and neurodegenerative diseases. At present, new compounds and novel routes of drug administration targeting insulin and its partners in the CNS, resulting in increased central efficiency of the hormone, and studies aiming to further elucidate central mechanisms of insulin action offer hope for novel ways of prevention and intervention in metabolic pathologies and complications.

K. Gerozissis
INSERM, Neuroendocrinologie Moléculaire de la Prise Alimentaire (NMPA), University Paris Sud-11, 15 rue Georges Clémenceau - Bat. 447, 91405 - Orsay Cedex, 91405 - Orsay Cedex, France
e-mail: gerozissis@yahoo.co.uk

S. Craft and Y. Christen (eds.), *Diabetes, Insulin and Alzheimer's Disease*, Research and Perspectives in Alzheimer's Disease
DOI 10.1007/978-3-642-04300-0_2, © Springer-Verlag Berlin Heidelberg 2010

# 1 Introduction

The central nervous system controls vital functions by efficiently coordinating peripheral and central cascades of signals and networks in an orchestrated manner. Historically, the brain was considered to be insulin independent. These earlier views have been challenged by findings demonstrating that insulin exerts multiple actions in the brain, regulating vital biological processes such as life span, neuronal survival, cognition, reproduction, feeding behavior, energy balance, and glucose and fat metabolism, and that inefficient central action of insulin contributes to the development of severe pathologies (Banks et al. 2000; Gerozissis 2003, 2004, 2008; Lustman and Clouse 2005; Okamoto et al. 2001; Park 2002; Perrin et al. 2004; Pocai et al. 2005; Reger et al. 2008; Schwartz and Porte, 2005; Schubert et al. 2004; van der Heide et al. 2005; Woods et al. 1979; Wrighten et al. 2008).

Insulin and specific insulin receptors are widely distributed in the networks of the central nervous system related mainly to feeding or cognition (Baskin et al. 1983; Bruning et al. 2000; Gerozissis 2003, 2008; Havrankova et al. 1978a, b; Schechter et al. 1996; Schulingkamp et al. 2000; Schwartz et al. 1992; Zhao et al. 2004). Insulin receptors located in the synapses of neurons and in astrocytes are present in high concentrations in the cerebral cortex, olfactory bulb, hippocampus, amygdala, cerebellum and hypothalamus (Abbott et al., 1999; Havrankova et al. 1978a; Unger et al. 1991; Zhao and Alkon, 2001). The major molecular structure and most of the properties of brain insulin receptors are identical to peripheral insulin receptors (Wozniak et al. 1993).

Additionally, the literature has demonstrated the regulation of insulin release or gene expression by glucose in hypothalamic cells in culture, in brain synaptosomes or in the hypothalamus (Gerozissis et al. 2001; Madadi et al. 2008a; Santos et al. 1999).

Insulin crosses the blood-brain barrier, and transport of blood insulin in the brain has been convincingly demonstrated. The hormone enters into circumventricular regions that lack a blood-brain barrier and can cross the blood-brain barrier via insulin receptor-mediated active transport (Banks 2004; Baura et al. 1993; Woods et al. 2003). However, several studies suggest that a portion of brain insulin is potentially de novo produced locally. Actually, reverse transcription-polymerase chain reaction (RT-PCR) and in situ hybridization approaches, for instance, clearly indicate insulin or preproinsulin gene expression in immortalized hypothalamic cells, the hypothalamus, the cortex or the hippocampus in the fetal, newborn and adult rodent brain (Banas et al. 2009; Clarke et al. 1986; de la Monte et al. 2005; Devaskar et al. 1994; Grunblatt et al. 2007; Hrytsenko et al. 2007; Madadi et al. 2008; Schechter et al. 1990, 1994, 1996; Steen et al. 2005).

## 2 Central Interplay of Insulin with Hormones, Neuropeptides and Neurotransmitters

By its action in the brain, insulin, together with various hormones, neuropeptides and transmitters, controls processes related to feeding behavior, energy and glucose homeostasis, and learning and memory and is potentially involved in the communication within brain structures, in particular the hypothalamus and the limbic system (Blundell 1984; Fehm et al. 2006; Fetissov et al. 2000; Feurte et al. 2000; Gerozissis 2003, 2004, 2008; Lam et al. 2005; Morton et al. 2006; Niswender and Schwartz, 2003; Obici et al. 2002a, b; Plum et al. 2006; Rother et al. 2008; Schwartz and Porte 2005; Woods et al. 2000; Yamada et al. 1997). So far, the most investigated field of central insulin action is the control of energy balance and the adjustment of both food intake and energy expenditure occurring mainly in the hypothalamus. However, extrahypothalamic structures, in particular the brain stem, the cortex and the hippocampus, also have an important role in those processes (Berthoud and Morrison 2008; Fehm et al. 2006; Tracy et al. 2001).

In the hypothalamus, there is a complex interplay of insulin with peptides and neurotransmitters, in particular serotonin (5-hydroxytryptamine, 5-HT), leptin, the brain-derived neurotrofic factor (BDNF), melanocortins, neuropeptide Y (NPY), agouti-related protein (AgRP), melatonin, ghrelin and potentially additional neuroregulators (Guillod-Maximin et al. 2009; Teccott et al. 1995). Insulin shares with those mediators, and in particular with leptin and serotonin, signaling pathways in the brain and in neuronal cells in culture, resulting in overall membrane and/or gene effects (Benomar et al. 2005; Gerozissis et al. 2007; Niswender and Schwartz 2003; Plum et al. 2006; Rother et al. 2008; reviewed by Gerozissis 2008; McNay 2007).

Abundant literature supports the hypothesis of an important role for insulin in processes related to memory. Numerous studies show that acute intravenous administration of the hormone facilitates cognition in humans (Kern et al. 2001; Park et al. 2000; Reger and Craft 2006; Watson and Craft 2004). Insulin's acute effects on memory follow a curvilinear dose–response pattern, such that very low and very high doses of insulin do not facilitate memory (Craft et al. 2003). In addition, chronically elevated peripheral levels of insulin appear to induce cognitive impairments (Craft 2007; Elias et al. 1997; Luchsinger et al. 2004). The mechanisms through which insulin affects memory are not well documented and require additional investigation. However, several studies demonstrate effects on cerebral energy metabolism, neurotransmitter levels and synaptic plasticity (Gispen and Biessels 2000; Reagan 2007). Although insulin does not appear to increase glucose transport into the brain, it stimulates glucose uptake in medial temporal lobe structures that support memory and has selective effects on cerebral glucose metabolism (Bingham et al. 2002; Doyle et al. 1995; Reger and Craft 2006).

In addition to insulin, the literature implicates leptin in the regulation of neuronal structure and function in the hippocampus, the cortex and other brain areas (Harvey and Aschford 2003; Huang et al. 1996; Shioda et al. 1998). Interestingly, insulin increases the expression of leptin receptors in the hippocampus. Furthermore, insulin modulates central levels of neurotransmitters known to affect cognition, such as acetylcholine, serotonin or norepinephrine (Figlewicz et al. 1993; Kopf and Baratti 1999; Orosco and Gerozissis 2001; Paulus et al. 2005). Neuronal signals activated by leptin may overlap with those generated by insulin, and this may explain the many similarities in their central effects. A direct central dialogue between insulin and leptin in particular at the PI3K level was convincingly demonstrated (Niswender and Schwartz 2003). Nevertheless, unrelated brain actions of the two hormones have also been reported.

Insulin and leptin further interact via other neuromodulators, in particular the anorexigenic neurotransmitter serotonin. We have demonstrated previously bidirectional effects of insulin and serotonin in the median hypothalamus, an interaction that seems to be a link in a larger cascade of events in the complex regulatory loop between hypothalamic neuromodulators and nutritional behavior (Gerozissis 2008; Orosco and Gerozissis 2001; Orosco et al. 2000). It has been suggested that the central serotonergic system acts on energy metabolism via leptin-responsive hypothalamic pathways (Calapai et al. 1999). In turn, leptin increases serotonin turnover and affects the acute feeding-induced hypothalamic serotonergic stimulation (Telles et al. 2003). The presence of leptin receptors on serotonin neurons in several cell groups suggests that leptin might exert some of its effects on energy balance through the serotonin system. Indeed, either depletion of serotonin or treatment with serotonin receptor antagonists attenuates the anorectic effect of leptin (Finn et al. 2001; Hay-Schmidt et al. 2001). Type 2c serotonin receptors, (5-HT2cR) and potentially 5-HT2bR are involved in the anorexic action of the transmitter (Gerozissis et al. 2007; Lam et al. 2008). Mice lacking functional 5-HT2cR, for instance, are hyperphagic and develop obesity (Tecott 1995).

As insulin and leptin are involved in the regulation of food intake and body weight and affect memory and learning processes, it appears likely that these hormones, together with serotonin, which is also involved in cognitive processes, might be a potential link between nutrition and cognition (Gerozissis 2003; Gerozissis et al. 2001; Morrison 2009). Our own data associated with the literature suggest that brain insulin, in interaction with serotonin, leptin, melanocortins and potentially additional regulators, might be involved in cognitive processes related to feeding, in particular in anticipation of meal time (Benoit et al. 2003; Drazen et al. 2005, 2006; Gerozissis 2003, 2008; Gerozissis et al. 2001).

## 3 Mechanisms of Action

As in peripheral tissues, in brain structures (mainly the hypothalamus and the hippocampus) and human neuroblastoma cells in culture, insulin activates several signaling pathways, such as insulin receptor substrate/phosphatidylinositol-3-kinase pathway (IRS/PI3K), Signal Transducer and Transcription Factor 3 (STAT3) and mitogen-activated protein kinase (MAPK; Benomar et al. 2005; de la Monte et al. 2005; Gerozissis et al. 2007; McNay 2007; Niswender et al. 2003; Plum et al. 2005,

2006; van der Heide et al. 2005). The IRS/PI3K signaling pathway operating in the hypothalamus is implicated in food intake and glucose regulation, whereas in the hippocampus it regulates cognition (Huang et al. 2004; Könner et al. 2007; Niswender and Schwartz 2003; Niswender et al. 2003a, 2004; van der Heide et al. 2005). Blockade of PI3K activation by intracerebroventricular administration of inhibitors of PI3K largely restores food intake that was decreased by either insulin, serotonin or leptin administration (Gerozissis et al. 2001, 2006; Niswender et al. 2003a, 2004; Niswender and Schwartz 2004), assigning to the PI3K pathway a key role in mediating the impact of the three partners on feeding behavior. PI3K activation induces both membrane and genomic effects. Insulin binds to its receptor on pro-opiomelanocortin (POMC) and AgRP neurons, stimulating receptor autophosphorylation and activating its signal cascade. Insulin receptor substrate proteins bind to the phosphorylated residues on the insulin receptor, recruit the regulatory subunit p85 of PI3K and thus activate PI3K, which phosphorylates phosphatidylinositol-4, 5-bisphosphate, generating phosphatidylinositol 3,4,5-trisphosphate (PIP3). The protein kinase B/AKT and phosphoinositide-dependent protein kinase 1 (PDK1) bind to PIP3. The phosphorylated AKT enters the nucleus, where it phosphorylates and inactivates forkhead box protein O1 (FOXO-1). Various other hormonal and nutrient-related signals regulate the activity of hypothalamic POMC neurons, in particular leptin and serotonin, acting via receptors that are structurally different from the insulin receptor, resulting in the release of melanocortins and regulation of feeding behavior (Ahima et al. 2000; Finn et al. 2001; Carvalheira et al., 2001; Gerozissis 2008; Heisler et al. 2002; Kalra et al. 1999; Könner et al. 2007; Lam et al. 2007; Niswender and Schwartz 2003; Rother et al. 2008; Zhou et al. 2005; Fig. 1).

In hypothalamic AgRP neurons, insulin treatment leads to membrane hyperpolarization and a decrease in action-potential frequency via activation of PI3K and $K_{ATP}$ channels and electrical silencing of these cells, resulting in reduced release of AgRP, NPY, and other transmitters from AgRP neurons, regulation of liver innervation and suppression of hepatic glucose production. Thus, acting through the central nervous system, insulin regulates hepatic interleukin-6 expression to control gluconeogenesis via STAT3 activation in liver parenchymal cells, thereby potentializing systemic insulin's direct suppressing effect on hepatic gluconeogenic gene expression via activation of PI3K signaling and subsequent export of FOXO-1 from the nucleus (Inoue et al. 2006; Könner et al. 2007; Rother et al. 2008). Leptin and serotonin are also involved in hepatic gluconeogenesis acting rather directly in the liver via specific receptors (see Könner et al. 2007; Rother et al. 2008).

Our own observations, obtained ex vivo in rat hypothalami or in a human neural cell culture, suggest that insulin, serotonin and leptin, in addition to PI3K, share MAPK and STAT3 pathways (Gerozissis et al. 2007).

The well-documented signaling pathway for insulin in the hypothalamus via activation of PI3K is also involved in many of insulin's actions within the hippocampus, such as memory enhancement. Some effects of insulin in the hippocampus may also be mediated through activation of MAPK (de la Monte et al. 2005; Huang et al. 2004; O'Malley and Harvey 2007; McNay 2007; van der Heide et al. 2005).

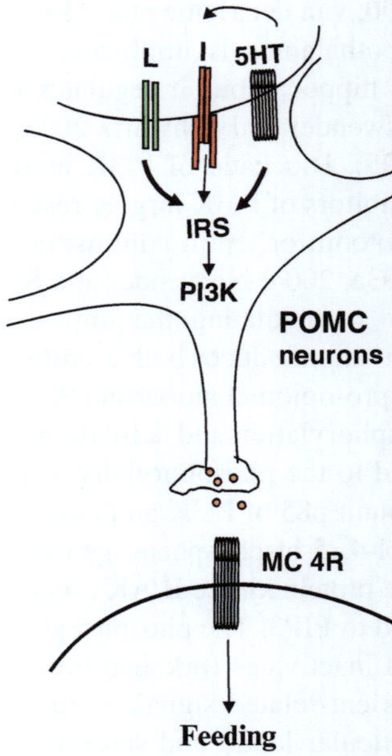


**Fig. 1** In pro-opiomelanocortin (POMC) neurons, insulin (I), leptin (L) and serotonin (5HT) modify the production or the release of melanocortins and thus regulate feeding behavior. The three partners might act via the PI3K pathway directly or/and indirectly. MC4R, melanocortin type 4 receptors

Interestingly, insulin and serotonin, together with BDNF, also cooperate in biological processes that influence aging and age-related diseases (Mattson et al. 2004a). Their signaling pathways activate one or more transcription factors that regulate expression of genes encoding proteins involved in neural plasticity, stress resistance and cell survival. They can also exert rapid transcription-independent effects on neurons and glial cells (Mattson et al. 2004a, b). The presence of both insulin and leptin receptors, the increased expression of the long form of the leptin receptor following application of insulin in the hippocampus and also the action of leptin via PI3K in this structure again suggest a possible cross-talk between insulin and leptin (Paulus et al. 2005; Shanley et al. 2002).

## 4 Metabolic, Mood and Neurodegenerative Pathologies

Impairment of brain insulin gene expression, brain insulin levels or responsiveness observed in aged subjects and rodents or humans with metabolic or neurodegenerative diseases raised the question of the importance of altered brain insulin efficiency in those pathologies. In type 1 diabetic models, insulinopenia impairs brain insulin signaling that is associated with the biochemical and behavioral characteristics of Alzheimer's disease (Jolivalt et al. 2008). In terms of brain insulin availability, the literature shows decreased cortical insulin concentration in post-mortem studies in humans, whereas our own studies in rats show an extracellular hypothalamic insulin decrease with age in normal weight animals and a worsening of hypothalamic insulin deficiency with age in genetically obese models (Frolich et al. 1998; Gerozissis et al. 1993, 2001). Chronic peripheral hyperinsulinemia downregulates blood–brain barrier receptors and reduces insulin transport into the brain (Kaiyala et al. 2000; Stein et al. 1987). Thus, conditions characterized by insulin resistance, such as type 2 diabetes, may reduce the efficiency of central nervous system insulin uptake and decrease brain insulin levels (Baskin et al. 1985; Israel et al. 1993; Kaiyala et al. 2000; reviewed by Reger and Craft 2006). The literature suggests that lack of insulin in the brain promotes neuronal differentiation affecting learning, memory, and Alzheimer's disease (Li and Holscher 2007; Schechter et al. 2005). However, even more than brain insulin deficiency, inappropriate insulin concentrations and impaired ratio of brain to peripheral insulin levels appear pathophysiologically important. Actually, brain insulinopenia associated with mild or high insulinemia, resulting in an important negative shift of brain to peripheral insulin ratio, is reported in aged or obese rodents and humans and in patients with Alzheimer's disease (Cohen and Dillin 2008; Craft et al. 1998; Kern et al. 2006; Steen et al. 2005; Stein et al. 1987; Fig. 2).

At present, the classic concept stating that only skeletal muscle, beta-cells and liver are involved in metabolic dysfunctions seems not satisfactory anymore. There is increasing evidence in favor of complex interactions among the various sites of insulin action and of redundant mechanisms for inter-organ communication. It appears likely that every tissue contributes to the onset of diabetes. The brain has a pivotal role in those processes (Accili 2003; Bruning et al. 2000; Dore et al. 1997; Fig. 3).

Intact insulin signaling in the brain is essential for energy and glucose homeostasis. The blockading of insulin action in the hypothalamic arcuate nucleus by insulin antibodies, decreasing insulin receptors by antisense oligonucleotides, or

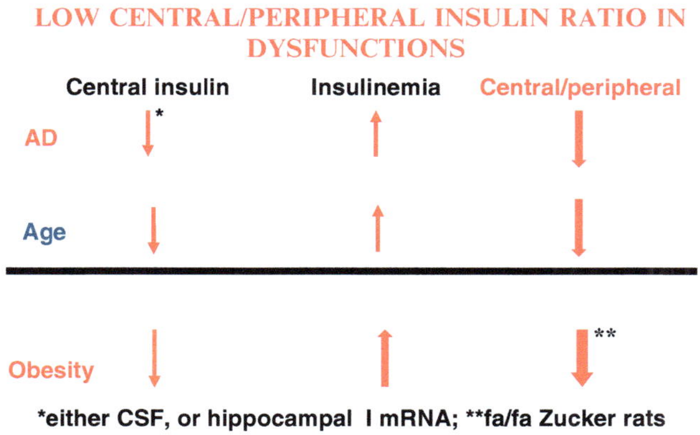


**Fig. 2** An important negative shift of brain-to-peripheral (insulinemia) insulin ratio was observed in the hypothalamus or cortex of aged or obese rodents or humans and in patients with Alzheimer's disease (AD). Central insulin was determined by measurements of the hormone or its gene expression (mRNA) in the cerebrospinal fluid (CSF), the hypothalamus, the hippocampus and the cortex in vivo or in post mortem studies

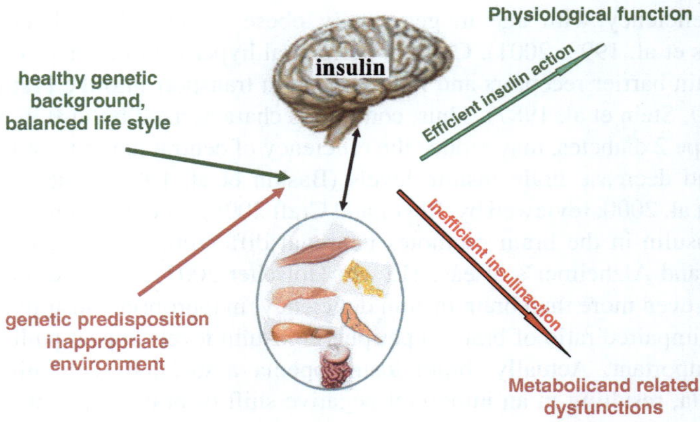


**Fig. 3** With a healthy genetic background and environment, a balanced lifestyle via an orchestrated, matched inter-organ communication maintains efficient brain insulin action and physiological functions. Inheritance, stress, inappropriate nutrition and physical exercise, and low mental activity disrupt brain insulin signaling and efficiency and, together with peripheral insulin resistance, deregulate normal functions, contributing to the onset of obesity, diabetes and related pathologies

inhibiting activation of PI3K lead to decreased ability of circulating insulin to suppress endogenous glucose production (Obici et al. 2002a, b; Prodi and Obici 2006). In humans, a lack of IRS-2 in the hypothalamus results in increased appetite and body weight, leading to insulin resistance and finally diabetes (Lin et al. 2004). Furthermore, mice with neuron-specific insulin receptor deletion show an increase in food intake and body weight (Bruning et al. 2000). The restoration of insulin receptors in the brains of mice with tissue-restricted insulin receptor expression maintains energy homeostasis and prevents diabetes (Okamoto et al. 2004). On the other hand, cerebrocortical insulin resistance is observed in either obese patients or individuals with the Gly972Arg polymorphism in IRS-1, a type 2 diabetes risk gene (Tschritter et al. 2006). Additional literature suggests that defective insulin signaling within key hypothalamic neuronal pathways, along with impaired leptin signaling, can be included in the potential mechanisms linking obesity to type 2 diabetes (Lin et al. 2004; Obici et al. 2002; Porte et al. 2005; Schwartz, 2001; Schwartz and Porte, 2005). Recent studies demonstrate that chronic brain leptin infusion in rats in vivo stimulates hypothalamic phosphotyrosine phosphatase (PTP-1B) expression, suggesting a potential progressive installation of leptin and insulin inefficiency since this phosphatase negatively affects both leptin and insulin signaling pathways (Benomar et al. 2009; Berthou et al. 2008; White et al. 2009).

Deficits in insulin receptor signaling and impairments in hypothalamic–pituitary–adrenal axis function contribute to the neurological complications of diabetic patients. The literature supports the hypothesis that a long-term consequence of diabetes and obesity is accelerated brain aging that results in neuropsychological deficits and increased vulnerability to co-morbidities such as Alzheimer's disease and depressive syndromes (Kodl and Seaquist 2008; Reagan 2007; Wrighten et al. 2009). On the other hand, administration of insulin improves cognitive performance in healthy subjects (Park et al. 2000) and aged subjects (Messier et al. 1997; Winocur and Gagnon, 1998) and in experimental models of insulin resistance and in Alzheimer's disease patients (Craft et al. 1999; Jolivalt et al. 2008; Reagan 2007).

The diminished central concentration of insulin in the cerebrospinal fluid of Alzheimer's disease patients is probably due to a lower rate of insulin transport in the brain, resulting from a down-regulation of insulin receptors at the level of blood-brain barrier (Reger and Craft 2006). Additionally, diminished local insulin production cannot be excluded, since lower insulin gene expression messenger RNA levels are reported in the hippocampus of Alzheimer's disease patients (Steen et al. 2005). The main pathophysiological characteristics of obesity and diabetes – impaired glucose homeostasis, central and peripheral insulin resistance, lack of insulin in the brain, chronic hyperinsulinemia and hyperleptinema – affect the hippocampus, where they induce insulin and/or leptin resistance, alter synaptogenesis and contribute to accelerated brain aging in diabetic patients (Craft 2007; Reagan 2007). Those hippocampal alterations increase vulnerability to the development of neurodegenerative disorders, including Alzheimer's disease, which is considered by some authors to be brain type 2 diabetes or type 3 diabetes (Elias et al.1997; Helkala et al. 1995; Hoyer 1998; Luchsinger et al. 2004; Strachan et al. 1997; Steen et al. 2005; reviewed by Reger and Craft 2006; Wrighten et al. 2008).

The obesity of the Zucker rat, a shorter-living animal than its lean congener, is associated with numerous metabolic and neurochemical disturbances in central transmitters regulating feeding behavior. In this model, both hypothalamic insulin levels and serotonin responsiveness to food intake are impaired with aging (Gerozissis et al. 2001; Lemierre et al. 1998). The alteration of serotonin responsiveness in this model appears of interest because of the crucial role of the serotonergic system in depression and its eventual importance in metabolic dysfunctions (Gerozissis 2008; Kalia 2005; Okamura et al. 2000).

Actually, abundant literature associates obesity/diabetes and depressive disorders. Epidemiologic studies report increased incidence of depression in diabetics, including adolescents, and vice versa (Lustman and Clouse 2005; Lustman et al. 2008; McIntyre et al. 2007; Stewart et al. 2005). In view of the above observations, some authors propose to reclassify depressive syndromes as "Metabolic syndrome type 2" (McIntyre et al. 2007).

The two processes, diabetes and depression, negatively interact, in that depression leads to poor metabolic control and hyperglycemia exacerbates depression. Depression constitutes a major risk factor in the development of type 2 diabetes and may accelerate the onset of diabetes complications. Interestingly, depressive, non-diabetic patients have several insulin- and glucose-metabolism disturbances. The pathophysiological similarities between diabetes and stress-related depression suggest that common mediators may be involved in the etiology and progression of the neurological complications of these disorders.

The activation of the serotonergic system modulates the function of the hippocampus, which receives a strong serotonergic projection from the raphe nuclei and expresses serotonin receptors at high density (Kalia 2005). Patients with clinical depression develop glucose intolerance and impaired insulin sensitivity. These abnormalities can be resolved after treatment with antidepressants and recovery from depression (Okamura et al. 2000). The metabolic syndrome is associated with suppressed neuroendocrine responses to serotonin. In a type 2 diabetic model, the Goto Kakizaki rat, for instance, our pilot studies demonstrate inefficient serotonergic responsiveness of the hypothalamus to food intake (Gerozissis et al. 2008). Literature based on a study of suicide victims suggests that the PI3K signaling pathway is involved in serotonergic action in the brain (Hsiung et al. 2003). However, the exact molecular mechanisms affected by the insulin-serotonin interrelated action in both metabolic and mood disorders remain elusive.

Concerning leptin, both leptin insufficiency and leptin resistance may contribute to the onset of depression. Low levels of leptin are associated with depressive behaviors in rodents and humans, whereas pharmacological studies indicate that leptin has antidepressant-like efficacy (Lu 2007).

There is now a growing body of literature supporting the implication of both leptin and serotonin in Alzheimer's disease. Leptin facilitates hippocampal synaptic plasticity and thus it may improve hippocampus-dependent behavioral performances (Shanley et al. 2001). Additional studies suggest that alterations in insulin and leptin signaling within non-hypothalamic brain areas are among potential mechanisms linking obesity and diabetes to impaired cognitive function (reviewed by Wrighten et al. 2008). Further studies indicate that inefficient leptin signaling, a well-described phenomenon in the hypothalamus, may also be observed in the hippocampus in obesity and diabetes. Thus leptin resistance might contribute to deficits in hippocampal synaptic plasticity and to enhanced neuronal susceptibility to damage (Shanley et al. 2001; Signore et al. 2008; Wrighten et al. 2008). As mentioned above, leptin and insulin share numerous signaling cascades that are also shared by other neuroprotective molecules. The signaling mechanism underlying neuroprotective effects involves the activation of PI3K, MAPK, and Src tyrosine kinases (Shanley et al. 2001). Finally, leptin has been proposed as a novel therapeutic agent for Alzheimer's disease (Fewlass et al. 2004; Greco et al. 2009).

Alzheimer's disease and depressive syndromes are closely related. In fact, a history of depression is a risk factor for patients with Alzheimer's disease, whereas neurochemical alterations in the brains of those patients include reduced levels of serotonin and other neurotransmitters. It appears that inefficient activation of serotonin signaling pathways might promote synaptic dysfunction and neuronal death in Alzheimer's disease (reviewed by Mattson et al. 2004a). Serotonin may enhance synaptic plasticity by up-regulating the expression of brain-derived neurotrophic factor (BDNF), a neuropeptide that shares signaling pathways with brain insulin. This signaling mechanism may be compromised in Alzheimer's disease (Tong et al. 2001). Interestingly, serotonin receptor agonists and serotonin-selective reuptake inhibitors (SSRI), widely prescribed for the treatment of clinical depression, may also be used as neuroprotective agents (Lezonalc'h 2007; Nelson et al. 2007; Sanchez et al. 2001).

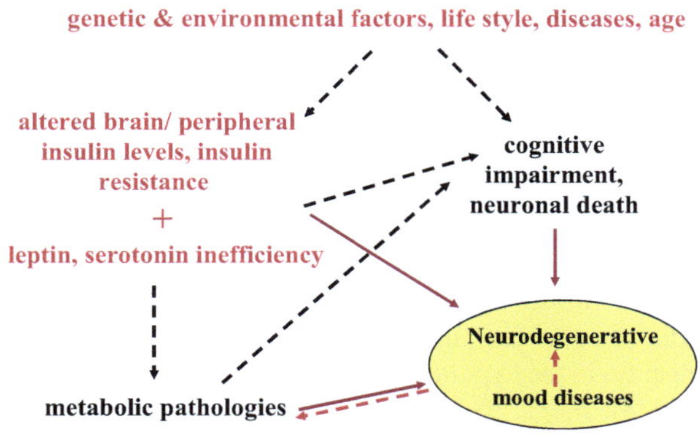


**Fig. 4** Genetic predisposition, environment, inappropriate nutrition, low mental and physical activity, illnesses and age induce cognitive impairment, alter the levels and efficiency of peripheral and central insulin, serotonin and leptin, and induce metabolic pathologies that contribute to the initiation and progress of mood and neurodegenerative diseases. In turn, mood and cognitive disorders have a negative impact on metabolic dysfunctions

Figure 4 points out some aspects of the obesity/diabetes interconnection with mood and neurodegenerative pathologies via multiple insulin, leptin and serotonin interactions.

Most of the initiating events that alter the control of metabolic homeostatic mechanisms appear to have an environmental origin. However, a genetic predisposition facilitates the initiation of these disturbances. The risk of metabolic disorders in subjects with genetic susceptibility is strongly increased by factors such as an abundant or an unbalanced energy-dense diet, combined with inappropriate physical activity and stress. Interesting information is coming from diet-induced obesity (DIO) models. Actually, high fat diets that do not affect the age-matched resistant controls induce obesity in DIO rats (Clegg et al. 2005; Irani et al. 2007).

High-fat diets are incriminated in mechanisms operating in both the periphery and the brain, disrupting energy balance, glucose homeostasis and cognitive performance (Gerozissis 2004, 2008; Greenwood and Winocur, 2001; Prada et al. 2005; Woods et al. 2004). Food rich in fats, in particular saturated, as well as diabetes, alter insulin and leptin levels and transport in the brain, modify energy and glucose metabolism, and induce cognitive impairments (Banks et al. 2006; Gerozissis et al. 1997,1999; Kayala et al. 2000; Rouch et al. 2005; Woods et al. 2003). In the central nervous system, those diets modify insulin gene expression and efficiency, leptin signaling and serotonin responsiveness to food intake (Banas et al. 2009; Koros et al. 2009; Lin et al. 2000). Given for a short period, during which body weight remains stable, high-fat diets induce in the hypothalamus a transitory suppression of the orexigenic peptides NPY and AgRP (Ziotopoulou et al. 2000) and a transitory increase of the gene expression of insulin, insulin receptors and insulin receptor substrates (Banas et al. 2009). All these modifications might represent adaptive mechanisms attempting to counteract the inefficiency of regulatory elements, such as diminished responsiveness of central serotonin, reduced release and efficiency of additional neuromodulators (Banas et al. 2009; Woods et al. 2004) and/or reduced peripheral insulin sensitivity (Cruciani-Guglielmacci et al. 2004; Griffin et al. 1999). Even if rats fed those diets for short periods develop more abundant fat tissue than the chow-fed controls (Banas et al. 2009), longer periods of fat ingestion are necessary to increase body weight gain and develop obesity (Drake et al. 2005; Ghibaudi et al. 2002; Kitraki et al. 2004; Soulis et al. 2005). A long-term, rich-in-fat regimen results in peripheral insulin resistance through an impairment of the ability of insulin to activate the IRS/PI3K pathway. A similar mechanism leading to combined insulin and leptin resistance during enriched-in-fat feeding could operate in the hypothalamus, in both animals and humans (for review, see Gerozissis 2008).

## 5 Current treatment trends

Current strategies for treating metabolic diseases and complications are not adequately effective. Most antidiabetic therapies induce many adverse effects and probably initiate a vicious cycle of problems (Fehm et al. 2006; Gerozissis 2008). Strict glucose control, for instance, promotes weight gain that appears to play a fundamental role in the pathophysiology of diabetes (Fehm et al. 2006; Looker et al. 2001). Further, although more than one quarter of diabetic patients show clinical depression, the pathology is recognized and treated in only about one third of these cases. Finally, pharmacological management of Alzheimer's disease is, at best, palliative and quasi-ineffective, inducing important adverse effects. However, early intervention in Alzheimer's disease is possible, because cognitive impairment is detectable years prior to diagnosis (Kidd 2008).

New perspectives for more efficient interventional approaches to metabolic diseases and complications have come from recent brain studies. Targeting the human brain for therapy is a particularly challenging problem because of the difficulty of delivering drugs within the brain and maintaining therapeutic local levels. Nonetheless, novel strategies are under active investigation.

One interesting approach focuses on insulin analogs, with enhanced transport from the periphery to the central nervous system and thus with increased efficiency in the brain. Interestingly, such analogs, which improve glycemic control, do not increase weight gain (Gerich 2002; Hennige et al. 2006).

Alternative techniques and routes of insulin administration are proposed. As insulin receptors are present in the olfactory bulb, the intranasal route is a practicable way to reach the brain while maintaining euglycemia (Born et al. 2002; Reger and Craft 2006; Reger et al. 2008a, 2008b). Actually, this route allows direct rapid access of the hormone to the cerebrospinal fluid compartment, and the effects observed in animals and humans correspond to the diverse actions of insulin. Intranasal administration of insulin enhances memory and mood and decreases body weight in healthy humans without causing hypoglycemia. Interestingly, a differential sensitivity of men and women to body weight- and memory-improving effects was observed in subjects receiving intranasal insulin. Women appear to be more resistant to the acute anorexigenic effect of central nervous insulin signaling and more sensitive than men to insulin's beneficial effect on hippocampus-dependent memory functions (Benedict et al. 2008; Hallschmidt et al. 2004).

Intranasal administration of centrally efficient insulin analogs, which may increase the number of molecules transported from the nasal cavity to the brain, is an interesting combination that appears to enhance insulin's efficiency on hippocampal memory processing, resulting to memory improvement (Benedict et al. 2007).

It appears that the stabilization and reversal of metabolic and related pathologies require integrative management: improvement of life style through appropriate diets, mental and physical activities, reduction of stress, and early initiation of efficient treatment of metabolic and related dysfunctions.

Associated with, or alternatively to, insulin therapy, multiple molecule combinations and approaches that simultaneously use more efficient routes of administration targeting both the periphery and the central nervous system, targeting insulin and/or its partners may be of considerable relevance for clinical applications in the treatment of metabolic and related pathologies (Banks et al. 2002; Bojanowska 2005; Cohen and Horton 2007; Kidd 2008; Lustman et al. 2008; Prodi and Obici, 2006; Gerozisssis 2008; Kidd 2008; Lustman et al. 2005; Perry and Greig, 2002; Porte et al. 2002; Pardridge et al. 2007; Reger and Craft 2006; Stockhorst et al. 2004; Wrighten et al. 2008).

Nonetheless, given the close interaction of insulin with numerous partners and the involvement of all those actors in multiple physiopathological situations, avoiding important side effects is crucial and has to be seriously considered in the design of an efficient therapy. Further investigations are necessary to determine the best strategies for long-term management of patients with these chronic pathologies.

# 6 Conclusion

A multitude of organs, networks and factors contribute to maintaining or altering insulin efficiency. Both central and peripheral insulinopenia and/or insulin resistance as well as central to peripheral insulin imbalance may contribute to the initiation and progress of metabolic, neurodegenerative and mood pathologies. Insulin's complex interaction with peptides and neurotransmitters in the brain appears to offer a largely unexplored area for novel efficient preventive and curative approaches. Together with maintaining a healthy life style through an appropriate diet, mental and physical exercise, avoidance of stress, early treatment and global management with multiple interventional combinations appear promising for a positive global outcome. Efficient treatments for metabolic and related pathologies will, we hope, benefit from further investigations into the mechanisms of insulin action in the brain.

## References


Abbott MA, Wells DG, Fallon JR (1999) The insulin receptor tyrosine kinase substrate p58/53 and the insulin receptor are components of CNS synapses. J Neurosci 19:7300–7308

Accili D (2004) Lilly lecture 2003: the struggle for mastery in insulin action: from triumvirate to republic. Diabetes 53:1633–1642

Ahima RS, Saper CB, Flier JS, Elmquist JK (2000) Leptin regulation of neuroendocrine systems. Front Neuroendocrinol 21:263–307

Banas SM, Rouch C, Kassis N, Markaki EM, Gerozissis K (2009) A dietary fat excess alters metabolic and neuroendocrine responses before the onset of metabolic diseases. Cell Mol Neurobiol 29:157–168

Banks WA (2004) The source of cerebral insulin. Eur J Pharmacol 490:5–12

Banks WA, Lebel CR (2002) Strategies for the delivery of leptin to the CNS. J Drug Target 10:297–308

Banks WA, Farr SA, Morley JE (2006) The effects of high fat diets on the blood-brain barrier transport of leptin: failure or adaptation? Physiol Behav 88:244–248

Baskin DG, Porte D, Jr., Guest K, Dorsa DM (1983) Regional concentrations of insulin in the rat brain. Endocrinology 112:898–903

Baskin DG, Stein LJ, Ikeda H, Woods SC, Figlewicz DP, Porte D, Jr., Greenwood MR, Dorsa DM (1985) Genetically obese Zucker rats have abnormally low brain insulin content. Life Sci 36:627–633

Baura GD, Foster DM, Porte D, Jr., Kahn SE, Bergman RN, Cobelli C, Schwartz MW (1993) Saturable transport of insulin from plasma into the central nervous system of dogs in vivo. A mechanism for regulated insulin delivery to the brain. J Clin Invest 92:1824–1830

Benedict C, Hallschmid M, Schultes B, Born J, Kern W (2007) Intranasal insulin to improve memory function in humans. Neuroendocrinology 86:136–142

Benedict C, Kern W, Schultes B, Born J, Hallschmid M (2008) Differential sensitivity of men and women to anorexigenic and memory-improving effects of intranasal insulin. J Clin Endocrinol Metab 93:1339–1344

Benoit SC, Clegg DJ, Barrera JG, Seeley RJ, Woods SC (2003) Learned meal initiation attenuates the anorexic effects of the melanocortin agonist MTII. Diabetes 52:2684–2688

Benomar Y, Roy AF, Aubourg A, Djiane J, Taouis M (2005) Cross down-regulation of leptin and insulin receptor expression and signalling in a human neuronal cell line. Biochem J 388:929–939

Benomar Y, Berthou F, Vacher CM, Bailleux V, Gertler A, Djiane J, Taouis M (2009) Leptin but not ciliary neurotrophic factor (CNTF) induces phosphotyrosine phosphatase-1B expression in human neuronal cells (SH-SY5Y): putative explanation of CNTF efficacy in leptin-resistant state. Endocrinology 150:1182–1191

Berthoud HR, Morrison C (2008) The brain, appetite, and obesity. Annu Rev Psychol 59:55–92

Bingham EM, Hopkins D, Smith D, Pernet A, Hallett W, Reed L, Marsden PK, Amiel SA (2002) The role of insulin in human brain glucose metabolism: an 18fluoro-deoxyglucose positron emission tomography study. Diabetes 51:3384–3390

Blundell JE (1984) Serotonin and appetite. Neuropharmacology 23:1537–1551

Bojanowska E (2005) Physiology and pathophysiology of glucagon-like peptide-1 (GLP-1): the role of GLP-1 in the pathogenesis of diabetes mellitus, obesity, and stress. Med Sci Monit 11: RA271–278

Born J, Lange T, Kern W, McGregor GP, Bickel U, Fehm HL (2002) Sniffing neuropeptides: a transnasal approach to the human brain. Nature Neurosci 5:514–516

Bruning JC, Gautam D, Burks DJ, Gillette J, Schubert M, Orban PC, Klein R, Krone W, Muller-Wieland D, Kahn CR (2000) Role of brain insulin receptor in control of body weight and reproduction. Science 289:2122–2125

Burks DJ, Font de Mora J, Schubert M, Withers DJ, Myers MG, Towery HH, Altamuro SL, Flint CL, White MF (2000) IRS-2 pathways integrate female reproduction and energy homeostasis. Nature 407:377–382

Calapai G, Corica F, Corsonello A, Sautebin L, Di Rosa M, Campo GM, Buemi M, Mauro VN, Caputi AP (1999) Leptin increases serotonin turnover by inhibition of brain nitric oxide synthesis. J Clin Invest 104:975–982

Carvalheira JB, Siloto RM, Ignacchitti I, Brenelli SL, Carvalho CR, Leite A, Velloso LA, Gontijo JA, Saad MJ (2001) Insulin modulates leptin-induced STAT3 activation in rat hypothalamus. FEBS Lett 500:119–124

Clarke DW, Mudd L, Boyd FT, Jr., Fields M, Raizada MK (1986) Insulin is released from rat brain neuronal cells in culture. J Neurochem 47:831–836

Clegg DJ, Benoit SC, Reed JA, Woods SC, Dunn-Meynell A, Levin BE (2005) Reduced anorexic effects of insulin in obesity-prone rats fed a moderate-fat diet. Am J Physiol Regul Integr Comp Physiol 288:R981–R986

Cohen A, Horton ES (2007) Progress in the treatment of type 2 diabetes: new pharmacologic approaches to improve glycemic control. Curr Med Res Opin 23:905–917

Cohen E, Dillin A (2008) The insulin paradox: aging, proteotoxicity and neurodegeneration. Nature Rev Neurosci 9:759–767

Craft S (2007) Insulin resistance and Alzheimer's disease pathogenesis: potential mechanisms and implications for treatment. Curr Alzheimer Res 4:147–152

Craft S, Peskind E, Schwartz MW, Schellenberg GD, Raskind M, Porte D, Jr. (1998) Cerebrospinal fluid and plasma insulin levels in Alzheimer's disease: relationship to severity of dementia and apolipoprotein E genotype. Neurology 50:164–168

Craft S, Asthana S, Newcomer JW, Wilkinson CW, Matos IT, Baker LD, Cherrier M, Lofgreen C, Latendresse S, Petrova A, Plymate S, Raskind M, Grimwood K, Veith RC (1999) Enhancement of memory in Alzheimer disease with insulin and somatostatin, but not glucose. Arch Gen Psychiat 56:1135–1140

Craft S, Asthana S, Cook DG, Baker LD, Cherrier M, Purganan K, Wait C, Petrova A, Latendresse S, Watson GS, Newcomer JW, Schellenberg GD, Krohn AJ (2003) Insulin dose-response effects on memory and plasma amyloid precursor protein in Alzheimer's disease: interactions with apolipoprotein E genotype. Psychoneuroendocrinology 28:809–822

Cruciani-Guglielmacci C, Vincent-Lamon M, Rouch C, Orosco M, Ktorza A, Magnan C (2005) Early changes in insulin secretion and action induced by high-fat diet are related to a decreased sympathetic tone. Am J Physiol Endocrinol Metab 288:E148–E154

de la Monte SM, Wands JR (2005) Review of insulin and insulin-like growth factor expression, signaling, and malfunction in the central nervous system: relevance to Alzheimer's disease. J Alzheimers Dis 7:45–61

Devaskar SU, Giddings SJ, Rajakumar PA, Carnaghi LR, Menon RK, Zahm DS (1994) Insulin gene expression and insulin synthesis in mammalian neuronal cells. J Biol Chem 269:8445–8454

Dore S, Kar S, Rowe W, Quirion R (1997) Distribution and levels of [125I]IGF-I, [125I]IGF-II and [125I]insulin receptor binding sites in the hippocampus of aged memory-unimpaired and –impaired rats. Neuroscience 80:1033–1040

Doyle P, Cusin I, Rohner-Jeanrenaud F, Jeanrenaud B (1995) Four-day hyperinsulinemia in euglycemic conditions alters local cerebral glucose utilization in specific brain nuclei of freely moving rats. Brain Res 684:47–55

Drake AJ, Livingstone DE, Andrew R, Seckl JR, Morton NM, Walker BR (2005) Reduced adipose glucocorticoid reactivation and increased hepatic glucocorticoid clearance as an early adaptation to high-fat feeding in Wistar rats. Endocrinology 146:913–919

Drazen DL, Wortman MD, Seeley RJ, Woods SC (2005) Neuropeptide Y prepares rats for scheduled feeding. Am J Physiol Regul Integr Comp Physiol 288:R1606–R1611

Drazen DL, Vahl TP, D'Alessio DA, Seeley RJ, Woods SC (2006) Effects of a fixed meal pattern on ghrelin secretion: evidence for a learned response independent of nutrient status. Endocrinology 147:23–30

Elias PK, Elias MF, D'Agostino RB, Cupples LA, Wilson PW, Silbershatz H, Wolf PA (1997) NIDDM and blood pressure as risk factors for poor cognitive performance. The Framingham Study. Diabetes Care 20:1388–1395

Fehm HL, Kern W, Peters A (2006) The selfish brain: competition for energy resources. Prog Brain Res 153:129–140

Fetissov SO, Gerozissis K, Orosco M, Nicolaidis S (2000) Synergistic effect of arcuate and raphe nuclei graft to alleviate insulinemia and obesity in Zucker rats. Acta Diabetol 37:65–70

Feurte S, Nicolaidis S, Gerozissis K (2000) Is the early increase in leptinemia one of the anorectic signals induced by an essential amino acid-deficient diet in the rat? Endocrinology 141:3916–3919

Fewlass DC, Noboa K, Pi-Sunyer FX, Johnston JM, Yan SD, Tezapsidis N (2004) Obesity-related leptin regulates Alzheimer's Abeta. Faseb J 18:1870–1878

Figlewicz DP, Szot P, Israel PA, Payne C, Dorsa DM (1993) Insulin reduces norepinephrine transporter mRNA in vivo in rat locus coeruleus. Brain Res 602:161–164

Finn PD, Cunningham MJ, Rickard DG, Clifton DK, Steiner RA (2001) Serotonergic neurons are targets for leptin in the monkey. J Clin Endocrinol Metab 86:422–426

Frolich L, Blum-Degen D, Bernstein HG, Engelsberger S, Humrich J, Laufer S, Muschner D, Thalheimer A, Turk A, Hoyer S, Zochling R, Boissl KW, Jellinger K, Riederer P (1998) Brain insulin and insulin receptors in aging and sporadic Alzheimer's disease. J Neural Transm 105:423–438

Gerich JE (2002) Novel insulins: expanding options in diabetes management. Am J Med 113:308–316

Gerozissis K (2003) Brain insulin: regulation, mechanisms of action and functions. Cell Mol Neurobiol 23:1-25

Gerozissis K (2004) Brain insulin and feeding: a bi-directional communication. Eur J Pharmacol 490:59–70

Gerozissis K (2008) Brain insulin, energy and glucose homeostasis; genes, environment and metabolic pathologies. Eur J Pharmacol 585:38–49

Gerozissis K, Banas SM, Markaki EM, Apazoglou K, Rouch C (2006) The serotonin connection in the insulin-leptin neuroregulation of food intake. 36$^{nd}$ SFN meeting, Atlanta, GA, USA

Gerozissis K, Markaki EM, Rouch C, Banas SM, Berthou F, Bailbe D, Taouis M (2008) Alteration of both hypothalamic insulin and serotonin in a type 2 diabetes model, Diabetologia 51 (Suppl1):A774

Gerozissis K, Orosco M, Rouch C, Nicolaidis S (1993) Basal and hyperinsulinemia-induced immunoreactive hypothalamic insulin changes in lean and genetically obese Zucker rats revealed by microdialysis. Brain Res 611:258–263

Gerozissis K, Orosco M, Rouch C, Nicolaidis S (1997) Insulin responses to a fat meal in hypothalamic microdialysates and plasma. Physiol Behav 62:767–772

Gerozissis K, Rouch C, Nicolaidis S, Orosco M (1999) Brain insulin response to feeding in the rat is both macronutrient and area specific. Physiol Behav 66:271–275

Gerozissis K, Rouch C, Lemierre S, Nicolaidis S, Orosco M (2001) A potential role of central insulin in learning and memory related to feeding. Cell Mol Neurobiol 21:389–401

Gerozissis K, Banas SM, Berthou F, Bailleux V, Rouch C, Taouis M (2007) Brain insulin signaling regulation by leptin and serotonin. Potential impact on the development of metabolic diseases. Diabetologia, 50 (Suppl 1):A0015

Ghibaudi L, Cook J, Farley C, van Heek M, Hwa JJ (2002) Fat intake affects adiposity, comorbidity factors, and energy metabolism of sprague-dawley rats. Obes Res 10:956–963

Gispen WH, Biessels GJ (2000) Cognition and synaptic plasticity in diabetes mellitus. Trends Neurosci 23:542–549

Greco SJ, Sarkar S, Johnston JM, Zhu X, Su B, Casadesus G, Ashford JW, Smith MA, Tezapsidis N (2008) Leptin reduces Alzheimer's disease-related tau phosphorylation in neuronal cells. Biochem Biophys Res Commun 376:536–541

Greenwood CE, Winocur G (2001) Glucose treatment reduces memory deficits in young adult rats fed high-fat diets. Neurobiol Learn Mem 75:179–189

Griffin ME, Marcucci MJ, Cline GW, Bell K, Barucci N, Lee D, Goodyear LJ, Kraegen EW, White MF, Shulman GI (1999) Free fatty acid-induced insulin resistance is associated with activation of protein kinase C theta and alterations in the insulin signaling cascade. Diabetes 48:1270–1274

Grunblatt E, Salkovic-Petrisic M, Osmanovic J, Riederer P, Hoyer S (2007) Brain insulin system dysfunction in streptozotocin intracerebroventricularly treated rats generates hyperphosphorylated tau protein. J Neurochem 101:757–770

Guillod-Maximin E, Roy AF, Vacher CM, Aubourg A, Bailleux V, Lorsignol A, Penicaud L, Parquet M, Taouis M (2009) Adiponectin receptors are expressed in hypothalamus and colocalized with proopiomelanocortin and neuropeptide Y in rodent arcuate neurons. J Endocrinol 200:93–105

Hallschmid M, Benedict C, Schultes B, Fehm HL, Born J, Kern W (2004) Intranasal insulin reduces body fat in men but not in women. Diabetes 53:3024–3029

Harvey J, Ashford ML (2003) Leptin in the CNS: much more than a satiety signal. Neuropharmacology 44:845–854

Havrankova J, Roth J, Brownstein M (1978a) Insulin receptors are widely distributed in the central nervous system of the rat. Nature 272:827–829

Havrankova J, Schmechel D, Roth J, Brownstein M (1978b) Identification of insulin in rat brain. Proc Natl Acad Sci USA 75:5737–5741

Hay-Schmidt A, Helboe L, Larsen PJ (2001) Leptin receptor immunoreactivity is present in ascending serotonergic and catecholaminergic neurons of the rat. Neuroendocrinology 73:215–226

Heisler LK, Cowley MA, Tecott LH, Fan W, Low MJ, Smart JL, Rubinstein M, Tatro JB, Marcus JN, Holstege H, Lee CE, Cone RD, Elmquist JK (2002) Activation of central melanocortin pathways by fenfluramine. Science 297:609–611

Helkala EL, Niskanen L, Viinamaki H, Partanen J, Uusitupa M (1995) Short-term and long-term memory in elderly patients with NIDDM. Diabetes Care 18:681–685

Hennige AM, Sartorius T, Tschritter O, Preissl H, Fritsche A, Ruth P, Haring HU (2006) Tissue selectivity of insulin detemir action in vivo. Diabetologia 49:1274–1282

Hoyer S (1998) Is sporadic Alzheimer disease the brain type of non-insulin dependent diabetes mellitus? A challenging hypothesis. J Neural Transm 105:415–422

Hrytsenko O, Wright JR, Jr., Morrison CM, Pohajdak B (2007) Insulin expression in the brain and pituitary cells of tilapia (Oreochromis niloticus). Brain Res 1135:31–40

Hsiung SC, Adlersberg M, Arango V, Mann JJ, Tamir H, Liu KP (2003) Attenuated 5-HT1A receptor signaling in brains of suicide victims: involvement of adenylyl cyclase, phosphatidylinositol 3-kinase, Akt and mitogen-activated protein kinase. J Neurochem 87:182–194

Huang CC, Lee CC, Hsu KS (2004) An investigation into signal transduction mechanisms involved in insulin-induced long-term depression in the CA1 region of the hippocampus. J Neurochem 89:217–231

Inoue H, Ogawa W, Asakawa A, Okamoto Y, Nishizawa A, Matsumoto M, Teshigawara K, Matsuki Y, Watanabe E, Hiramatsu R, Notohara K, Katayose K, Okamura H, Kahn CR, Noda T, Takeda K, Akira S, Inui A, Kasuga M (2006) Role of hepatic STAT3 in brain-insulin action on hepatic glucose production. Cell Metab 3:267–275

Irani BG, Dunn-Meynell AA, Levin BE (2007) Altered hypothalamic leptin, insulin, and melanocortin binding associated with moderate-fat diet and predisposition to obesity. Endocrinology 148:310–316

Israel PA, Park CR, Schwartz MW, Green PK, Sipols AJ, Woods SC, Porte D, Jr., Figlewicz DP (1993) Effect of diet-induced obesity and experimental hyperinsulinemia on insulin uptake into CSF of the rat. Brain Res Bull 30:571–575

Jolivalt CG, Lee CA, Beiswenger KK, Smith JL, Orlov M, Torrance MA, Masliah E (2008) Defective insulin signaling pathway and increased glycogen synthase kinase-3 activity in the

brain of diabetic mice: parallels with Alzheimer's disease and correction by insulin. J Neurosci Res 86:3265–3274
Kaiyala KJ, Prigeon RL, Kahn SE, Woods SC, Schwartz MW (2000) Obesity induced by a high-fat diet is associated with reduced brain insulin transport in dogs. Diabetes 49:1525–1533
Kalia M (2005) Neurobiological basis of depression: an update. Metabolism 54:24–27
Kalra SP, Dube MG, Pu S, Xu B, Horvath TL, Kalra PS (1999) Interacting appetite-regulating pathways in the hypothalamic regulation of body weight. Endocr Rev 20:68–100
Kern W, Benedict C, Schultes B, Plohr F, Moser A, Born J, Fehm HL, Hallschmid M (2006) Low cerebrospinal fluid insulin levels in obese humans. Diabetologia 49:2790–2792
Kidd PM (2008) Alzheimer's disease, amnestic mild cognitive impairment, and age-associated memory impairment: current understanding and progress toward integrative prevention. Altern Med Rev 13:85–115
Kitraki E, Soulis G, Gerozissis K (2004) Impaired neuroendocrine response to stress following a short-term fat-enriched diet. Neuroendocrinology 79:338–345
Kodl CT, Seaquist ER (2008) Cognitive dysfunction and diabetes mellitus. Endocr Rev 29:494–511
Konner AC, Janoschek R, Plum L, Jordan SD, Rother E, Ma X, Xu C, Enriori P, Hampel B, Barsh GS, Kahn CR, Cowley MA, Ashcroft FM, Bruning JC (2007) Insulin action in AgRP-expressing neurons is required for suppression of hepatic glucose production. Cell Metab 5:438–449
Kopf SR, Baratti CM (1999) Effects of posttraining administration of insulin on retention of a habituation response in mice: participation of a central cholinergic mechanism. Neurobiol Learn Mem 71:50–61
Koros C, Boukouvalas G, Gerozissis K, Kitraki E (2009) Fat diet affects leptin receptor levels in the rat cerebellum. Nutrition 25:85–87
Lam DD, Farooqi IS, Heisler LK (2007) Melanocortin receptors as targets in the treatment of obesity. Curr Top Med Chem 7:1085–1097
Lam DD, Przydzial MJ, Ridley SH, Yeo GS, Rochford JJ, O'Rahilly S, Heisler LK (2008) Serotonin 5-HT2C receptor agonist promotes hypophagia via downstream activation of melanocortin 4 receptors. Endocrinology 149:1323–1328
Lam TK, Pocai A, Gutierrez-Juarez R, Obici S, Bryan J, Aguilar-Bryan L, Schwartz GJ, Rossetti L (2005) Hypothalamic sensing of circulating fatty acids is required for glucose homeostasis. Nat Med 11:320–327
Lemierre S, Rouch C, Nicolaidis S, Orosco M (1998) Combined effect of obesity and aging on feeding-induced monoamine release in the rostromedial hypothalamus of the Zucker rat. Int J Obes Relat Metab Disord 22:993–999
Lezoualc'h F (2007) 5-HT4 receptor and Alzheimer's disease: the amyloid connection. Exp Neurol 205:325–329
Li L, Holscher C (2007) Common pathological processes in Alzheimer disease and type 2 diabetes: a review. Brain Res Rev 56:384–402
Lin S, Storlien LH, Huang XF (2000) Leptin receptor, NPY, POMC mRNA expression in the diet-induced obese mouse brain. Brain Res 875:89–95
Lin X, Taguchi A, Park S, Kushner JA, Li F, Li Y, White MF (2004) Dysregulation of insulin receptor substrate 2 in β cells and brain causes obesity and diabetes. J Clin Invest 114:908–916
Looker HC, Knowler WC, Hanson RL (2001) Changes in BMI and weight before and after the development of type 2 diabetes. Diabetes Care 24:1917–1922
Lu XY (2007) The leptin hypothesis of depression: a potential link between mood disorders and obesity? Curr Opin Pharmacol 7:648–652
Luchsinger JA, Tang MX, Shea S, Mayeux R (2004) Hyperinsulinemia and risk of Alzheimer disease. Neurology 63:1187–1192
Lustman PJ, Clouse RE (2005) Depression in diabetic patients: the relationship between mood and glycemic control. J Diabetes Complications 19:113–122

Lustman PJ, Penckofer SM, Clouse RE (2008) Recent advances in understanding depression in adults with diabetes. Curr Psychiatry Rep 10:495–502
Madadi G, Dalvi PS, Belsham DD (2008) Regulation of brain insulin mRNA by glucose and glucagon-like peptide 1. Biochem Biophys Res Commun 376:694–699
Mattson MP, Maudsley S, Martin B (2004a) BDNF and 5-HT: a dynamic duo in age-related neuronal plasticity and neurodegenerative disorders. Trends Neurosci 27:589–594
Mattson MP, Maudsley S, Martin B (2004b) A neural signaling triumvirate that influences ageing and age-related disease: insulin/IGF-1, BDNF and serotonin. Ageing Res Rev 3:445–464
McIntyre RS, Soczynska JK, Konarski JZ, Woldeyohannes HO, Law CW, Miranda A, Fulgosi D, Kennedy SH (2007) Should depressive syndromes be reclassified as "Metabolic Syndrome Type II"? Ann Clin Psychiat 19:257–264
McNay EC (2007) Insulin and ghrelin: peripheral hormones modulating memory and hippocampal function. Curr Opin Pharmacol 7:628–632
Messier C, Gagnon M (1996) Glucose regulation and cognitive functions: relation to Alzheimer's disease and diabetes. Behav Brain Res 75:1–11
Messier C, Gagnon M, Knott V (1997) Effect of glucose and peripheral glucose regulation on memory in the elderly. Neurobiol Aging 18:297–304
Morrison CD (2009) Leptin signaling in brain: A link between nutrition and cognition? Biochim Biophys Acta 1792:401–408
Morton GJ, Cummings DE, Baskin DG, Barsh GS, Schwartz MW (2006) Central nervous system control of food intake and body weight. Nature 443:289–295
Nakazato M, Murakami N, Date Y, Kojima M, Matsuo H, Kangawa K, Matsukura S (2001) A role for ghrelin in the central regulation of feeding. Nature 409:194–198
Nelson RL, Guo Z, Halagappa VM, Pearson M, Gray AJ, Matsuoka Y, Brown M, Martin B, Iyun T, Maudsley S, Clark RF, Mattson MP (2007) Prophylactic treatment with paroxetine ameliorates behavioral deficits and retards the development of amyloid and tau pathologies in 3xTgAD mice. Exp Neurol 205:166–176
Niswender KD, Schwartz MW (2003) Insulin and leptin revisited: adiposity signals with overlapping physiological and intracellular signaling capabilities. Front Neuroendocrinol 24:1–10
Niswender KD, Gallis B, Blevins JE, Corson MA, Schwartz MW, Baskin DG (2003a) Immunocytochemical detection of phosphatidylinositol 3-kinase activation by insulin and leptin. J Histochem Cytochem 51:275–283
Niswender KD, Morrison CD, Clegg DJ, Olson R, Baskin DG, Myers MG, Jr., Seeley RJ, Schwartz MW (2003b) Insulin activation of phosphatidylinositol 3-kinase in the hypothalamic arcuate nucleus: a key mediator of insulin-induced anorexia. Diabetes 52:227–231
Niswender KD, Baskin DG, Schwartz MW (2004) Insulin and its evolving partnership with leptin in the hypothalamic control of energy homeostasis. Trends Endocrinol Metab 15:362–369
Obici S, Feng Z, Karkanias G, Baskin DG, Rossetti L (2002a) Decreasing hypothalamic insulin receptors causes hyperphagia and insulin resistance in rats. Nature Neurosci 5:566–572
Obici S, Zhang BB, Karkanias G, Rossetti L (2002b) Hypothalamic insulin signaling is required for inhibition of glucose production. Nature Med 8:1376–1382
Okamoto H, Nakae J, Kitamura T, Park BC, Dragatsis I, Accili D (2004) Transgenic rescue of insulin receptor-deficient mice. J Clin Invest 114:214–223
Okamoto H, Obici S, Accili D, Rossetti L (2005) Restoration of liver insulin signaling in Insr knockout mice fails to normalize hepatic insulin action. J Clin Invest 115:1314–1322
Okamura F, Tashiro A, Utumi A, Imai T, Suchi T, Tamura D, Sato Y, Suzuki S, Hongo M (2000) Insulin resistance in patients with depression and its changes during the clinical course of depression: minimal model analysis. Metabolism 49:1255–1260
O'Malley D, Harvey J (2007) MAPK-dependent actin cytoskeletal reorganization underlies BK channel activation by insulin. Eur J Neurosci 25:673–682

Orosco M, Gerozissis K (2001) Macronutrient-induced cascade of events leading to parallel changes in hypothalamic serotonin and insulin. Neurosci Biobehav Rev 25:167–174

Orosco M, Gerozissis K, Rouch C, Meile MJ, Nicolaidis S (1995) Hypothalamic monoamines and insulin in relation to feeding in the genetically obese Zucker rat as revealed by microdialysis. Obes Res 3 Suppl 5:655S–665S

Orosco M, Rouch C, Gerozissis K (2000) Activation of hypothalamic insulin by serotonin is the primary event of the insulin-serotonin interaction involved in the control of feeding. Brain Res 872:64–70

Pardridge WM (2007) Drug targeting to the brain. Pharm Res 24:1733–1744

Park CR (2001) Cognitive effects of insulin in the central nervous system. Neurosci Biobehav Rev 25:311–323

Park CR, Seeley RJ, Craft S, Woods SC (2000) Intracerebroventricular insulin enhances memory in a passive-avoidance task. Physiol Behav 68:509–514

Paulus K, Schulz C, Lehnert H (2005) Central nervous effects of leptin and insulin on hippocampal leptin and insulin receptor expression following a learning task in Wistar rats. Neuropsychobiology 51:100–106

Perrin C, Knauf C, Burcelin R (2004) Intracerebroventricular infusion of glucose, insulin, and the adenosine monophosphate-activated kinase activator, 5-aminoimidazole-4-carboxamide-1-beta-D-ribofuranoside, controls muscle glycogen synthesis. Endocrinology 145:4025–4033

Perry T, Greig NH (2002) The glucagon-like peptides: a new genre in therapeutic targets for intervention in Alzheimer's disease. J Alzheimers Dis 4:487–496

Plum L, Schubert M, Bruning JC (2005) The role of insulin receptor signaling in the brain. Trends Endocrinol Metab 16:59–65

Plum L, Belgardt BF, Bruning JC (2006) Central insulin action in energy and glucose homeostasis. J Clin Invest 116:1761–1766

Pocai A, Lam TK, Gutierrez-Juarez R, Obici S, Schwartz GJ, Bryan J, Aguilar-Bryan L, Rossetti L (2005) Hypothalamic K(ATP) channels control hepatic glucose production. Nature 434:1026–1031

Porte D, Jr., Baskin DG, Schwartz MW (2005) Insulin signaling in the central nervous system: a critical role in metabolic homeostasis and disease from C. elegans to humans. Diabetes 54:1264–1276

Prada PO, Zecchin HG, Gasparetti AL, Torsoni MA, Ueno M, Hirata AE, Corezola do Amaral ME, Hoer NF, Boschero AC, Saad MJ (2005) Western diet modulates insulin signaling, c-Jun N-terminal kinase activity, and insulin receptor substrate-1ser307 phosphorylation in a tissue-specific fashion. Endocrinology 146:1576–1587

Prodi E, Obici S (2006) Minireview: the brain as a molecular target for diabetic therapy. Endocrinology 147:2664–2669

Reagan LP (2007) Insulin signaling effects on memory and mood. Curr Opin Pharmacol 7:633–637

Reger MA, Craft S (2006) Intranasal insulin administration: a method for dissociating central and peripheral effects of insulin. Drugs Today (Barc) 42:729–739

Reger MA, Watson GS, Green PS, Baker LD, Cholerton B, Fishel MA, Plymate SR, Cherrier MM, Schellenberg GD, Frey WH, 2nd, Craft S (2008a) Intranasal insulin administration dose-dependently modulates verbal memory and plasma amyloid-β in memory-impaired older adults. J Alzheimers Dis 13:323–331

Reger MA, Watson GS, Green PS, Wilkinson CW, Baker LD, Cholerton B, Fishel MA, Plymate SR, Breitner JC, DeGroodt W, Mehta P, Craft S (2008b) Intranasal insulin improves cognition and modulates β-amyloid in early AD. Neurology 70:440–448

Rother E, Konner AC, Bruning JC (2008) Neurocircuits integrating hormone and nutrient signaling in control of glucose metabolism. Am J Physiol Endocrinol Metab 294:E810–E816

Rouch C, Meile MJ, Gerozissis K (2005) Persisting neural and endocrine modifications induced by a single fat meal. Cell Mol Neurobiol 25:995–1008

Sanchez V, Camarero J, Esteban B, Peter MJ, Green AR, Colado MI (2001) The mechanisms involved in the long-lasting neuroprotective effect of fluoxetine against MDMA ('ecstasy')-induced degeneration of 5-HT nerve endings in rat brain. Br J Pharmacol 134:46–57

Santos MS, Pereira EM, Carvaho AP (1999) Stimulation of immunoreactive insulin release by glucose in rat brain synaptosomes. Neurochem Res 24:33–36

Schechter R, Sadiq HF, Devaskar SU (1990) Insulin and insulin mRNA are detected in neuronal cell cultures maintained in an insulin-free/serum-free medium. J Histochem Cytochem 38:829–836

Schechter R, Whitmire J, Wheet GS, Beju D, Jackson KW, Harlow R, Gavin JR, 3rd (1994) Immunohistochemical and in situ hybridization study of an insulin-like substance in fetal neuron cell cultures. Brain Res 636:9–27

Schechter R, Beju D, Gaffney T, Schaefer F, Whetsell L (1996) Preproinsulin I and II mRNAs and insulin electron microscopic immunoreaction are present within the rat fetal nervous system. Brain Res 736:16–27

Schechter R, Beju D, Miller KE (2005) The effect of insulin deficiency on tau and neurofilament in the insulin knockout mouse. Biochem Biophys Res Commun 334:979–986

Schubert M, Gautam D, Surjo D, Ueki K, Baudler S, Schubert D, Kondo T, Alber J, Galldiks N, Kustermann E, Arndt S, Jacobs AH, Krone W, Kahn CR, Bruning JC (2004) Role for neuronal insulin resistance in neurodegenerative diseases. Proc Natl Acad Sci USA 101:3100–3105

Schulingkamp RJ, Pagano TC, Hung D, Raffa RB (2000) Insulin receptors and insulin action in the brain: review and clinical implications. Neurosci Biobehav Rev 24:855–872

Schwartz MW (2001) Progress in the search for neuronal mechanisms coupling type 2 diabetes to obesity. J Clin Invest 108:963–964

Schwartz MW, Porte D, Jr. (2005) Diabetes, obesity, and the brain. Science 307:375–379

Schwartz MW, Figlewicz DP, Baskin DG, Woods SC, Porte D, Jr. (1992) Insulin in the brain: a hormonal regulator of energy balance. Endocr Rev 13:387–414

Shanley LJ, Irving AJ, Harvey J (2001) Leptin enhances NMDA receptor function and modulates hippocampal synaptic plasticity. J Neurosci 21:RC186

Shanley LJ, O'Malley D, Irving AJ, Ashford ML, Harvey J (2002) Leptin inhibits epileptiform-like activity in rat hippocampal neurones via PI 3-kinase-driven activation of BK channels. J Physiol 545:933–944

Shioda S, Funahashi H, Nakajo S, Yada T, Maruta O, Nakai Y (1998) Immunohistochemical localization of leptin receptor in the rat brain. Neurosci Lett 243:41–44

Signore AP, Zhang F, Weng Z, Gao Y, Chen J (2008) Leptin neuroprotection in the CNS: mechanisms and therapeutic potentials. J Neurochem 106:1977–1990

Sipols AJ, Baskin DG, Schwartz MW (1995) Effect of intracerebroventricular insulin infusion on diabetic hyperphagia and hypothalamic neuropeptide gene expression. Diabetes 44:147–151

Soulis G, Kitraki E, Gerozissis K (2005) Early neuroendocrine alterations in female rats following a diet moderately enriched in fat. Cell Mol Neurobiol 25:869–880

Steen E, Terry BM, Rivera EJ, Cannon JL, Neely TR, Tavares R, Xu XJ, Wands JR, de la Monte SM (2005) Impaired insulin and insulin-like growth factor expression and signaling mechanisms in Alzheimer's disease–is this type 3 diabetes? J Alzheimers Dis 7:63–80

Stein LJ, Dorsa DM, Baskin DG, Figlewicz DP, Porte D, Jr., Woods SC (1987) Reduced effect of experimental peripheral hyperinsulinemia to elevate cerebrospinal fluid insulin concentrations of obese Zucker rats. Endocrinology 121:1611–1615

Stewart SM, Rao U, White P (2005) Depression and diabetes in children and adolescents. Curr Opin Pediatr 17:626–631

Stockhorst U, de Fries D, Steingrueber HJ, Scherbaum WA (2004) Insulin and the CNS: effects on food intake, memory, and endocrine parameters and the role of intranasal insulin administration in humans. Physiol Behav 83:47–54

Strachan MW, Deary IJ, Ewing FM, Frier BM (1997) Is type II diabetes associated with an increased risk of cognitive dysfunction? A critical review of published studies. Diabetes Care 20:438–445

Tecott LH, Sun LM, Akana SF, Strack AM, Lowenstein DH, Dallman MF, Julius D (1995) Eating disorder and epilepsy in mice lacking 5-HT2c serotonin receptors. Nature 374:542–546
Telles MM, Guimaraes RB, Ribeiro EB (2003) Effect of leptin on the acute feeding-induced hypothalamic serotonergic stimulation in normal rats. Regul Pept 115:11–18
Tong L, Thornton PL, Balazs R, Cotman CW (2001) Beta -amyloid-(1-42) impairs activity-dependent cAMP-response element-binding protein signaling in neurons at concentrations in which cell survival Is not compromised. J Biol Chem 276:17301–17306
Tracy AL, Jarrard LE, Davidson TL (2001) The hippocampus and motivation revisited: appetite and activity. Behav Brain Res 127:13–23
Tschritter O, Preissl H, Hennige AM, Stumvoll M, Porubska K, Frost R, Marx H, Klosel B, Lutzenberger W, Birbaumer N, Haring HU, Fritsche A (2006) The cerebrocortical response to hyperinsulinemia is reduced in overweight humans: a magnetoencephalographic study. Proc Natl Acad Sci USA 103:12103–12108
Unger JW, Livingston JN, Moss AM (1991) Insulin receptors in the central nervous system: localization, signalling mechanisms and functional aspects. Prog Neurobiol 36:343–362
van der Heide LP, Kamal A, Artola A, Gispen WH, Ramakers GM (2005) Insulin modulates hippocampal activity-dependent synaptic plasticity in a N-methyl-d-aspartate receptor and phosphatidyl-inositol-3-kinase-dependent manner. J Neurochem 94:1158–1166
Watson GS, Craft S (2004) Modulation of memory by insulin and glucose: neuropsychological observations in Alzheimer's disease. Eur J Pharmacol 490:97–113
White CL, Whittington A, Barnes MJ, Wang Z, Bray GA, Morrison CD (2009) HF diets increase hypothalamic PTP1B and induce leptin resistance through both leptin-dependent and -independent mechanisms. Am J Physiol Endocrinol Metab 296:E291–E299
Winocur G, Gagnon S (1998) Glucose treatment attenuates spatial learning and memory deficits of aged rats on tests of hippocampal function. Neurobiol Aging 19:233–241
Woods SC, Lotter EC, McKay LD, Porte D, Jr. (1979) Chronic intracerebroventricular infusion of insulin reduces food intake and body weight of baboons. Nature 282:503–505
Woods SC, Schwartz MW, Baskin DG, Seeley RJ (2000) Food intake and the regulation of body weight. Annu Rev Psychol 51:255–277
Woods SC, Seeley RJ, Baskin DG, Schwartz MW (2003) Insulin and the blood-brain barrier. Curr Pharm Des 9:795–800
Woods SC, D'Alessio DA, Tso P, Rushing PA, Clegg DJ, Benoit SC, Gotoh K, Liu M, Seeley RJ (2004) Consumption of a high-fat diet alters the homeostatic regulation of energy balance. Physiol Behav 83:573–578
Wozniak M, Rydzewski B, Baker SP, Raizada MK (1993) The cellular and physiological actions of insulin in the central nervous system. Neurochem Int 22:1–10
Wrighten SA, Piroli GG, Grillo CA, Reagan LP (2009) A look inside the diabetic brain: Contributors to diabetes-induced brain aging. Biochim Biophys Acta 1799:444–453
Yamada M, Ohnishi H, Sano S, Nakatani A, Ikeuchi T, Hatanaka H (1997) Insulin receptor substrate (IRS)-1 and IRS-2 are tyrosine-phosphorylated and associated with phosphatidylinositol 3-kinase in response to brain-derived neurotrophic factor in cultured cerebral cortical neurons. J Biol Chem 272:30334–30339
Zhao WQ, Alkon DL (2001) Role of insulin and insulin receptor in learning and memory. Mol Cell Endocrinol 177:125–134
Zhao WQ, Chen H, Quon MJ, Alkon DL (2004) Insulin and the insulin receptor in experimental models of learning and memory. Eur J Pharmacol 490:71–81
Zhou L, Williams T, Lachey JL, Kishi T, Cowley MA, Heisler LK (2005) Serotonergic pathways converge upon central melanocortin systems to regulate energy balance. Peptides 26:1728–1732
Ziotopoulou M, Mantzoros CS, Hileman SM, Flier JS (2000) Differential expression of hypothalamic neuropeptides in the early phase of diet-induced obesity in mice. Am J Physiol Endocrinol Metab 279:E838–E845

# Insulin-Mediated Neuroplasticity in the Central Nervous System

**Lawrence P. Reagan**

**Abstract** Neuroplasticity is a concept that can be broadly defined as the ability of the central nervous system (CNS) to respond and adapt to the surrounding milieu. Neuroplasticity occurs in a variety of ways, including changes in the structural and functional properties of neuronal and non-neuronal cells as well as alterations in receptor pharmacology and neurochemical profiles. Ultimately these changes affect cognitive performance. Emerging evidence from clinical and preclinical studies suggests that insulin is an important mediator and facilitator of neuroplasticity in the CNS. Insulin administration improves cognitive performance in a variety of clinical settings ranging from normal healthy volunteers to patients with mild cognitive impairment (MCI) and Alzheimer's disease (AD). Since the cognition-enhancing actions of insulin are likely mediated by insulin receptors (IR), these data suggest that impairments in the functional activities of insulin in the hippocampus may contribute to cognitive deficits observed in diabetes patients. The clinical and epidemiological data illustrate that diabetes patients have an increased risk of developing age-related disorders like AD, suggesting that deficits in IR signaling may be a key initiating factor in the development and progression of cognitive decline. An important question that remains to be addressed is the identification of the mechanisms through which insulin enhances cognitive performance and, conversely, how impairments in IR signaling may contribute to neuroplasticity deficits associated with diabetes and age-related disorders. This aim of this review is to discuss the literature that supports the hypothesis that insulin is a trophic factor in the CNS that supports neuroplasticity, particularly as it relates to how deficits in IR signaling may be a mechanistic link between neurological co-morbidities like AD and diabetes.

---

L.P. Reagan
Department of Pharmacology, Physiology and Neuroscience, School of Medicine, University of South Carolina, 6439 Garners Ferry Rd, Columbia, SC 29208, USA
e-mail: lpreagan@uscmed.sc.edu

S. Craft and Y. Christen (eds.), *Diabetes, Insulin and Alzheimer's Disease*, Research and Perspectives in Alzheimer's Disease
DOI 10.1007/978-3-642-04300-0_3, 

## 1 Introduction

Diabetes mellitus is an endocrine disorder of carbohydrate metabolism resulting from inadequate insulin release (type 1 diabetes; T1DM) or insulin insensitivity (type 2 diabetes; T2DM). T1DM is believed to occur in response to an autoimmune destruction of insulin-producing pancreatic β cells, whereas T2DM may be triggered or worsened by a number of factors, including obesity, hypertension, and other features of the metabolic syndrome. For many years it has been well accepted that diabetes results in peripheral complications such as retinopathy, peripheral neuropathy, stroke and coronary heart disease. There is now evidence from clinical and preclinical studies illustrating that the complications of diabetes extend to the central nervous system (CNS) and include structural, neurochemical and functional changes (Gispen and Biessels 2000). More simply, diabetes elicits deficits in neuroplasticity in the CNS. One brain region in which neuroplasticity is adversely affected in diabetes phenotypes is the hippocampus, an important integration center for learning and memory in the mammalian brain (McEwen 1999). Diabetes-mediated neuroplasticity deficits in the hippocampus may include neuronal atrophy, decreased hippocampal formation volumes, synaptic reorganization, decreases in neurogenesis/cell proliferation, decreases in synaptic transmission, elevations in oxidative stress mediators and alterations in receptor pharmacology, among others. -Ultimately, a long-term consequence of diabetes-induced neuroplasticity deficits may be cognitive impairments. While controversy remains regarding the magnitude and significance of cognitive deficits in diabetes patients, clinical and epidemiological data clearly demonstrate that diabetic patients are at increased risk of developing neurological co-morbidities, including mood disorders (Reagan 2007) and age-related disorders (Luchsinger et al. 2004, 2007). Accordingly, the aim of this chapter will be to review our current understanding of neuroplasticity deficits in diabetes, especially as they relate to cognitive performance and neurological comorbidities.

## 2 Morphological Plasticity

The hippocampus exhibits a dynamic range of structural changes depending upon the neuronal milieu. For example, changes in neuronal spine density may be stimulated in the hippocampus under physiological conditions and contribute to the facilitation of learning and memory (Bourne and Harris 2008). The birth of new dentate gyrus granule neurons (i.e., neurogenesis) has also been implicated as an important facilitator of hippocampus-dependent learning and memory (Aimone et al. 2009; Leuner et al. 2006). Conversely, deficits in the structural integrity of the hippocampus are intimately associated with cognitive decline observed in neurological disorders, including T1DM and T2DM. In experimental models, short-term hyperglycemia (i.e, 7 to 21 days) produces rapid structural changes in the rodent hippocampus. In this regard, streptozotocin (STZ) diabetic rats, an experimental model of T1DM, exhibit dendritic remodeling in the CA3 region of the rat hippocampus (Magariños and McEwen 2000). Our subsequent studies determined that hyperglycemia-mediated morphological changes are more widespread in the hippocampus of STZ rats and include redistribution of synaptic proteins that may affect neurotransmission and plasticity (Grillo et al. 2005). These morphological changes in response to short-term hyperglycemia occur in the absence of neuronal degeneration in the hippocampus of STZ rats (Reagan 2002). Suppression of cell proliferation and neurogenesis is also observed in STZ diabetic rodents (Kim et al. 2003; Saravia et al. 2006) and the NOD mouse (Beauquis et al. 2008). Interestingly, diabetes-induced decreases in neurogenesis/cell proliferation are inhibited by treatments that have previously been shown to increase cell proliferation in the dentate gyrus, including exercise (Kim et al. 2003), estrogen treatment (Saravia et al. 2006) and antidepressant treatment (Beauquis et al. 2006).

Additional studies have identified histopathological complications in diabetic rodents indicative of accelerated brain aging. In a study of Alzheimer-associated pathologies in animal models of diabetes, both T1DM and T2DM rats exhibited decreases in phosphorylated Akt and phosphorylated GSK-3β (Li et al. 2007), which may have contributed to the development of histological AD features. Additionally, tau hyperphosphorylation is observed in the hippocampus of STZ diabetic mice (Clodfelder-Miller et al. 2006; Planel et al. 2007; Zhao et al. 2003), as well as in the hippocampus of fatty Zucker rats, an experimental model of T2DM (Wrighten et al. 2008). It is interesting to speculate that morphological deficits and increased AD-like histopathology observed in experimental models of diabetes represent initiating factors in the long-term complications in diabetes, including increased risk of co-morbid, age-related disorders.

An important consideration, especially in relation to hippocampal function, is whether these neuroanatomical changes represent the initiation of irreversible neuronal damage in diabetic subjects. Following chronic hyperglycemia, neuronal apoptosis and decreases in neuronal density are observed in the hippocampus of diabetic rodents (Li et al. 2002). A recent magnetic resonance imaging study identified reduced T2 values in the CA3 region of the hippocampus, as well as decreases in hippocampal volume following chronic hyperglycemia (Toth et al. 2006). One caveat associated with these chronic hyperglycemia studies is that the relationship of these findings to the clinical situation remains to be determined. In this regard, clinical studies support these observations in that structural abnormalities are observed in the hippocampus of T1DM and T2DM patients. As such, the collective observations from clinical and preclinical studies illustrate that structural deficits are a neurological complication of diabetes.

## 3 Oxidative Stress

In physiological conditions, there is a delicate balance between the synthesis of oxygen free radicals and the activities of anti-oxidant pathways. Oxidative stress, lipid peroxidation and increased production of reactive oxygen species reduce the activity of proteins that are critical to neuronal homeostasis (Mattson 1998). Oxidative stress and reactive oxygen species are increased in diabetes (Baynes 1991; Wolff 1993) and are proposed to contribute to the development of diabetic encephalopathy (Gispen and Biessels 2000). In support of this hypothesis, superoxide production is increased in the serum of T1DM patients, increases that are reduced with improved glycemic control (Ceriello et al. 1991). Studies from animal models have strengthened the relationship between disrupted oxidative stress balance and diabetes. Lipid peroxidation products such as 4-hydroxynonenal (HNE) and malondialdehyde (MDA) are increased in the brains of diabetic rodents (Kumar and Menon 1993; Traverso et al. 1998) and more specifically in the hippocampus of STZ rats (Grillo et al. 2003; Reagan et al. 2000; Reagan 2002; Tuzcu and Baydas 2006). HNE has been shown to mediate β-amyloid toxicity (Mark et al. 1997) and oxidative stress-induced apoptosis in hippocampal primary cultures (Kruman et al. 1997). Specific protein targets of oxidative stress have been identified that may contribute to diabetes-mediated neuroplasticity deficits. In this regard, the glial glutamate transporter GLT-1 is a target of HNE protein conjugation, which may be responsible for HNE-mediated decreases in glutamate transport in primary astrocytic cultures (Blanc et al. 1998) and in rat cortical synaptosomal fractions (Keller et al. 1997). Our previous studies have identified the neuron-specific glucose transporter GLUT3 as a target of HNE protein conjugation in the hippocampus of diabetic rats (Reagan et al. 2000). Functionally, this post-translational modification of GLUT3 may be responsible for decreases in glucose uptake observed in primary neuronal cultures treated with HNE (Keller et al. 1997).

In addition to increases in pro-oxidant pathways, hyperglycemia decreases anti-oxidant pathways in the diabetic brain. For example, glutathione levels (Tuzcu and Baydas 2006), as well as the expression and activity of glyceraldehyde-3-phosphate dehydrogenase (Aragno et al. 2005), are reduced in the hippocampus of STZ rats. Interestingly, anti-oxidant treatments such as melatonin, vitamin E (Tuzcu and Baydas 2006) and dehydroepiandrosterone (DHEA; Aragno et al. 2000, 2005; Reagan et al. 2008) reverse the imbalances in anti-oxidant/pro-oxidant ratios by increasing anti-oxidant expression and activity. Metformin also restores neuronal oxidant/anti-oxidant balance in T2DM rats (Correia et al. 2008). These results illustrate that hyperglycemia-mediated increases in oxidative stress adversely affect neuronal metabolism and neurochemistry but also indicate that oxidative stress status may be re-established in the diabetic brain.

## 4 Hypothalamic-Pituitary-Adrenal Axis Dysfunction

Many pathological conditions are associated with hypothalamic-pituitary-adrenal (HPA) axis dysregulation, including diabetes (Reagan et al. 2008). Elevated plasma glucocorticoid (GC) levels and/or increased reactivity of the HPA axis contribute to the development of insulin resistance in obesity and in T2DM. These peripheral actions of GCs extend to the CNS. For example, short-term corticosterone (CORT) administration elicits peripheral insulin resistance and also decreases insulin sensitivity in the hippocampus, including decreases in insulin-stimulated translocation of the insulin sensitive glucose transporter, GLUT4 (Piroli et al. 2007). Diabetic animals exhibit increases in basal levels of corticosterone and increased sensitivity to stress (Leedom et al. 1998; Magariños and McEwen 2000; Oster et al. 1988; Scribner et al. 1991), which has lead to the suggestion that hyperglycemia may serve as a model of chronic stress (Scribner et al. 1993). Such observations raise the question of whether the actions of stress and/or glucocorticoids may be accelerated in the hippocampus of diabetic rats. In support of this hypothesis, our previous studies have shown that exposure to short-term restraint stress exacerbates diabetes-mediated complications, including dendritic atrophy (Magariños and McEwen 2000), impairments in IR signaling (Piroli et al. 2004), diabetes-mediated increases in oxidative stress (Grillo et al. 2003) and alterations in glutamate transporter expression (Reagan et al. 2008). Interestingly, adrenalectomized STZ rats and db/db mice provided low-dose CORT replacement exhibit improved spatial memory and novel object recognition when compared to sham-operated animals and adrenalectomized animals given high levels of CORT replacement (Stranahan et al. 2008). Such results support the notion that HPA axis dysfunction and/or exposure to elevated glucocorticoid levels contributes to the cognitive deficits observed in diabetic animals. Although it is challenging to dissociate cause and effects relationships, these findings may place impairments in HPA axis function at the center of the neurological complications of diabetes, including accelerated brain aging.

## 5 Insulin and the Glutamate System

The emerging literature regarding IR signaling in the hippocampus suggests that insulin modulates synaptic plasticity via a variety of mechanistic pathways. One target of IR signaling in the CNS is the glutamate system, a well-recognized mediator of promoting and maintaining neuroplasticity in the CNS. For example, insulin elicits surface expression of NMDA receptors (Skeberdis et al. 2001) and stimulates the phosphorylation of NR2A and NR2B NMDA receptor subunits in the hippocampus (Christie et al. 1999). Insulin also stimulates the phosphorylation and endocytosis of GluR2 in hippocampal slice preparations (Ahmadian et al. 2004). These insulin signaling events are PI3-kinase dependent and have been proposed to contribute to insulin-induced long-term depression (LTD) in hippocampal slices (Ahmadian et al. 2004; Huang et al. 2004; van der Heide et al. 2005). In diabetes phenotypes, IR-glutamate interactions are suppressed and may thereby contribute to diabetes-mediated deficits in hippocampal synaptic plasticity. For example, AMPA receptor binding activity (Gagne et al. 1997) and the functional activities of AMPA receptors (Chabot et al. 1997; Kamal et al. 2006) are reduced in the hippocampus of STZ diabetic rats. Regarding NMDA receptor expression and activity, NR2B mRNA and protein are decreased in the hippocampus of STZ rats

(Di Luca et al. 1999; Gardoni et al. 2002). Moreover, $Ca^{+}$/CaM-stimulated phosphorylation of hippocampal NR2A and NR2B subunits expressed in the postsynaptic density is reduced in STZ rats. As described above, redistribution and re-organization of PSD-95 that may be indicative of ongoing synaptogenesis in the hippocampus of STZ rats (Grillo et al. 2005) may also modulate gluatamatergic tone by modulating surface expression of glutamate receptors.

Functionally, these transcriptional, translational and post-translational modifications of glutamate receptors observed in diabetic rodents may adversely affect synaptic transmission and the electrophysiological properties of hippocampal neurons. In this regard, T1DM rodents exhibit impairments in hippocampal LTP and enhancement of LTD (Artola et al. 2005; Biessels et al. 1996; Izumi et al. 2003; Kamal et al. 1999; Valastro et al. 2002). Electrophysiological studies in experimental models of T2DM have failed to reach a consensus, with some studies stating that T2DM animals exhibit deficits in LTP (Gerges et al. 2003) whereas others have failed to observe electrophysiological changes (Belanger et al. 2004). A potential explanation for these discrepancies is that the physiological characteristics of the T2DM animals used in these studies may be dissimilar; such considerations may also be particularly relevant when considering the equivocal findings from clinical studies that examined structural and functional plasticity in diabetes patients.

# 6 Leptin

Ongoing epidemiological studies by the Centers for Disease Control estimate that more than 60% of the adult US population may be categorized as either overweight or obese (Ogden et al. 2006). Obesity-related CNS abnormalities may result from obesity-induced insulin resistance and/or impaired glucose tolerance (Pi-Sunyer 2002; Ronnemaa et al. 1997). An additional consideration in obesity phenotypes is the potential for increases in plasma levels of adipocyte-derived hormones, such as leptin. Leptin is synthesized and secreted by adipocytes and is transported across the blood-brain barrier (BBB) via a saturable transport system (Banks 2004). The actions of leptin in the hypothalamus are well described, especially in relation to normal metabolism and in pathophysiological settings such as diabetes and obesity phenotypes (Woods et al. 1998). In the hippocampus, a growing literature supports a role for leptin in the facilitation of neuronal structure and function (Harvey 2007). For example, leptin enhances hippocampal excitability via NMDA receptor-mediated mechanisms and regulates hippocampal plasticity by converting short-term potentiation into LTP (Shanley et al. 2001). Confocal immunoflurorescence analyses also determined that leptin regulates the morphological features of primary hippocampal cultures, including the motility of dendritic filopodia (O'Malley et al. 2007). These structural and functional enhancements of hippocampal plasticity may contribute to leptin's ability to improve hippocampus-dependent behavioral performance (Farr et al. 2006; Harvey 2007; Oomura et al. 2006). Conversely, genetic mutations that result in disrupted leptin signaling, such as in the *db/db* mouse and the Zucker *fa/fa* rat, are associated with reductions in LTP (Gerges et al. 2003; Li et al. 2002) and impaired performance of hippocampal-dependent tasks (Li et al. 2002; Winocur et al. 2005). Such results indicate that decreases in leptin signaling contribute to deficits in hippocampal synaptic plasticity and suggest that leptin resistance, a well-described phenomenon in the hypothalamus, may also be observed in the hippocampus in diabetes/obesity phenotypes.

## 7 Insulin

The insulin receptor (IR) is expressed in discrete neuronal populations in the CNS, including the hippocampus (Doré et al. 1997; Marks et al. 1991), where it is proposed to facilitate cognitive function (Park 2001). Insulin improves cognitive performance in humans and animals in a wide variety of settings, including healthy subjects (Kopf and Baratti 1994; Park et al. 2000; Parkes and White 2000), aged subjects (Manning et al. 1998; Messier et al. 1997; Winocur and Gagnon 1998), AD patients (Craft et al. 1999; Manning et al. 1993; Messier and Gagnon 1996) and in experimental models of insulin resistance (Greenwood and Winocur 2001). Additionally, behavioral training increases IR expression and strengthens IR signaling in the hippocampus (Zhao et al. 1999). More recent studies employing innovative delivery of insulin to the CNS, namely intranasal insulin treatment, further support the hypothesis that insulin enhances cognitive performance. For example, chronic intranasal insulin administration improves cognitive performance in both AD and non-demented individuals (Benedict et al. 2007a, b; Hallschmid et al. 2008; Reger et al. 2008; Sabayan et al. 2008), and acute insulin administration has been shown to improve declarative memory in AD patients (Craft et al. 1996). Moreover, administration of the insulin sensitizer, rosiglitazone, ameliorates cognitive decline in AD patients (Watson et al. 2005). These data support the hypothesis that activation of IR signaling cascades improves cognitive/behavioral performance.

The relationship between IR activity and impaired behavioral performance has also been examined in experimental models of diabetes (Belanger et al. 2004; Biessels et al. 1996; Choeiri et al. 2005; Li et al. 2002; Winocur et al. 2005). The take-home message from these pre-clinical studies is that a variety of factors may impact the outcome of behavior in diabetic animals, ranging from the physiological/pathophysiological characteristics of the animal model to the selection and analysis of the particular behavioral tests. However, prior studies have illustrated that insulin replacement reverses diabetes-mediated plasticity deficits in the rat hippocampus (Biessels et al. 1998; Magariños et al. 2001). More recent studies revealed that intranasal insulin replacement effectively inhibits structural and functional deficits in the CNS of STZ mice (Francis et al. 2008). These data suggest that deficits in insulin signaling are a key mechanistic mediator of diabetes-mediated neuroplasticity deficits.

One way to further validate this hypothesis would be to selectively impair IR signaling in the CNS. In the absence of receptor-specific ligands, recent studies have utilized molecular approaches to more selectively examine the functional activities of central insulin receptors in global IR knockout mice (Bruning et al. 2000) and in antisense oligonucleotide-treated rats (Obici et al. 2002). An emerging technology that provides an alternative to these approaches is virus-mediated gene transfer (Wilson and Yeomans 2002). Using this approach, we recently developed a lentivirus vector that contains an antisense sequence selective for the IR (LV-IRAS; Grillo et al. 2007). Following injections into the third ventricle to target IRs expressed in the arcuate nucleus, LV-IRAS rats exhibit significant decreases in IR expression and signaling in the hypothalamus when compared to rats treated with the control virus (LV-Con). LV-IRAS rats have increased body weight, greater subcutaneous adiposity and increased plasma leptin levels when compared to LV-Con rats. These hypothalamus-specific decreases in IR expression do not affect peripheral glucose metabolism or neuroendocrine responses to stress. While IR expression and signaling are unchanged in the hippocampus of LV-IRAS rats, downregulation of hypothalamic IRs elicits changes in the morphological and electrophysiological properties of hippocampal neurons and also elicits impairments in hippocampus-dependent behaviors. These deficits in hippocampal synaptic plasticity are strikingly similar to those observed in age-related disorders and mood disorders such as depressive illness (Reagan et al. 2008). As such, the use of the LV-IRAS construct provides a unique and innovative approach to examine how selective downregulation of IRs in one brain region may impact neuroplasticity throughout the CNS. Moreover, such an approach may identify common mechanisms and pathologies that provide etiological links between neurological co-morbidities observed in diabetes phenotypes.

# 8 Conclusions

In view of the expanding diabetes/obesity epidemic, the long-term neurological consequences of diabetes represent a significant threat to the stability of health care systems around the world. While some may argue that cognitive impairments are not among the neurological complications of diabetes, the clinical and epidemiological data clearly indicate that diabetes patients exhibit cognitive deficits (Cukierman-Yaffe et al. 2009). One take-home message from our provocative discussions during this meeting is that it is unlikely that a single factor is responsible for the neuroplasticity deficits observed in diabetes. It is more probable that a variety of mechanistic mediators interact in an additive or synergistic fashion to increase neuronal vulnerability and accelerate brain aging (see Fig. 1). Could these putative mechanistic mediators ultimately increase the risk of the development of dementia and AD in diabetes patients? A key step forward would be verification that these mechanistic mediators serve as biomarkers to identify individuals that

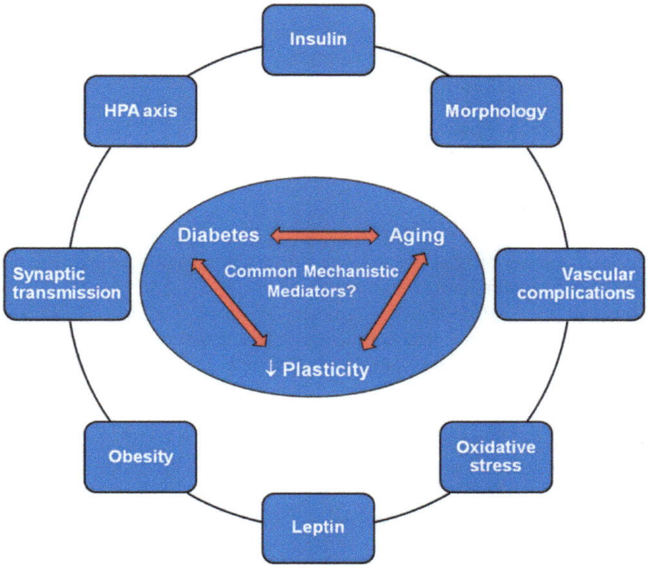


**Fig. 1** Potential mechanistic mediators of neuroplasticity deficits in diabetes. Since a variety of factors may adversely affect the CNS and contribute to accelerated brain aging in diabetes patients, the identification of a 'keystone' mediator appears unlikely. More realistically, the pathophysiological features of diabetes interact in an additive or synergistic fashion to impair neuronal plasticity and increase the risk for age-related disorders such as dementia and AD. See text for details

are susceptible to age-related co-morbidities. Continued integration of clinical, epidemiological and pre-clinical studies is essential to address this important health care issue.

**Acknowledgments** Research in the author's laboratory is supported in part by Juvenile Diabetes Research Foundation grant number 2-03-675, NIH grant number R01NS047728 and the University of South Carolina Research Foundation.

## References

Ahmadian G, Ju W, Liu L, Wyszynski M, Lee SH, Dunah AW, Taghibiglou C, Wang Y, Lu J, Wong TP, Sheng M, Wang YT (2004) Tyrosine phosphorylation of GluR2 is required for insulin-stimulated AMPA receptor endocytosis and LTD. EMBO J 23:1040–1050

Aimone JB, Wiles J, Gage FH (2009) Computational influence of adult neurogenesis on memory encoding. Neuron 61:187–202

Aragno M, Parola S, Tamagno E, Brignardello E, Manti R, Danni O, Boccuzzi G (2000) Oxidative derangement in rat synaptosomes induced by hyperglycemia: restorative effect of dehydroepiandrosterone treatment. Biochem Pharmacol 60:389–395

Aragno M, Mastrocola R, Medana C, Restivo F, Catalano MG, Pons N, Danni O, Boccuzzi G (2005) Up-regulation of advanced glycated products receptors in the brain of diabetic rats is prevented by antioxidant treatment. Endocrinology 146:5561–5567

Artola A, Kamal A, Ramakers GM, Biessels GJ, Gispen WH (2005) Diabetes mellitus concomitantly facilitates the induction of long-term depression and inhibits that of long-term potentiation in hippocampus. Eur J Neurosci 22:169–178

Banks WA (2004) The many lives of leptin. Peptides 25:331–338

Baynes JW (1991) Role of oxidative stress in development of complications in diabetes. Diabetes 40:405–412

Beauquis J, Roig P, Homo-Delarche F, De Nicola A, Saravia F (2006) Reduced hippocampal neurogenesis and number of hilar neurones in streptozotocin-induced diabetic mice: reversion by antidepressant treatment. Eur J Neurosci 23:1539–1546

Beauquis J, Saravia F, Coulaud J, Roig P, Dardenne M, Homo-Delarche F, De Nicola A (2008) Prominently decreased hippocampal neurogenesis in a spontaneous model of type 1 diabetes, the nonobese diabetic mouse. Exp Neurol 210:359–367

Belanger A, Lavoie N, Trudeau F, Massicotte G, Gagnon S (2004) Preserved LTP and water maze learning in hyperglycaemic-hyperinsulinemic ZDF rats. Physiol Behav 83:483–494

Benedict C, Hallschmid M, Schmitz K, Schultes B, Ratter F, Fehm HL, Born J, Kern W (2007a) Intranasal insulin improves memory in humans: superiority of insulin aspart. Neuropsychopharmacology 32:239–243

Benedict C, Hallschmid M, Schultes B, Born J, Kern W (2007b) Intranasal insulin to improve memory function in humans. Neuroendocrinology 86:136–142

Biessels G-J, Kamal A, Ramakers GM, Urban IJ, Spruijt BM, Erkelens DW, Gispen WH (1996) Place learning and hippocampal synaptic plasticity in streptozotocin-induced diabetic rats. Diabetes 45:1259–1266

Biessels G-J, Kamal A, Urban IJ, Spruijt BM, Erkelens DW, Gispen WH (1998) Water maze learning and hippocampal synaptic plasticity in streptozotocin-diabetic rats: effects of insulin treatment. Brain Res 800:125–135

Blanc EM, Keller JN, Fernandez S, Mattson MP (1998) 4-hydroxynonenal, a lipid peroxidation product, impairs glutamate transport in cortical astrocytes.Glia 22:149–160

Bourne JN, Harris KM (2008) Balancing structure and function at hippocampal dendritic spines. Annu Rev Neurosci 31:47–67

Bruning JC, Gautam D, Burks DJ, Gillette J, Schubert M, Orban PC, Klein R, Krone W, Muller-Wieland D, Kahn CR (2000) Role of brain insulin receptor in control of body weight and reproduction. Science 289:2122–2125

Ceriello A, Giugliano D, Quatraro A, Dello Russo P, Lefèbvre PJ (1991) Metabolic control may influence the increased superoxide generation in diabetic serum. Diabetic Med 8:540-542

Chabot C, Massicotte G, Milot M, Trudeau F, Gagne J (1997) Impaired modulation of AMPA receptors by calcium-dependent processes in streptozotocin-induced diabetic rats. Brain Res 768:249–256

Choeiri C, Hewitt K, Durkin J, Simard CJ, Renaud JM, Messier C (2005) Longitudinal evaluation of memory performance and peripheral neuropathy in the Ins2C96Y Akita mice. Behav Brain Res 157:31–38

Christie JM, Wenthold RJ, Monaghan DT (1999) Insulin causes a transient tyrosine phosphorylation of NR2A and NR2B NMDA receptor subunits in rat hippocampus. J Neurochem 72:1523-1528

Clodfelder-Miller BJ, Zmijewska AA, Johnson GV, Jope RS (2006) Tau is hyperphosphorylated at multiple sites in mouse brain in vivo after streptozotocin-induced insulin deficiency. Diabetes 55:3320–3325

Correia S, Carvalho C, Santos MS, Proenca T, Nunes E, Duarte AI, Monteiro P, Seica R, Oliveira CR, Moreira PI (2008) Metformin protects the brain against the oxidative imbalance promoted by type 2 diabetes. Med Chem 4:358–364

Craft S, Newcomer J, Kanne S, Dagogo-Jack S, Cryer P, Sheline Y, Luby J, Dagogo-Jack A, Alderson A (1996) Memory improvement following induced hyperinsulinemia in Alzheimer's disease. Neurobiol Aging 17:123–130

Craft S, Asthana S, Newcomer JW, Wilkinson CW, Matos IT, Baker LD, Cherrier M, Lofgreen C, Latendresse S, Petrova A, Plymate S, Raskind M, Grimwood K, Veith RC (1999) Enhancement of memory in Alzheimer Disease with insulin and somatostatin, but not glucose. Arch Gen Psychiat 56:1135–1140

Cukierman-Yaffe T, Gerstein HC, Williamson JD, Lazar RM, Lovato L, Miller ME, Coker LH, Murray A, Sullivan MD, Marcovina SM, Launer LJ (2009) Relationship between baseline glycemic control and cognitive function in individuals with type 2 diabetes and other cardiovascular risk factors: the action to control cardiovascular risk in diabetes-memory in diabetes (ACCORD-MIND) trial. Diabetes Care 32:221–226

Di Luca M, Ruts L, Gardoni F, Cattabeni F, Biessels G-J, Gispen WH (1999) NMDA receptor subunits are modified transcriptionally and post-translationally in the brain of streptozotocin-diabetic rats. Diabetologia 42:693–701

Doré S, Kar S, Rowe W, Quirion R (1997) Distribution and levels of [125I]IGF-I, [125I]IGF-II and [125I]Insulin receptor binding sites in the hippocampus of aged memory-unimpaired and –impaired rats. Neuroscience 80:1033–1040

Farr SA, Banks WA, Morley JE (2006) Effects of leptin on memory processing. Peptides 27:1420–1425

Francis GJ, Martinez JA, Liu WQ, Xu K, Ayer A, Fine J, Tuor UI, Glazner G, Hanson LR, Frey WH, Toth C (2008) Intranasal insulin prevents cognitive decline, cerebral atrophy and white matter changes in murine type I diabetic encephalopathy. Brain 131:3311–3334

Gagne J, Milot M, Gelinas S, Lahsaini A, Trudeau F, Martinoli MG, Massicotte G (1997) Binding properties of glutamate receptors in streptozotocin-induced diabetes in rats. Diabetes 46: 841–846

Gardoni F, Kamal A, Bellone C, Biessels GJ, Ramakers GM, Cattabeni F, Gispent WH, Di Luca M (2002) Effects of streptozotocin-diabetes on the hippocampal NMDA receptor complex in rats. J Neurochem 80:438–447

Gerges NZ, Aleisa AM, Alkadhi KA (2003) Impaired long-term potentiation in obese zucker rats: possible involvement of presynaptic mechanism. Neuroscience 120:535–539

Gispen WH, Biessels G-J (2000) Cognition and synaptic plasticity in diabetes mellitus. Trends Neurosci 23:542–549

Greenwood CE, Winocur G (2001) Glucose treatment reduces memory deficits in young adult rats fed high-fed diets. Neurobiol Learn Mem 75:179–189

Grillo CA, Piroli GG, Rosell DR, Hoskin EK, McEwen BS, Reagan LP (2003) Region specific increases in oxidative stress and superoxide dismutase in the hippocampus of diabetic rats subjected to stress. Neuroscience 121:133–140

Grillo CA, Piroli GG, Wood GE, Reznikov LR, McEwen BS, Reagan LP (2005) Immunocytochemical analysis of synaptic proteins provides new insights into diabetes-mediated plasticity in the rat hippocampus. Neuroscience 136:477–486

Grillo CA, Tamashiro KL, Piroli GG, Melhorn S, Gass JT, Newsom RJ, Reznikov LR, Smith A, Wilson SP, Sakai RR, Reagan LP (2007) Lentivirus-mediated downregulation of hypothalamic insulin receptor expression. Physiol Behav 92:691–701

Hallschmid M, Benedict C, Schultes B, Born J, Kern W (2008) Obese men respond to cognitive but not to catabolic brain insulin signaling. Int J Obes (Lond) 32:275–282

Harvey J (2007) Leptin regulation of neuronal excitability and cognitive function. Curr Opin Pharmacol 7:643–647

Huang CC, Lee CC, Hsu KS (2004) An investigation into signal transduction mechanisms involved in insulin-induced long-term depression in the CA1 region of the hippocampus. J Neurochem 89:217–231

Izumi Y, Yamada KA, Matsukawa M, Zorumski CF (2003) Effects of insulin on long-term potentiation in hippocampal slices from diabetic rats. Diabetologia 46:1007–1012

Kamal A, Biessels G-J, Urban IJA, Gispen WH (1999) Hippocampal synaptic plasticity in streptozotocin-diabetic rats: impairment of long-term potentiation and facilitation of long-term depression. Neuroscience 90:737–745

Kamal A, Biessels GJ, Gispen WH, Ramakers GM (2006) Synaptic transmission changes in the pyramidal cells of the hippocampus in streptozotocin-induced diabetes mellitus in rats. Brain Res 1073–1074:276–280

Keller JN, Pang Z, Geddes JW, Begley JG, Germeyer A, Waeg G, Mattson MP (1997) Impairment of glucose and glutamate transport and induction of mitochondrial oxidative stress and dysfunction in synaptosomes by amyloid b-peptide: role of the lipid peroxidation product 4-hydroxynonenal. J Neurochem 69:273–284

Kim HB, Jang MH, Shin MC, Lim BV, Kim YP, Kim KJ, Kim EH, Kim CJ (2003) Treadmill exercise increases cell proliferation in dentate gyrus of rats with streptozotocin-induced diabetes. J Diabetes Complications 17:29–33

Kopf SR, Baratti CM (1994) Memory-improving actions of glucose: involvement of a central cholinergic muscarinic mechanism. Behav Neur Biol 62:237–243

Kruman I, Bruce-Keller AJ, Bredesen D, Waeg G, Mattson MP (1997) Evidence that 4-hydroxynonenal mediates oxidative stress-induced neuronal apoptosis. J Neurosci 17:5089–5100

Kumar JSS, Menon VP (1993) Effect of diabetes on levels of lipid peroxides and glycolipids in rat brain. Metabolism 42:1435–1439

Leedom LJ, Meehan WP, Zeidler A (1987) Avoidance responding in mice with diabetes mellitus. Physiol Behav 40:447–51

Leuner B, Gould E, Shors TJ (2006) Is there a link between adult neurogenesis and learning? Hippocampus 16:216–224

Li XL, Aou S, Oomura Y, Hori N, Fukunaga K, Hori T (2002) Impairment of long-term potentiation and spatial memory in leptin receptor-deficient rodents. Neuroscience 113:607–615

Li ZG, Zhang W, Grunberger G, Sima AA (2002) Hippocampal neuronal apoptosis in type 1 diabetes. Brain Res 946:221–231

Li ZG, Mang W, Sima AA (2007) Alzheimer-like changes in rat models of spontaneous diabetes. Diabetes 56:2650

Luchsinger JA, Tang MX, Shea S, Mayeux R (2004) Hyperinsulinemia and risk of Alzheimer disease. Neurology 63:1187–1192

Luchsinger JA, Reitz C, Patel B, Tang MX, Manly JJ, Mayeux R (2007) Relation of diabetes to mild cognitive impairment. Arch Neurol 64:570–575

Magariños AM, McEwen BS (2000) Experimental diabetes in rats causes hippocampal dendritic and synaptic reorganization and increased glucocorticoid reactivity to stress. Proc Natl Acad Sci USA 97:11056–11061

Magariños AM, Jain K, Blount ED, Reagan L, Smith BH, McEwen BS (2001) Peritoneal implantation of microencapsulated porcine pancreatic islets in diabetic rats ameliorates severe hyperglycemia and prevents retraction and simplification of hippocampal dendrites. Brain Res 902:282–287

Manning CA, Ragozzino ME, Gold PE (1993) Glucose enhancement of memory in patients with probable senile dementia of the Alzheimer's type. Neurobiol Aging 14:523–528

Manning CA, Stone WS, Korol DL, Gold PE (1998) Glucose enhancement of 24-h memory retrival in healthy elderly humans. Behav Brain Res 93:71–76

Mark RJ, Lovell MA, Markesbery WR, Uchida K, Mattson MP (1997) A role for 4-hydroxynonenal, an aldehyde product of lipid peroxidation, in disruption of ion homeostasis and neuronal death induced by amyloid b-peptide. J Neurochem 68:255–264

Marks JL, Porte DJr, Stahl WL, Baskin DG (1991) Localization of insulin receptor mRNA in rat brain by in situ hybridization. Endocrinology 127:3234–3236

Mattson MP (1998) Modification of ion homeostasis by lipid peroxidation: roles in neuronal degeneration and adaptive plasticity. Trends Neurosci 21:53–57

McEwen BS (1999) Stress and hippocampal plasticity. Annu Rev Neurosci 22:105–122

Messier C, Gagnon M (1996) Glucose regulation and cognitive functions: relation to Alzheimer's disease and diabetes. Behav Brain Res 75:1–11

Messier C, Gagnon M, Knott V (1997) Effect of glucose and peripheral glucose regulation on memory in the ederly. Neurobiol Aging 18:297–304

O'Malley D, MacDonald N, Mizielinska S, Connolly CN, Irving AJ, Harvey J (2007) Leptin promotes rapid dynamic changes in hippocampal dendritic morphology. Mol Cell Neurosci 35:559–572

Obici S, Feng Z, Karkanias G, Baskin DG, Rossetti L (2002) Decreasing hypothalamic insulin receptors causes hyperphagia and insulin resistance in rats. Nature Neurosci 5: 566–572

Ogden CL, Carroll MD, Curtin LR, McDowell MA, Tabak CJ, Flegal KM (2006) Prevalence of overweight and obesity in the United States, 1999–2004. JAMA 295:1549–1555

Oomura Y, Hori N, Shiraishi T, Fukunaga K, Takeda H, Tsuji M, Matsumiya T, Ishibashi M, Aou S, Li XL, Kohno D, Uramura K, Sougawa H, Yada T, Wayner MJ, Sasaki K (2006) Leptin facilitates learning and memory performance and enhances hippocampal CA1 long-term potentiation and CaMK II phosphorylation in rats. Peptides 27:2738–2749

Oster MH, Castonguay TM, Keen CL, Stern JS (1988) Circadian rhythm of corticosterone in diabetic rats. Life Sci 43:1643–1645

Park CR (2001) Cognitive effects of insulin in the central nervous system. Neurosci Biobehav Rev 25:311–323

Park CR, Seely RJ, Craft S, Woods SC (2000) Intracerebroventricular insulin enhances memory in a passive-avoidance task. Physiol Behav 68:509–514

Parkes M, White KG (2000) Glucose attenuation of memory impairments. Behav Neurosci 114:307–319

Pi-Sunyer FX (2002) The obesity epidemic: Pathophysiology and consequences of obesity. Obesity Res 10:97S–104S

Piroli GG, Grillo CA, Charron MJ, McEwen BS, Reagan LP (2004) Biphasic effects of stress upon GLUT8 glucose transporter expression and trafficking in the diabetic rat hippocampus. Brain Res 1006:28–35

Piroli GG, Grillo CA, Reznikov LR, Adams S, Mcewen BS, Charron MJ, Reagan LP (2007) Corticosterone impairs insulin-stimulated translocation of GLUT4 in the rat hippocampus. Neuroendocrinology 85:71–80

Planel E, Tatebayashi Y, Miyasaka T, Liu L, Wang L, Herman M, Yu WH, Luchsinger JA, Wadzinski B, Duff KE, Takashima A (2007) Insulin dysfunction induces in vivo tau hyperphosphorylation through distinct mechanisms. J Neurosci 27:13635–13648

Reagan LP (2002) Glucose, stress and hippocampal neuronal vulnerability. Int Rev Neurobiol 51:289–324

Reagan LP (2007) Insulin signaling effects on memory and mood. Curr Opin Pharmacol 7: 633–637

Reagan LP, Magariños AM, Yee DK, Szweda LI, Van Bueren A, McCall AL, McEwen BS (2000) Oxidative stress and HNE conjugation of GLUT3 are increased in the hippocampus of diabetic rats subjected to stress. Brain Res 862:292–300

Reagan LP, Grillo CA, Piroli GG (2008) The As and Ds of stress: metabolic, morphological and behavioral consequences. Eur J Pharmacol 585:64–75

Reger MA, Watson GS, Green PS, Wilkinson CW, Baker LD, Cholerton B, Fishel MA, Plymate SR, Breitner JCS, DeGroodt W, Mehta P, Craft S (2008) Intranasal insulin improves cognition and modulates beta-amyloid in early AD. Neurology 70:440–448

Ronnemaa T, Koskenvuo M, Marniemi J, Koivunen T, Sajantila A, Rissanen A, Kaitsaari M, Bouchard C, Kaprio J (1997) Glucose metabolism in identical twins discordant for obesity. The critical role of visceral fat. J Clin Endocrinol Metab 82:383–387

Sabayan B, Foroughinia F, Mowla A, Borhanihaghighi A (2008) Role of Insulin Metabolism Disturbances in the Development of Alzheimer Disease: Mini Review. Am J Alzheimers Dis Other Demen 23:192–199

Saravia FE, Beauquis J, Revsin Y, Homo-Delarche F, de Kloet ER, De Nicola AF (2006) Hippocampal neuropathology of diabetes mellitus is relieved by estrogen treatment. Cell Mol Neurobiol 26:943–957

Scribner KA, Walker CD, Cascio CS, Dallman MF (1991) Chronic streptozotocin diabetes in rats facilitates the acute stress response without altering pituitary or adrenal responsiveness to secretagogues. Endocrinology 129:99–108

Scribner KA, Akana SF, Walker CD, Dallman MF (1993) Streptozotocin-diabetic rats exhibit facilitated adrenocorticotropin responses to acute stress, but normal sensitivity to feedback by corticosteroids. Endocrinology 133:2667–2674

Shanley LJ, Irving AJ, Harvey J (2001) Leptin enhances NMDA receptor function and modulates hippocampal synaptic plasticity. Journal of Neuroscience 21:art-RC186

Skeberdis VA, Lan J, Zheng X, Zukin RS, Bennett MV (2001) Insulin promotes rapid delivery of N-methyl-D- aspartate receptors to the cell surface by exocytosis. Proc Natl Acad Sci USA 98:3561–3566

Stranahan AM, Arumugam TV, Cutler RG, Lee K, Egan JM, Mattson MP (2008) Diabetes impairs hippocampal function through glucocorticoid-mediated effects on new and mature neurons. Nature Neurosci 11:309–317

Toth C, Schmidt AM, Tuor UI, Francis G, Foniok T, Brussee V, Kaur J, Yan SF, Martinez JA, Barber PA, Buchan A, Zochodne DW (2006) Diabetes, leukoencephalopathy and rage. Neurobiol Dis 23:445–461

Traverso N, Menini S, Cosso L, Odetti E, Albano E, Pronzato MA, Marinari UM (1998) Immunological evidence for increased oxidative stress in diabetic rats. Diabetologia 41:265-270

Tuzcu M, Baydas G (2006) Effect of melatonin and vitamin E on diabetes-induced learning and memory impairment in rats. Eur J Pharmacol 537:106–110

Valastro B, Cossette J, Lavoie N, Gagnon S, Trudeau F, Massicotte G (2002) Up-regulation of glutamate receptors is associated with LTP defects in the early stages of diabetes mellitus. Diabetologia 45:642–650

van der Heide LP, Kamal A, Artola A, Gispen WH, Ramakers GM (2005) Insulin modulates hippocampal activity-dependent synaptic plasticity in a N-methyl-d-aspartate receptor and phosphatidyl-inositol-3-kinase-dependent manner. J Neurochem 94:1158–1166

Watson GS, Cholerton BA, Reger MA, Baker LD, Plymate SR, Asthana S, Fishel MA, Kulstad JJ, Green PS, Cook DG, Kahn SE, Keeling ML, Craft S (2005) Preserved cognition in patients with early Alzheimer disease and amnestic mild cognitive impairment during treatment with rosiglitazone: a preliminary study. Am J Geriatr Psychiat 13:950–958

Wilson SP, Yeomans DC (2002) Virally mediated delivery of enkephalin and other neuropeptide transgenes in experimental pain models. Ann NY Acad Sci 971:521

Winocur G, Gagnon S (1998) Glucose treatment attenuates spatial learning and memory deficts of aged rats on tests of hippocampal function. Neurobiol Aging 19:233–241

Winocur G, Greenwood CE, Piroli GG, Grillo CA, Reznikov LR, Reagan LP, McEwen BS 2005) Memory Impairment in Obese Zucker Rats: An Investigation of Cognitive Function in an Animal Model of Insulin Resistance and Obesity. Behav Neurosci 119:1389–1395

Wolff SP (1993) Diabetes mellitus and free radicals. Brit Med Bull 49:642–652

Woods SC, Seeley RJ, Porte D, Jr., Schwartz MW (1998) Signals that regulate food intake and energy homeostasis. Science 280:1378–1383

Wrighten SA, Piroli GG, Grillo CA, Reagan LP (2009) A look inside the diabetic brain: Contributors to diabetes-induced brain aging. Biochim Biophys Acta 1792:444–453

Zhao W, Chen H, Moore E, Meiri N, Quon MJ, Alkon DL (1999) Brain insulin receptors and spatial memory. J Biol Chem 274:34893–34902

Zhao YM, Pei JJ, Ji ZJ, Zhao ZW, Qian YY, Sheng SL (2003) Effect of amyloid precursor protein 17mer peptide on microtubule structure and tau protein hyperphosphorylation in hippocampal neurons of experimental diabetic mice. Neuroreport 14:61–66

# Stress Hormones and Neuroplasticity in the Diabetic Brain

**Alexis M. Stranahan and Mark P. Mattson**

**Abstract** Diabetes is associated with metabolic dysfunction across multiple organ systems, and the central nervous system is no exception. Neurons in the diabetic brain exhibit functional alterations that may increase the risk of cognitive decline and Alzheimer's disease. Diabetes is associated with changes in the hypothalamic-pituitary-adrenal (HPA) axis, but the relationship between HPA axis function and cognitive dysfunction in diabetes is still being elucidated. Here we review evidence for and against HPA axis dysfunction in diabetes, and its consequences for neuroplasticity in the hippocampus, a brain region that mediates certain aspects of learning and memory. The tripartite relationship between diabetes, HPA axis alterations, and cognitive impairment will be discussed. The evidence favors a role for adrenal steroid hormones as central and peripheral mediators of diabetes-induced cellular dysfunction. In the hippocampus, adrenal corticosteroids may perturb neurotrophic factor signaling, resulting in impaired neurogenesis, synaptic plasticity and cognitive function. The adverse effects of diabetes on hippocampal plasticity may be allayed by exercise and dietary energy restriction.

## 1 Introduction

Diabetes is a metabolic disorder characterized by deficits in insulin production (Type 1) or impairment of insulin sensitivity (Type 2). Poorly controlled diabetes has deleterious consequences for multiple organ systems, including the brain. In humans, diabetes increases the risk for depression and dementia (Messier 2005; Greenwood and Winocur 2005). Both Type 1 and Type 2 diabetes are associated with cognitive deficits, relative to age-matched non-diabetic subjects. Functionally,

A.M. Stranahan (✉)
National Institutes of Health, National Institute on Aging, Biomedical Research Center, Laboratory of Neurosciences, 251 Bayview Blvd, Baltimore, MD 21224-0000, USA
e-mail: alexis.stranahan@jhu.edu

S. Craft and Y. Christen (eds.), *Diabetes, Insulin and Alzheimer's Disease*, Research and Perspectives in Alzheimer's Disease,
DOI 10.1007/978-3-642-04300-0_4, 

insulin-deficient individuals perform poorly on the Wechsler Abbreviated Intelligence test relative to healthy control subjects (Northam et al. 2009). Insulin resistant diabetes may have a particularly prominent negative impact on temporal lobe regions, including the hippocampus. Diabetes was associated with reduced cerebral blood volume in the dentate gyrus of the hippocampus and entorhinal cortex (Wu et al. 2008). Diabetes reduces hippocampal volume, and this structural alteration is associated with impaired performance in behavioral tests of hippocampal function (Convit et al. 2003; Gold et al. 2007; den Heijer et al. 2003; Korf et al. 2006; but see Burns et al. 2007). Overall, the human data support the idea that diabetes is associated with temporal lobe dysfunction.

## 2 Deficits in Hippocampal Structure and Function in Rodent Models of Diabetes

Cognitive deficits have also been documented in rodent models of diabetes. The best-characterized rodent model of diabetes is the streptozocin (STZ)-treated rat. STZ specifically targets the beta-cells of the pancreas, resulting in their death, insulin deficiency and chronically elevated glucose levels. Functionally, STZ-treated rodents exhibit impaired performance on tests of spatial learning (Biessels et al. 1996) and object recognition (Stranahan et al. 2008a). Learning involves changes in the strength of synapses between groups of neurons; strengthening of synapses during learning is associated with a phenomenon known as long-term potentiation (LTP). LTP is impaired in both the perforant path input to the dentate gyrus (Fig. 1; Kamal et al. 1999, Stranahan et al. 2008a) and the Schaffer collateral input to hippocampal area CA1 (Biessels et al. 1996) in STZ-induced diabetes.

Neurons exhibit changes in their morphology with synaptic activity, particularly among small, highly motile protrusions known as dendritic spines. In hippocampal area CA1, the density of spines is reduced following the induction of diabetes with STZ (Martinez-Tellez et al. 2005). In the hippocampal dentate gyrus, new neurons are continually generated, and this form of structural plasticity is also impaired in STZ diabetes (Stranahan et al. 2008a; Zhang et al. 2008; Revsin et al. 2009). Overall, it is clear that untreated insulin-deficient diabetes leads to changes in neuronal structure and function in the hippocampus.

Similar deficits have been reported in leptin-deficient and leptin receptor-deficient rodents. The leptin-deficient mouse (*ob*/*ob*), along with leptin receptor-deficient mice (*db*/*db*) and rats (Zucker), are the most commonly used models for insulin-resistant diabetes. Functionally, *db*/*db* mice and Zucker rats perform poorly on tests of hippocampus-dependent memory (Stranahan et al. 2008a; Winocur et al. 2005; Li et al. 2002; but see Bélanger et al. 2004). At the synaptic level, *db*/*db* mice exhibit LTP impairment at perforant path synapses in the dentate gyrus (Fig. 1; Stranahan et al. 2008a) and Schaffer collateral synapses in hippocampal area CA1 (Li et al. 2002). In contrast, Zucker rats show LTP impairment in the CA1 subfield

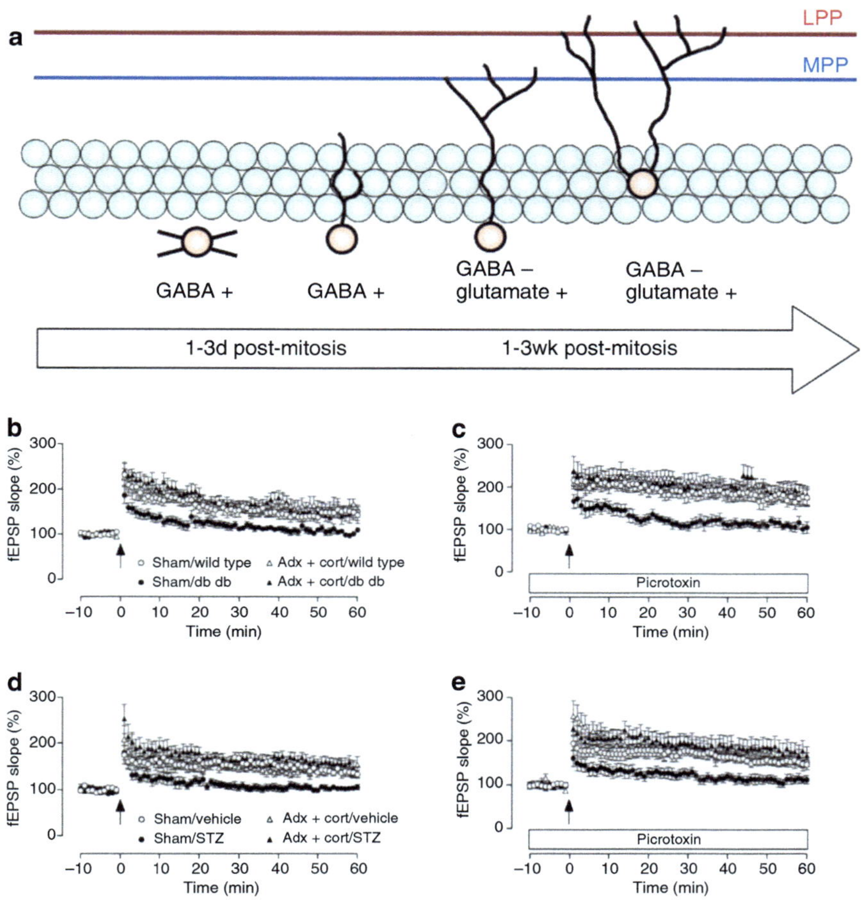


**Fig. 1** Deficits in perforant path synaptic plasticity are reversed when normal levels of corticosterone are maintained under pharmacological conditions that activate both new and mature dentate gyrus granule neurons. (**a**) Time course of granule cell maturation during adult neurogenesis. Initially, the inhibitory neurotransmitter GABA is excitatory in newly generated neurons (Ge et al. 2006). As the neuron matures and begins to make synaptic contacts, the cell switches from GABAergic excitation to glutamatergic excitation. (**b**) In *db*/*db* mice, medial perforant path LTP is impaired under recording conditions that permit activation of new and mature neurons. Adrenalectomy (Adx) and low-dose corticosterone (cort) replacement reverse this deficit. (**c**) The GABA A antagonist picrotoxin blocks GABAergic excitation on new neurons (Saxe et al. 2006; Stranahan et al. 2008a). Under this pharmacological condition, which restricts activation to mature neurons, glucocorticoids mediate the impairment of medial perforant path synaptic plasticity in *db*/*db* mice. (**d**) Medial perforant path LTP was impaired in insulin-deficient STZ-diabetic rats under recording conditions that activate new and mature neurons. This deficit was preventable by maintaining normal physiological levels of corticosterone. (**e**) Under recording conditions that restrict activation to mature neurons, STZ diabetes was again associated with deficits in synaptic plasticity that were reversible by lowering corticosterone levels

of the hippocampus (Gerges et al. 2003; Alzoubi et al. 2005; but see Bélanger et al. 2004) without detectable impairment in the dentate gyrus (Alzoubi et al. 2005).

Structurally, *db*/*db* mice show reduced dendritic spine density in the hippocampal dentate gyrus (Stranahan et al. 2009). Both *db*/*db* mice and Zucker rats show reduced levels of adult neurogenesis in the dentate gyrus (Stranahan et al. 2008a; Hwang et al. 2008). Levels of dendritic spine density (Stranahan et al. 2008c) and adult neurogenesis (Lindqvist et al. 2006) are also reduced in rats with diet-induced insulin resistance and obesity. Because parallel observations have been reported in human subjects with either Type 1 or Type 2 diabetes, and in animal models of both types of diabetes, it is unlikely that global changes in insulin levels are directly responsible for impaired hippocampal plasticity.

## 3 Diabetes and the HPA Axis

What is the peripheral metabolic relationship between changes in insulin, glucose, and glucocorticoids? Changes in glucocorticoid signaling contribute to hyperglycemia in diabetes. Specifically, restricting glucocorticoid signaling normalizes fasting glucose levels in insulin-resistant Zucker rats (Watts et al. 2005) and *db*/*db* mice (Stranahan et al. 2008a). In contrast with effects on fasting glucose levels, there was no change in postprandial glucose concentrations following adrenalectomy and low-dose corticosterone replacement in the *ob*/*ob* mouse model of insulin resistance (Tokuyama and Himms–Hagen 1987). Similarly, there was no effect of interference with glucocorticoid receptor signaling on postprandial glucose levels in Zucker rats (Langley and York 1990). Overall, these studies indicate that interference with glucocorticoid signaling regulates fasting but not fed glucose levels in rodent models of Type 2 diabetes.

How do glucocorticoids and diabetes interact in the hippocampus? Studies in rodent models support the idea that exposure to chronically elevated glucocorticoids mediates deficits in cognition following psychological stress (Oitzl et al. 1998; Wright et al. 2006). Long-term exposure to stress levels of glucocorticoids also impairs synaptic plasticity (Alfarez et al. 2003; Kerr et al. 1989; Korz and Frey 2003; Pavlides et al. 1993). Structurally, prolonged stress decreases hippocampal dendritic spine and synapse density (Hajszan et al. 2009) and impairs hippocampal neurogenesis (Gould et al. 1997), opening the possibility that diabetes-induced elevations in glucocorticoids may mediate impairments in hippocampal plasticity.

In humans, poor glycemic control is associated with hyperactivation of the HPA axis, leading to elevated levels of circulating cortisol (Bruehl et al. 2007). Levels of adrenal glucocorticoids are also increased in rodents with experimental diabetes (Stranahan et al. 2008a; Magariños and McEwen 2001; Shimomura et al. 1987). Because elevated glucocorticoids are thought to contribute to cognitive impairment following psychological stress (Oei et al. 2006), aging (MacLullich et al. 2005), and in Alzheimer's disease (AD; Elgh et al. 2006), changes in glucocorticoid production may be relevant for cognitive dysfunction in diabetes.

# 4 Effect of Diabetes on Hippocampal Structure and Function: A Role for Glucocorticoids

Diabetes exerts deleterious effects on the brain that may be mediated in part by chronic hyperglycemia and deficient insulin signaling. Emerging evidence suggests that adrenal steroid hormones also contribute to deficits in neurogenesis, synaptic plasticity, and cognition. In rodent models, adrenalectomy and low-dose corticosterone replacement reverse the impairment of learning and memory in both insulin-deficient and insulin-resistant animals. Restoring normal physiological levels of corticosterone also prevents deficits in synaptic plasticity and adult hippocampal neurogenesis. In terms of the cellular interaction between adult neurogenesis and synaptic plasticity, both newly generated neurons and pre-existing, mature dentate gyrus granule neurons respond to diabetes with impaired synaptic plasticity (Fig. 1). Lowering corticosterone levels ameliorates LTP deficits in both neuronal populations (Stranahan et al. 2008a).

Prolonged exposure to elevated levels of glucocorticoids (corticosterone in rodents and cortisol in humans) has a negative impact on learning in non-diabetic rodents (Wright et al. 2006) and humans (Grillon et al. 2004). Because there have been reports of cognitive and synaptic deficits in both insulin-resistant and insulin-deficient diabetes models, it is likely that global changes in insulin levels do not account for these effects. Moreover, levels of insulin and glucose in the hippocampus have not been shown to change with insulin deficiency or insulin resistance (Stranahan et al. 2008a). This evidence does not support a role for local changes in insulin and glucose levels as mediators of diabetes-induced cognitive deficits. However, the absence of local changes in the concentration of glucose and insulin levels in the hippocampus does not preclude changes in the expression of glucose and insulin receptors. Diabetes is associated with changes in the expression of glucose transporters (Reagan et al. 2001) and insulin signaling transcripts (Clodfelder-Miller et al. 2005) in the hippocampus. It is intriguing to consider the possibility that changes in insulin signaling and glucose transporter expression in the diabetic brain may be secondary to changes in corticosterone levels (Fig. 2).

The effects of insulin on cognition are opposite to the effects of prolonged corticosterone administration. Exposure to elevated corticosterone levels reduces translocation of the insulin-sensitive glucose transporter GLUT4 in the hippocampus of rats (Piroli et al. 2007). Intrahippocampal insulin administration (Moosavi et al. 2007) or pharmacological activation of insulin signaling pathways (Revest et al. 2005) protects against stress-induced deficits in learning and memory, suggesting that the suppressive effects of diabetes on hippocampal function may be attributable to an interaction between elevated glucocorticoids and insulin receptor signaling.

Restoring normal physiological levels of corticosterone in diabetes also reverses impairments on tasks that activate new and mature neurons differently. Although performance in the water maze is unaffected following pharmacological inhibition of new neuron production (Shors et al. 2002), genetic inhibition of adult

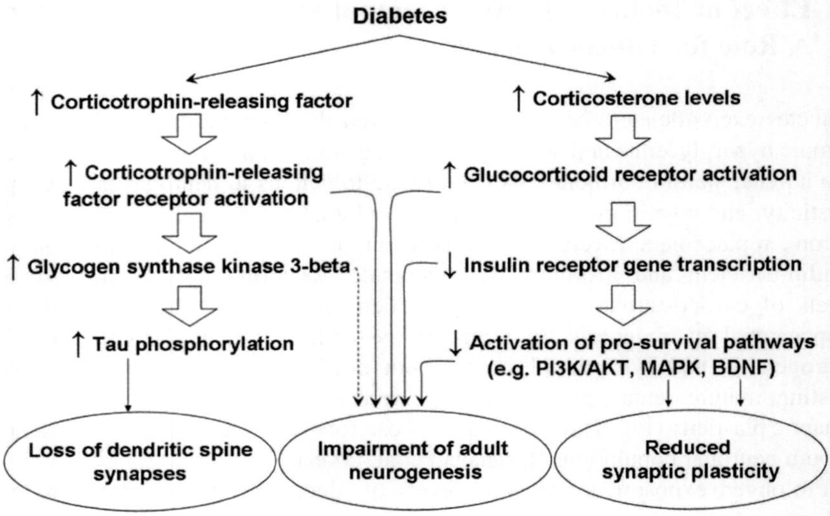


**Fig. 2** Model for molecular changes and biological processes disrupted by HPA axis hormones in diabetes. Type 2 diabetes is associated with increased corticotrophin-releasing factor concentrations and alterations in corticotrophin-releasing factor receptor expression (Jöhren et al. 2007). Stress increases tau phosphorylation through corticotrophin-releasing factor-mediated activation of glycogen synthase kinase 3-beta (Rissman et al. 2007). Abnormal accumulation of tau may lead to loss of spines and synapses (Thies and Mandelkow 2007). Diabetes is also associated with increases in circulating corticosterone levels (Stranahan et al. 2008a, b). Exposure to elevated corticosterone levels (Gould et al. 1992) and corticotrophin-releasing factor receptor activation (Alonso et al. 2004) both impair adult neurogenesis. Similarly, glucocorticoid-mediated down-regulation of insulin receptor expression could also disrupt adult neurogenesis. Downstream of glucocorticoid-mediated changes in insulin receptor expression is the reduced activation of pro-survival pathways, which could contribute both to the impairment of adult neurogenesis and reductions in synaptic plasticity

neurogenesis leads to deficits in this task (Zhang et al. 2008). Similarly, recognition memory was impaired following genetic ablation of adult neurogenesis (Jessberger et al. 2009) but unaffected following focal cranial irradiation (Saxe et al., 2006). While the contribution of adult-generated neurons to learning and memory is still being elucidated, the therapeutically relevant question is whether new neurons can influence cognition in disease models, such as diabetes. It is apparent that, in diabetes models, both newly generated neurons and pre-existing mature neurons exhibit dynamic regulation of synaptic plasticity by glucocorticoids.

## 5 Diabetes and Elevated Glucocorticoids in Human Cognition

Studies in humans also suggest that diabetes adversely affects learning and memory. However, not all aspects of cognition are equally affected by diabetes. Diabetic individuals show impairments on tasks that require episodic memory, whereas attention and language faculties are unaffected (Messier 2005). Because episodic memory recruits temporal lobe structures, whereas attention and language involve other cortical and prefrontal regions, these data can be interpreted to suggest a particular vulnerability of the hippocampus and associated cortical regions to the adverse consequences of metabolic impairment in diabetes.

Additional studies have begun to elucidate the role of glucocorticoid signaling in cognitive deficits in diabetic humans. Specifically, inhibition of the enzyme 11–β–hydroxysteroid dehydrogenase 1 (11βHSD1), which locally regulates glucocorticoid actions in the hippocampus by reactivating cortisol from inactive cortisone, reverses cognitive deficits in humans with insulin-resistant diabetes (Sandeep et al. 2005). The specific nature of cognitive impairment in diabetes, and the observation of improved cognition following treatments that limit the actions of cortisol, indicates that elevated cortisol levels may also contribute to the impairment of hippocampal function in humans.

## 6 Interventions that Attenuate Diabetes also Reverse Central Diabetic Encephalopathy

Diabetes induces a neurodegenerative behavioral phenotype that is reminiscent of aging and AD (Sandeep et al. 2004; Biessels et al. 1996; Li et al. 2002; Stranahan et al. 2008a; Martínez-Tellez et al. 2005; Zhou et al. 2007; Winocur et al. 2005). Recently, we demonstrated that attenuation of the peripheral metabolic characteristics of insulin resistance also restores hippocampal neurotrophin levels and dendritic spine density. *db/db* mice exhibit reduced dendritic spine density, an effect that is partially mitigated following exercise and dietary restriction (Stranahan et al. 2009). These structural findings may be related to hippocampal function, based on the extensive literature linking plasticity among spines and synapses with hippocampus-dependent memory (Leuner and Shors 2004).

While previous studies have shown reductions in hippocampal spine and synapse density in insulin-deficient diabetes (Martínez-Tellez et al. 2005; Zhou et al. 2007), fewer studies have addressed the possibility of similar changes in insulin-resistant diabetes. One study reported deficits in presynaptic marker expression in whole-hippocampal homogenates from *db/db* and *ob/ob* mice (Ahima et al. 1999). Using a diet-induced insulin resistance model, we have observed reduced dendritic spine density and impaired LTP in the hippocampal CA1 subfield (Stranahan et al. 2008c). However, no studies had previously investigated changes in dentate gyrus granule neuron morphology in insulin resistance.

Brain-derived neurotrophic factor (BDNF) is a pleiotropic growth factor that contributes to neuronal survival, differentiation, and synaptic integration. Emerging data suggest that BDNF also influences cellular energy metabolism (Burkhalter et al. 2003; Yeo et al. 2004). The coincident roles of BDNF in cellular metabolism and synaptogenesis led us to investigate the regulation of BDNF protein levels and dendritic spine density in *db/db* mice, which have deficient metabolic function. We chose to investigate the dentate gyrus of the hippocampus because BDNF is particularly abundant in the dentate gyrus, relative to the CA1 subfield (Friedman et al. 1991). To follow up on our observation of reduced BDNF and loss of dendritic spines in *db/db* mice, and to model potential human interventions, we assessed the consequences of voluntary wheel running and caloric restriction. These manipulations enhanced dendritic spine density and hippocampal BDNF expression in wild type mice and partially reversed abnormalities in *db/db* mice. These findings suggest that the adverse effects of diabetes on hippocampal structural plasticity can be ameliorated by increasing energy expenditure and decreasing energy intake.

## 7 Brain-derived Neurotrophic Factor and Diabetes

Voluntary wheel running and caloric restriction, which have anti-diabetic effects, also increase levels of BDNF in the hippocampus (Mattson et al. 2004a, b; Neeper et al. 1996; Ding et al. 2006). Emerging evidence suggests that the enhancement of metabolic efficiency is not unique to peripheral tissues and may occur also in the brain (Vaynman et al. 2006; Gomez-Pinilla et al. 2008). This "metabotrophic hypothesis" for the effects of exercise and caloric restriction on hippocampal structure and biochemistry has potential relevance for the treatment and prevention of neurodegenerative disease.

The effects of diabetes on neuronal structure in the hippocampus are qualitatively similar to the consequences of restricting BDNF signaling. Hippocampal BDNF levels fluctuate in an inverse relationship with fasting glucose levels (Anson et al. 2003; Duan et al. 2003). BDNF hetereozygous knockout mice are obese and insulin resistant (Duan et al. 2003), and this phenotype extends to the reported case of a mutation in the gene coding for BDNF in humans (Gray et al. 2006). In the dentate gyrus, intact BDNF signaling is a prerequisite for the actions of antidepressants, suggestive of a role in anxiety and mood regulation (Adachi et al. 2008). This finding indicates that correlated alterations in dentate gyrus BDNF signaling and neuronal structure may be associated with the changes in anxiety-like behavior that have been reported in rodent models of insulin resistance (Asakawa et al. 2003).

Intracerebroventricular infusion of BDNF enhances peripheral glucose sensitivity in *db/db* mice (Nakagawa et al. 2000), opening the possibility that increases in central BDNF could be related to reduced fasting glucose levels in serum. However, based on the data thus far, we cannot conclude that changes in BDNF are driving changes in peripheral metabolic markers and central dendritic spines, or vice versa. An alternative hypothesis would suggest that alterations in the cerebral vasculature could be driving diabetes-induced changes in spines, and alterations in BDNF could be unrelated to dendritic alterations. Future studies will be needed to determine the contributions of changes in hippocampal BDNF levels to alterations in dendritic spine density in *db/db* mice.

## 8 Stress, Diabetes, and AD Pathology

Prolonged administration of stress levels of corticosterone increases Abeta and tau pathology in a triple-transgenic mouse model of AD (Green et al. 2006). Social isolation stress also exacerbates Abeta deposition in Tg2576 mice (Dong et al. 2008). In non-human primates, glucocorticoid treatment reduces levels of insulin-degrading enzyme, which clears Abeta from the brain (Kulstadt et al. 2005). Stress is sufficient to increase tau phosphorylation (Stein-Behrens et al. 1994; Rissman et al. 2007), and this effect of stress may occur as the result of glucocorticoid-dependent and glucocorticoid-independent mechanisms. The stress-induced increase in phospho-tau is mediated in part by corticotrophin-releasing factor (CRF). Stress-induced elevations in amyloid beta in Tg2576 were also dependent on the actions of CRF rather than corticosterone (Kang et al. 2007).

STZ-induced diabetes is associated with hyperphosphorylation of tau (Planel et al. 2007). Diet-induced insulin resistance exacerbates Abeta pathology in the brains of Tg2576 mice (Ho et al. 2004). However, obesity and insulin resistance are not sufficient to cause AD pathology (Moroz et al. 2008). No studies to date have addressed the possibility that diabetes-induced increases in tau phosphorylation might be dependent on the actions of corticosterone or CRF.

It should be noted that exercise and caloric restriction, which have antidiabetic effects, also attenuate pathology in AD models. Caloric restriction ameliorates hippocampus-dependent learning deficits and reduces Abeta accumulation and tau phosphorylation in a triple-transgenic mouse model of AD (Halagappa et al. 2007). Similarly, voluntary wheel running attenuates cognitive deficits and reduces Abeta accumulation in the TgCRND8 mouse model of AD (Adlard et al. 2005). Both exercise and caloric restriction enhance levels of BDNF in the hippocampus (Mattson et al. 2004a, b; Neeper et al. 1996; Ding et al. 2006). Levels of hippocampal BDNF are reduced in humans with AD (Phillips et al. 1991), raising the possibility that exercise and caloric restriction might exert protective effects by increasing levels of BDNF. However, the mechanistic relationship between BDNF and Abeta deposition is as yet unknown.

## 9 Epidemiology of Diabetes, Stress, and AD in Humans

Diabetes is more frequent among populations with stress-related disorders, such as depression (Knol et al. 2006). The reverse is also true; among individuals with either Type 1 or Type 2 diabetes, depression and anxiety are more common (Gendelman et al. 2009; Collins et al. 2009). Similarly, AD occurs more frequently among individuals with depression (Geerlings et al. 2008), and symptoms of depression are commonly reported among AD patients (Amieva et al. 2008). Indeed, this relationship has been upheld at the level of plaque deposition, with individuals suffering from comorbid AD and depression exhibiting a higher plaque burden (Rapp et al. 2008).

Most studies have supported the association between diabetes and AD (Rönnemaa et al. 2008; for review see Messier 2005; Greenwood and Winocur 2005). This relationship was upheld at the level of brain structure, with coincident AD and insulin resistance leading to greater cortical atrophy (Biessels et al. 2006). Taken together, diabetes, stress, and AD form an epidemiological triumvirate of interrelated processes that are deleterious for brain function.

## 10 Summary and Conclusion

Diabetes has a negative impact on the structure and function of the temporal lobe. In humans, diabetes is associated with poor learning and memory and hippocampal atrophy (Messier et al. 2005; Greenwood and Winocur 2005; Biessels et al. 2006). Inhibiting the actions of glucocorticoids ameliorates cognitive deficits in diabetic humans (Sandeep et al. 2005) but it remains to be seen what neuroimaging or structural correlates might occur in diabetic humans following anti-glucorticoid treatments. In rodent models, diabetes impairs neuronal function at the behavioral and synaptic levels, in both the hippocampal dentate gyrus (Kamal et al. 1999; Stranahan et al. 2008a) and CA1 fields (Biessels et al. 1996; Stranahan et al. 2008c; Kamal et al. 1999). Structurally, untreated diabetes is associated with reduced dendritic spine density in the hippocampal dentate gyrus of genetically insulin-resistant mice (Stranahan et al. 2009) and in the CA1 subfield of diet-induced insulin-resistant rats (Stranahan et al. 2008c). The production of newly generated neurons is also reduced across pharmacological (Zhang et al. 2008; Stranahan et al. 2008a; Revsin et al. 2009), genetic (Stranahan et al. 2008a; Hwang et al. 2008), and dietary (Lindqvist et al. 2006) models of insulin resistance and insulin deficiency.

Some of the negative consequences of diabetes are mediated by elevated glucocorticoids. Lowering corticosterone levels through adrenalectomy and corticosterone replacement (Stranahan et al. 2008a) or restricting glucocorticoid signaling through the use of corticosterone receptor antagonists (Revsin et al. 2009) prevents diabetes-induced alterations in adult neurogenesis in the hippocampus. Maintaining normal physiological levels of corticosterone also restores synaptic plasticity and hippocampus-dependent memory (Stranahan et al. 2008a). Peripherally, lowering corticosterone levels (Stranahan et al. 2008a) or restricting glucocorticoid signaling (Watts et al. 2005) also attenuates the metabolic sequelae of Type 2 diabetes. Future studies will be needed to evaluate a possible role for BDNF in these effects. However, behavioral interventions that restore hippocampal BDNF also attenuate the loss of dendritic spines on dentate gyrus granule neurons in the brains of Type 2 diabetic mice (Stranahan et al. 2009).

Because the effects of diabetes and AD on the structure and function of the hippocampus are qualitatively similar, it is possible that they share common mechanisms. Molecular targets for the actions of diabetes, elevated HPA axis hormones, and AD include insulin-degrading enzyme and tau phosphorylation. It will be exciting to identify novel shared mechanisms and elucidate potential therapeutics for the prevention of diabetes, depression, and AD.

## References


Adachi M, Barrot M, Autry AE, Theobald D, Monteggia LM (2008) Selective loss of brain-derived neurotrophic factor in the dentate gyrus attenuates antidepressant efficacy. Biol Psychiat 63:642–649

Adlard PA, Perreau VM, Pop V, Cotman CW (2005) Voluntary exercise decreases amyloid load in a transgenic model of Alzheimer's disease. J Neurosci 25:4217–4221

Ahima RS, Bjorbaek C, Osei S, Flier JS (1999) Regulation of neuronal and glial proteins by leptin: implications for brain development. Endocrinology 140:2755–2762

Alfarez DN, Joels M, Krugers HJ (2003) Chronic unpredictable stress impairs long–term potentiation in rat hippocampal CA1 area and dentate gyrus in vitro. Eur J Neurosci 17:1928–1934

Alonso R, Griebel G, Pavone G, Stemmelin J, Le Fur G, Soubrié P (2004) Blockade of CRF(1) or V(1b) receptors reverses stress-induced suppression of neurogenesis in a mouse model of depression. Mol Psychiat 9:278–286, 224

Alzoubi KH, Aleisa AM, Alkadhi KA (2005) Impairment of long-term potentiation in the CA1, but not dentate gyrus, of the hippocampus in Obese Zucker rats: role of calcineurin and phosphorylated CaMKII. J Mol Neurosci 27:337–346

Amieva H, Le Goff M, Millet X, Orgogozo JM, Pérès K, Barberger-Gateau P, Jacqmin-Gadda H, Dartigues JF (2008) Prodromal Alzheimer's disease: successive emergence of the clinical symptoms. Ann Neurol 64:492–498

Anson RM, Guo Z, de Cabo R, Iyun T, Rios M, Hagepanos A, Ingram DK, Lane MA, Mattson MP (2003) Intermittent fasting dissociates beneficial effects of dietary restriction on glucose metabolism and neuronal resistance to injury from calorie intake. Proc Natl Acad Sci USA 100:6216–6220

Asakawa A, Inui A, Inui T, Katsuura G, Fujino MA, Kasuga M (2003) Leptin treatment ameliorates anxiety in ob/ob obese mice. J Diabetes Complications 17:105–107

Bélanger A, Lavoie N, Trudeau F, Massicotte G, Gagnon S (2004) Preserved LTP and water maze learning in hyperglycaemic-hyperinsulinemic ZDF rats. Physiol Behav 83:483–494

Biessels GJ, Kamal A, Ramakers GM, Urban IJ, Spruijt BM, Erkelens DW, Gispen WH (1996) Place learning and hippocampal synaptic plasticity in streptozotocin-induced diabetic rats. Diabetes 45:1259–1266

Biessels GJ, De Leeuw FE, Lindeboom J, Barkhof F, Scheltens P (2006) Increased cortical atrophy in patients with Alzheimer's disease and type 2 diabetes mellitus. J Neurol Neurosurg Psychiat 77:304–307

Bruehl H, Rueger M, Dziobek I, Sweat V, Tirsi A, Javier E, Arentoft A, Wolf OT, Convit A (2007) Hypothalamic-pituitary-adrenal axis dysregulation and memory impairments in type 2 diabetes. J Clin Endocrinol Metab 92:2439–2445

Burkhalter J, Fiumelli H, Allaman I, Chatton JY, Martin JL (2003) Brain-derived neurotrophic factor stimulates energy metabolism in developing cortical neurons. J Neurosci 23:8212–8220

Burns JM, Donnelly JE, Anderson HS, Mayo MS, Spencer-Gardner L, Thomas G, Cronk BB, Haddad Z, Klima D, Hansen D, Brooks WM (2007) Peripheral insulin and brain structure in early Alzheimer disease. Neurology 69:1094–1104

Clodfelder-Miller B, De Sarno P, Zmijewska AA, Song L, Jope RS (2005) Physiological and pathological changes in glucose regulate brain Akt and glycogen synthase kinase-3. J Biol Chem 280:39723–39731

Collins MM, Corcoran P, Perry IJ (2009) Anxiety and depression symptoms in patients with diabetes Diabet Med 26:153–161
Convit A, Wolf OT, Tarshish C, de Leon MJ (2003) Reduced glucose tolerance is associated with poor memory performance and hippocampal atrophy among normal elderly. Proc Natl Acad Sci USA 100:2019–2022
den Heijer T, Vermeer SE, van Dijk EJ, Prins ND, Koudstaal PJ, Hofman A, Breteler MM (2003) Type 2 diabetes and atrophy of medial temporal lobe structures on brain. MRI. Diabetologia 46:1604–1610
Ding Q, Vaynman S, Akhavan M, Ying Z, Gomez-Pinilla F (2006) Insulin-like growth factor I interfaces with brain-derived neurotrophic factor-mediated synaptic plasticity to modulate aspects of exercise-induced cognitive function. Neuroscience 140:823–833
Dong H, Yuede CM, Yoo HS, Martin MV, Deal C, Mace AG, Csernansky JG (2008) Corticosterone and related receptor expression are associated with increased beta-amyloid plaques in isolated Tg2576 mice Neuroscience 155:154–163
Duan W, Guo Z, Jiang H, Ware M, Mattson MP (2003) Reversal of behavioral and metabolic abnormalities and insulin resistance syndrome by dietary restriction in mice deficient in brain-derived neurotrophic factor. Endocrinology 144:2446–2453
Elgh E, Lindqvist Astot A, Fagerlund M, Eriksson S, Olsson T, Näsman B (2006) Cognitive dysfunction, hippocampal atrophy and glucocorticoid feedback in Alzheimer's disease. Biol Psychiatry 59:155–161
Friedman WJ, Olson L, Persson H. (1991) Cells that express brain-derived neurotrophic factor mRNA in the developing postnatal rat brain. Eur J Neurosci 3: 688–697
Ge S, Goh EL, Sailor KA, Kitabatake Y, Ming GL, Song H (2006) GABA regulates synaptic integration of newly generated neurons in the adult brain. Nature 439:589–593
Geerlings MI, den Heijer T, Koudstaal PJ, Hofman A, Breteler MM (2008) History of depression, depressive symptoms, and medial temporal lobe atrophy and the risk of Alzheimer disease. Neurology 70:1258–1264
Gendelman N, Snell-Bergeon JK, McFann K, Kinney G, Wadwa RP, Bishop F, Rewers M, Maahs DM (2009) Prevalence and Correlates of Depression in Persons with and without Type 1 Diabetes. Diabetes Care 32:575–579
Gerges NZ, Aleisa AM, Alkadhi KA (2003) Impaired long-term potentiation in obese zucker rats: possible involvement of presynaptic mechanism. Neuroscience 120:535–539
Gold SM, Dziobek I, Sweat V, Tirsi A, Rogers K, Bruehl H, Tsui W, Richardson S, Javier E, Convit A (2007) Hippocampal damage and memory impairments as possible early brain complications of type 2 diabetes. Diabetologia 50:711–719
Gomez-Pinilla F, Vaynman S, Ying Z (2008) Brain-derived neurotrophic factor functions as a metabotrophin to mediate the effects of exercise on cognition. Eur J Neurosci 28:2278–2287
Gould E, Cameron HA, Daniels DC, Woolley CS, McEwen BS (1992) Adrenal hormones suppress cell division in the adult rat dentate gyrus. J Neurosci 12:3642–3650
Gould E, McEwen BS, Tanapat P, Galea LA, Fuchs E (1997) Neurogenesis in the dentate gyrus of the adult tree shrew is regulated by psychosocial stress and NMDA receptor activation. J Neurosci 17:2492–2498
Gray J, Yeo GS, Cox JJ, Morton J, Adlam AL, Keogh JM, Yanovski JA, El Gharbawy A, Han JC, Tung YC, Hodges JR, Raymond FL, O'rahilly S, Farooqi IS (2006) Hyperphagia, severe obesity, impaired cognitive function, and hyperactivity associated with functional loss of one copy of the brain-derived neurotrophic factor (BDNF) gene. Diabetes 55:3366–3371
Green KN, Billings LM, Roozendaal B, McGaugh JL, LaFerla FM (2006) Glucocorticoids increase amyloid-beta and tau pathology in a mouse model of Alzheimer's disease. J Neurosci 26:9047–9056
Greenwood CE, Winocur G (2005) High–fat diets, insulin resistance, and declining cognitive function. Neurobiol Aging 26 (Suppl 1):S42–45
Grillon C, Smith K, Haynos A, Nieman LK (2004) Deficits in hippocampus-mediated Pavlovian conditioning in endogenous hypercortisolism. Biol Psychiatry 56:837–843

Hajszan T, Dow A, Warner-Schmidt JL, Szigeti-Buck K, Sallam NL, Parducz A, Leranth C, Duman RS (2009) Remodeling of hippocampal spine synapses in the rat learned helplessness model of depression. Biol Psychiat 65:392–400

Halagappa VK, Guo Z, Pearson M, Matsuoka Y, Cutler RG, Laferla FM, Mattson MP (2007) Intermittent fasting and caloric restriction ameliorate age-related behavioral deficits in the triple-transgenic mouse model of Alzheimer's disease. Neurobiol Dis 26:212–220

Ho L, Qin W, Pompl PN, Xiang Z, Wang J, Zhao Z, Peng Y, Cambareri G, Rocher A, Mobbs CV, Hof PR, Pasinetti GM (2004) Diet-induced insulin resistance promotes amyloidosis in a transgenic mouse model of Alzheimer's disease FASEB J 18:902–904

Hwang IK, Yi SS, Kim YN, Kim IY, Lee IS, Yoon YS, Seong JK (2008) Reduced hippocampal cell differentiation in the subgranular zone of the dentate gyrus in a rat model of type II diabetes. Neurochem Res 33:394–400

Jessberger S, Clark RE, Broadbent NJ, Clemenson GD Jr, Consiglio A, Lie DC, Squire LR, Gage FH (2009) Dentate gyrus-specific knockdown of adult neurogenesis impairs spatial and object recognition memory in adult rats. Learn Mem 16:147–154

Jöhren O, Dendorfer A, Dominiak P, Raasch W (2007) Gene expression of mineralocorticoid and glucocorticoid receptors in the limbic system is related to type-2 like diabetes in leptin-resistant rats. Brain Res 1184:160–167

Kamal A, Biessels GJ, Urban IJ, Gispen WH (1999) Hippocampal synaptic plasticity in streptozotocin-diabetic rats: impairment of long-term potentiation and facilitation of long-term depression. Neuroscience 90:737–745

Kang JE, Cirrito JR, Dong H, Csernansky JG, Holtzman DM (2007) Acute stress increases interstitial fluid amyloid-beta via corticotropin-releasing factor and neuronal activity. Proc Natl Acad Sci USA 104:10673–10678

Kerr DS, Campbell LW, Hao SY, Landfield PW (1989) Corticosteroid modulation of hippocampal potentials: increased effect with aging. Science 245:1505–1509

Knol MJ, Twisk JW, Beekman AT, Heine RJ, Snoek FJ, Pouwer F (2006) Depression as a risk factor for the onset of type 2 diabetes mellitus: a meta-analysis. Diabetologia 49:837–845

Korf ES, White LR, Scheltens P, Launer LJ (2006) Brain aging in very old men with type 2 diabetes: the Honolulu-Asia Aging Study. Diabetes Care 29:2268–2274

Korz V, Frey JU (2003) Stress–related modulation of hippocampal long–term potentiation in rats: Involvement of adrenal steroid receptors. J Neurosci 23:7281–7287

Kulstad JJ, McMillan PJ, Leverenz JB, Cook DG, Green PS, Peskind ER, Wilkinson CW, Farris W, Mehta PD, Craft S (2005) Effects of chronic glucocorticoid administration on insulin-degrading enzyme and amyloid-beta peptide in the aged macaque. J Neuropathol Exp Neurol 64:139–146

Langley SC, York DA (1990) Effects of antiglucocorticoid RU486 on development of obesity in obese fa/fa Zucker rats. Am J Physiol 259:539–544

Leuner B, Shors TJ (2004) New spines, new memories. Mol Neurobiol 29:117–130

Li XL, Aou S, Oomura Y, Hori N, Fukunaga K, Hori T (2002) Impairment of long-term potentiation and spatial memory in leptin receptor-deficient rodents. Neuroscience 113:607–615

Lindqvist A, Mohapel P, Bouter B, Frielingsdorf H, Pizzo D, Brundin P, Erlanson-Albertsson C (2006) High-fat diet impairs hippocampal neurogenesis in male rats. Eur J Neurol 13:1385–1388

MacLullich AM, Deary IJ, Starr JM, Ferguson KJ, Wardlaw JM, Seckl JR (2005) Plasma cortisol levels, brain volumes and cognition in healthy elderly men. Psychoneuroendocrinology 30:505–515

Magariños AM, McEwen BS (2000) Experimental diabetes in rats causes hippocampal dendritic and synaptic reorganization and increased glucocorticoid reactivity to stress. Proc Natl Acad Sci USA 97:11056–11061

Martínez-Tellez R, Gómez-Villalobos J, Flores G (2005) Alteration in dendritic morphology of cortical neurons in rats with diabetes mellitus induced by streptozotocin. Brain Res 1048:108–115

Mattson MP, Maudsley S, Martin B (2004a) A neural signaling triumvirate that influences ageing and age-related disease: insulin/IGF-1, BDNF and serotonin. Ageing Res Rev 3:445–464

Mattson MP, Duan W, Wan R, Guo Z (2004b) Prophylactic activation of neuroprotective stress response pathways by dietary and behavioral manipulations. NeuroRx 1:111–116

Messier C (2005) Impact of impaired glucose tolerance and type 2 diabetes on cognitive aging. Neurobol Aging 26 (Suppl 1):S26–S30

Moosavi M, Naghdi N, Maghsoudi N, Zahedi Asl S (2007) Insulin protects against stress–induced impairments in water maze performance. Behav Brain Res 176:230–236

Moroz N, Tong M, Longato L, Xu H, de la Monte SM (2008) Limited Alzheimer-type neurodegeneration in experimental obesity and type 2 diabetes mellitus. J Alzheimers Dis 15:29–44

Nakagawa T, Tsuchida A, Itakura Y, Nonomura T, Ono M, Hirota F, Inoue T, Nakayama C, Taiji M, Noguchi H (2000) Brain-derived neurotrophic factor regulates glucose metabolism by modulating energy balance in diabetic mice. Diabetes 49:436–444

Neeper SA, Gómez-Pinilla F, Choi J, Cotman CW (1996) Physical activity increases mRNA for brain-derived neurotrophic factor and nerve growth factor in rat brain. Brain Res 726:49–56

Northam EA, Rankins D, Lin A, Wellard RM, Pell GS, Finch SJ, Werther GA, Cameron FJ (2009) Central nervous system function in youth with type 1 diabetes 12 years after disease onset. Diabetes Care 32:445–450

Oei NY, Everaerd WT, Elzinga BM, van Well S, Bermond B (2006) Psychosocial stress impairs working memory at high loads: an association with cortisol levels and memory retrieval. Stress 9:133–141

Oitzl MS, Fluttert M, Sutanto W, de Kloet ER (1998) Continuous blockade of brain glucocorticoid receptors facilitates spatial learning and memory in rats. Eur J Neurosci 10:3759–3766

Pavlides C, Watanabe Y, McEwen BS (1993) Effects of glucocorticoids on hippocampal long-term potentiation. Hippocampus 3:183–192

Phillips HS, Hains JM, Armanini M, Laramee GR, Johnson SA, Winslow JW (1991) BDNF mRNA is decreased in the hippocampus of individuals with Alzheimer's disease. Neuron 7:695–702

Piroli GG, Grillo CA, Reznikov LR, Adams S, McEwen BS, Charron MJ, Reagan LP (2007) Corticosterone impairs insulin-stimulated translocation of GLUT4 in the rat hippocampus. Neuroendocrinology 85:71–80

Planel E, Tatebayashi Y, Miyasaka T, Liu L, Wang L, Herman M, Yu WH, Luchsinger JA, Wadzinski B, Duff KE, Takashima A (2007) Insulin dysfunction induces in vivo tau hyperphosphorylation through distinct mechanisms. J Neurosci 27:13635–13648

Rapp MA, Schnaider-Beeri M, Purohit DP, Perl DP, Haroutunian V, Sano M (2008) Increased neurofibrillary tangles in patients with Alzheimer disease with comorbid depression. Am J Geriatr Psychiat 16:168–174

Reagan LP, Gorovits N, Hoskin EK, Alves SE, Katz EB, Grillo CA, Piroli GG, McEwen BS, Charron MJ (2001) Localization and regulation of GLUTx1 glucose transporter in the hippocampus of streptozotocin diabetic rats Proc Natl Acad Sci USA 98:2820–2825

Revest JM, Di Blasi F, Kitchener P, Rougé-Pont F, Desmedt A, Turiault M, Tronche F, Piazza PV (2005) The MAPK pathway and Egr-1 mediate stress-related behavioral effects of glucocorticoids. Nature Neurosci 8:664–672

Revsin Y, Rekers NV, Louwe MC, Saravia FE, De Nicola AF de Kloet ER, Oitzl MS (2009) Glucocorticoid receptor blockade normalizes hippocampal alterations and cognitive impairment in streptozotocin-induced type 1 diabetes mice. Neuropsychopharmacology 34:747–758

Rissman RA, Lee KF, Vale W, Sawchenko PE (2007) Corticotropin-releasing factor receptors differentially regulate stress-induced tau phosphorylation. J Neurosci 27:6552–6562

Rönnemaa E, Zethelius B, Sundelöf J, Sundström J, Degerman-Gunnarsson M, Berne C, Lannfelt L, Kilander L (2008) Impaired insulin secretion increases the risk of Alzheimer disease. Neurology 71:1065–1071

Sandeep TC, Yau JL, MacLullich AM, Noble J, Deary IJ, Walker BR, Seckl JR (2005) 11Beta-hydroxysteroid dehydrogenase inhibition improves cognitive function in healthy elderly men and type 2 diabetics. Proc Natl Acad Sci USA 101:6734–6739

Saxe MD, Battaglia F, Wang JW, Malleret G, David DJ, Monckton JE, Garcia AD, Sofroniew MV, Kandel ER, Santarelli L, Hen R, Drew MR (2006) Ablation of hippocampal neurogenesis impairs contextual fear conditioning and synaptic plasticity in the dentate gyrus. Proc Natl Acad Sci USA 103:17501–17506

Shimomura Y, Bray GA, Lee M (1987) Adrenalectomy and steroid treatment in obese (ob/ob) and diabetic (db/db) mice. Horm Metab Res 19:295–299

Shors TJ, Townsend DA, Zhao M, Kozorovitskiy Y, Gould E (2002) Neurogenesis may relate to some but not all types of hippocampal-dependent learning. Hippocampus 12:578–584

Stein-Behrens B, Mattson MP, Chang I, Yeh M, Sapolsky R (1994) Stress exacerbates neuron loss and cytoskeletal pathology in the hippocampus. J Neurosci 14:5373–5380

Stranahan AM, Arumugam TV, Cutler RG, Lee K, Egan JM, Mattson MP (2008a) Diabetes impairs hippocampal function through glucocorticoid-mediated effects on new and mature neurons. Nature Neurosci 11:309–317

Stranahan AM, Lee K, Pistell PJ, Nelson CM, Readal N, Miller MG, Spangler EL, Ingram DK, Mattson MP (2008b) Accelerated cognitive aging in diabetic rats is prevented by lowering corticosterone levels. Neurobiol Learn Mem 90:479–483

Stranahan AM, Norman ED, Lee K, Cutler RG, Telljohann RS, Egan JM, Mattson MP (2008c) Diet-induced insulin resistance impairs hippocampal synaptic plasticity and cognition in middle-aged rats. Hippocampus 18:1085–1088

Stranahan AM, Lee K, Martin B, Maudsley S, Golden E, Cutler RG, Mattson MP (2009) Voluntary exercise and caloric restriction enhance hippocampal dendritic spine density and BDNF levels in diabetic mice. Hippocampus 19:951–961

Thies E, Mandelkow EM (2007) Missorting of tau in neurons causes degeneration of synapses that can be rescued by the kinase MARK2/Par-1. J Neurosci 27:2896–2907

Tokuyama K, Himms–Hagen J (1987) Increased sensitivity of the genetically obese mouse to corticosterone. Am J Physiol 252:202–208

Vaynman S, Ying Z, Wu A, Gomez-Pinilla F (2006) Coupling energy metabolism with a mechanism to support brain-derived neurotrophic factor-mediated synaptic plasticity. Neuroscience 139:1221–1234

Watts LM, Manchem VP, Leedom TA, Rivard AL, McKay RA, Bao D, Neroladakis T, Monia BP, Bodenmiller DM, Cao JX, Zhang HY, Cox AL, Jacobs SJ, Michael MD, Sloop KW, Bhanot S (2005) Reduction of hepatic and adipose tissue glucocorticoid receptor expression with antisense oligonucleotides improves hyperglycemia and hyperlipidemia in diabetic rodents without causing systemic glucocorticoid antagonism. Diabetes 54:1846–1853

Winocur G, Greenwood CE, Piroli GG, Grillo CA, Reznikov LR, Reagan LP, McEwen BS (2005) Memory impairment in obese Zucker rats: an investigation of cognitive function in an animal model of insulin resistance and obesity. Behav Neurosci 119:1389–1395

Wright RL, Lightner EN, Harman JS, Meijer OC, Conrad CD (2006) Attenuating corticosterone levels on the day of memory assessment prevents chronic stress–induced impairments in spatial memory Eur J Neurosci 24:595–605

Wu W, Brickman AM, Luchsinger J, Ferrazzano P, Pichiule P, Yoshita M, Brown T, DeCarli C, Barnes CA, Mayeux R, Vannucci SJ, Small SA (2008) The brain in the age of old: the hippocampal formation is targeted differentially by diseases of late life. Ann Neurol 64:698–706

Yeo GS, Connie Hung CC, Rochford J, Keogh J, Gray J, Sivaramakrishnan S, O'Rahilly S, Farooqi IS (2004) A de novo mutation affecting human TrkB associated with severe obesity and developmental delay. Nature Neurosci 7:1187–1189

Zhang CL, Zou Y, He W, Gage FH, Evans RM (2008) A role for adult TLX-positive neural stem cells in learning and behaviour. Nature 451:1004–1007

Zhang WJ, Tan YF, Yue JT, Vranic M, Wojtowicz JM (2008) Impairment of hippocampal neurogenesis in streptozotocin-treated diabetic rats Acta Neurol Scand 117:205–210

Zhou J, Wang L, Ling S, Zhang X (2007) Expression changes of growth-associated protein-43 (GAP-43) and mitogen-activated protein kinase phosphatase-1 (MKP-1) and in hippocampus of streptozotocin-induced diabetic cognitive impairment rats. Exp Neurol 206:201–208

Saxe MD, Battaglia F, Wang JW, Malleret G, David DJ, Monckton JE, Garcia AD, Sofroniew MV, Kandel ER, Santarelli L, Hen R, Drew MR (2006) Ablation of hippocampal neurogenesis impairs contextual fear conditioning and synaptic plasticity in the dentate gyrus. Proc Natl Acad Sci USA 103:17501–17506

Sh[illegible] Y, [illegible] (2007) [illegible] Neurosci Res 19:[illegible]

Shors TJ, Townsend DA, Zhao M, Kozorovitskiy Y, Gould E (2002) Neurogenesis may relate to some but not all types of hippocampal-dependent learning. Hippocampus 12:578–584

Stein-Behrens B, Mattson MP, Chang I, Yeh M, Sapolsky R (1994) Stress exacerbates neuron loss and cytoskeletal pathology in the hippocampus. J Neurosci 14:5373–5380

Stranahan AM, Arumugam TV, Cutler RG, Lee K, Egan JM, Mattson MP (2008) Diabetes impairs hippocampal function through glucocorticoid-mediated effects on new and mature neurons. Nat Neurosci 11:309–317

Stranahan AM, Lee K, Pistell PJ, Nelson CM, Readal N, Miller MG, Spangler EL, Ingram DK, Mattson MP (2008) Accelerated cognitive aging in diabetic rats is prevented by lowering corticosterone levels. Neurobiol Learn Mem 90:479–483

Stranahan AM, Norman ED, Lee K, Cutler RG, Telljohann RS, Egan JM, Mattson MP (2008) Diet-induced insulin resistance impairs hippocampal synaptic plasticity and cognition in middle-aged rats. Hippocampus 18:1085–1088

Stranahan AM, Lee K, Martin B, Maudsley S, Golden E, Cutler RG, Mattson MP (2009) Voluntary exercise and caloric restriction enhance hippocampal dendritic spine density and BDNF levels in diabetic mice. Hippocampus 19:951–961

Thi[illegible] E, [illegible] (20[illegible]) [illegible] can be rescued by the kinase MAPK [illegible]

Tokuyama K, Himms-Hagen J (1987) Increased sensitivity of the genetically obese mouse to corticosterone. Am J Physiol 252:E202–E208

Vaynman S, Ying Z, Wu A, Gomez-Pinilla F (2006) Coupling energy metabolism with a mechanism to support brain-derived neurotrophic factor-mediated synaptic plasticity. Neuroscience 139:1221–1234

Wa[illegible] (2005) [illegible]

Winocur G, Greenwood CE, Piroli GG, Grillo CA, Reznikov LR, Reagan LP, McEwen BS (2005) Memory impairment in obese Zucker rats: an investigation of cognitive function in an animal model of insulin resistance and obesity. Behav Neurosci 119:1389–1395

Wright RL, Lightner EN, Harman JS, Meijer OC, Conrad CD (2006) Attenuating corticosterone levels on the day of memory assessment prevents chronic stress-induced impairments in spatial memory. Eur J Neurosci 24:595–605

Wu W, Brickman AM, Luchsinger J, Ferrazzano P, Pichiule P, Yoshita M, Brown T, DeCarli C, Barnes CA, Mayeux R, Vannucci SJ, Small SA (2008) The brain in the age of old: the hippocampal formation is targeted differentially by diseases of late life. Ann Neurol 64:698–706

Yeo GS, Connie Hung CC, Rochford J, Keogh J, Gray J, Sivaramakrishnan S, O'Rahilly S, Farooqi IS (2004) A de novo mutation affecting human TrkB associated with severe obesity and developmental delay. Nat Neurosci 7:1187–1189

Zhang CL, Zou Y, He W, Gage FH, Evans RM (2008) A role for adult TLX-positive neural stem cells in learning and behaviour. Nature 451:1004–1007

Zhang W, Tan YF, Yue JT, Vranic M, Wojtowicz JM (2008) Impairment of hippocampal neurogenesis in streptozotocin-treated diabetic rats. Acta Neurol Scand 117:205–210

Zhou J, Wang L, Ling S, Zhang X (2007) Expression changes of growth-associated protein-43 (GAP-43) and mitogen-activated protein kinase phosphatase-1 (MKP-1) and in hippocampus of streptozotocin-induced diabetic cognitive impairment rats. Exp Neurol 206:201–208

# Diabetes and the Brain – An Epidemiologic Perspective

**Lenore J. Launer**

Type 2 diabetes (T2D) and cognitive impairment are two of the most common chronic conditions found in persons 60 years and older. After that age, studies suggest approximately 18%–20% of older persons have diabetes (Harris et al. 1988), about 19% are mildly cognitively impaired (MCI) in multiple domains (Lopez et al. 2003), and about 6% of community dwelling individuals have some dementia (Lobo et al. 2000). The prevalence of MCI and dementia increases with age as does the prevalence of diabetes; there is also an alarming trend towards a younger age of diabetes onset (Chaturvedi 2007). Several lines of investigation suggest a link between diabetes and disorders of cognitive function. Thus, the age-related trends in diabetes and cognitive disorders indicate there may be an even greater increase in the number of persons with MCI and dementia, in excess of the increase that is expected based on the age structure of the population.

## 1 Brief Description of Evidence Linking Diabetes to Late-age Cognitive Disorders

The metabolic and hemodynamic profile of diabetes, including co-morbidities such as hypertension, hyperinsulinemia and obesity, modulate vascular health and neuronal survival through multiple mechanisms. Pathophysiologic mechanisms that have been identified include inflammation, oxidative stress, energy imbalance, protein misfolding, glucocorticoid-mediated effects, and differences in gene expression (Klein and Waxman 2003; Baumbach 1994; Sasaki et al. 1998; Craft and Watson 2004; Stranahan et al. 2008). More recently several endocrine proteins

L.J. Launer
Laboratory of Epidemiology, Demography and Biometry, National Institute on Aging, Gateway Building, 3C309, 7201 Wisconsin Avenue, Bethesda, MD 20892, USA

S. Craft and Y. Christen (eds.), *Diabetes, Insulin and Alzheimer's Disease*,
Reserach and Perspectives in Alzheimer's Disease,
DOI 10.1007/978-3-642-04300-0_5, © Springer-Verlag Berlin Heidelberg 2010

( i.e., angioneurins such as VEGF), have been shown to modify both vascular health and neuronal survival (Zacchigna et al. 2008). Finally, genetic findings may identify new pathways contributing to diabetes that may also increase the susceptibility to cerebral disease (Diabetes Genetic Initiative et al. 2007).

The main threat of T2D to brain health is considered to be macrovascular clinical stroke (Buse et al. 2007). However, there is evidence suggesting diabetes can have a critical role in the build-up of pervasive and mixed cerebral brain pathology. Further, there are few population studies (van Harten et al. 2006), that address questions about the prevalence, incidence, and consequences of T2D on clinically silent diffuse and focal brain changes in smaller vessels. For example, it is not known whether changes in the microvasculature of the retina, kidney, and peripheral nervous system are also common in the brain. Microvascular damage is of interest because it is highly prevalent in neuropathologic samples of demented individuals (Fernando and Ince 2004), and there is good clinical trial evidence that it can be reduced with intensive glycemic control (ADA/ACCF/AHA 2009).

## 2 Integrated Community-Based Studies of Diabetes and the Brain

Despite gaps in our knowledge of the interaction between diabetes and cerebral disease, existing data are consistent with the hypothesis that diabetic pathologies lead to both Alzheimer's disease (AD) type neurodegeneration and vascular damage, and it is the mix of these pathologies that is the anatomical basis for clinical and sub-clinical cognitive impairment in diabetes. To test this hypothesis, it is helpful to take an integrated vertical approach based on different measures of brain structure/function. With this approach, it is possible to check for consistency among correlated phenotypes or develop new hypotheses based on newly identified associations. This approach is taken using findings from the Honolulu Asia Aging Study (HAAS; White et al. 1996), the Age Gene/Environment Susceptibility – Reykjavik Study (AGES-Reykjavik Study; Harris et al. 2007), and the Memory in Diabetes (MIND; Williamson et al. 2007) sub-study embedded in the ACCORD trial (Action to Control Risk in Diabetes Study Group 2008).

The Honolulu-Asia Aging Study (HAAS) began in 1991 as a continuation of the Honolulu Heart Program, a population-based longitudinal study of Japanese-American men born between 1900 and 1919 and living in Oahu, Hawaii, when the study began in 1965. Participants were seen at three mid-life exams (1965–68, 1968–70, 1971–74), and, as reported here, at four exams in late-life (1991–93, 1994–96, 1997–99; 2001–02); follow-up is on-going. Clinical measurements, demographic, and medical information were collected at each exam. Starting in 1991, global cognitive function was measured in the total sample and cases of dementia ascertained. An autopsy study nested within the cohort was also started in 1991; a MRI sub-study of 575 men was performed in 1995–1996. HAAS has provided valuable insights on the association of diabetes and related risk factors, high blood pressure and hyperinsulinemia, to clinical disease and brain pathology. (Launer et al. 2000; Petrovitch et al. 2000; Havlik et al. 2002; Peila et al. 2004).

AGES-Reykjavik Study is a population-based follow up study of men and women born 1907–1934 who participated in the Reykjavik Study, established in 1967 by the Icelandic Heart Association. AGES-Reykjavik focuses on studying environmental and genetic factors contributing to disease in the neurocognitive, vascular, musculoskeletal, and metabolic systems. From 2002–2006, 5,764 cohort members were re-examined with state-of-art imaging technology, questionnaires, and clinical measures. All participants were administered a battery of cognitive tests of speed, memory and working memory, and all eligible participants underwent a brain MRI. Retinal photos, which provide a measure of microvessels (Harris et al. 2007), were also obtained.

In addition to studying the possible physiologic contribution of diabetes to late age brain pathology, prevention and treatment strategies are needed to reduce the risk of cognitive impairment in persons with diabetes. To test strategies to reduce brain changes due to hyperglycemia, hypertension, and dyslipidemia, ACCORD MIND (Williamson et al. 2007) was designed as a sub-study embedded in the Action to Control Cardiovascular Risk in Diabetes (ACCORD) clinical trial (ACCORD Study Group 2008). ACCORD is a randomized controlled trial of 10,251 persons with T2D and a screening A1C (glycaded hemoglobin) of 7.5% or higher and and who are at high risk for cardiovascular disease. The trial aims to compare the effect on the rate of macrovascular cardiovascular events of standard vs. intensive therapeutic strategies to lower A1C (goal $<6\%$), lower systolic BP ($<120$ mmHG goal) or increase HDL (double blind placebo trial). The Memory in Diabetes (MIND) sub-study of ACCORD compares the effect of these interventions on cognitive function (memory, psychomotor speed, executive and global function) measured at baseline, 20 and 40 months after randomization in a subset of 2,977 subjects, and structural brain changes measured at baseline and 40 months in 620 of the MIND sample. At baseline, participating MIND patients were mean age 62 yrs (range 54–79 yrs old).

# 3 Cognitive Function

Clinical studies have shown that patients with T2D have impaired neuropsychologic functioning (Coker and Schumaker 2003). In community-dwelling samples, compared to normoglycemic persons, those with diabetes have a higher prevalence of global cognitive impairment (Kalmijn et al. 1995) and a higher incidence of cognitive decline (Gregg et al. 2000). Several large epidemiologic studies have reported an association of diabetes with dementia, including AD (Leibson et al. 1997; Luchsinger et al. 2001; Ott et al. 1999). Dementia and global cognitive dysfunction have been reported in persons with diabetes, but it is still unclear as to whether there are specific deficits in T2D or what factors may modulate the risk for cognitive disorders in T2D.

In the AGES-Reykjavik Study, we recently examined the association of specific cognitive functions measured by composite scores of memory, processing speed (PS), and executive function derived from a neuropsychological test battery. Compared to persons with no diabetes, those diagnosed with diabetes had significantly slower PS than normoglycemics (beta $= -0.12$; $P < 0.05$); those with impaired fasting glucose (IGF) performed similar to normotensive subjects. We found that the duration of diabetes was an important modulator of risk, and cognitive function (PS and executive function) declined significantly as duration of diabetes increased (Saczynski et al. 2008). A finding of interest is that those who were detected as being diabetic based solely on the fasting glucose level (thus duration is not known) performed significantly more poorly than normoglycemic subjects in memory and PS. Genetic susceptibility may also modulate the risk for cognitive disorders in T2D. In the HAAS (Peila et al. 2002) and the Cardiovascular Health Study (Irie 2008), we found that those with diabetes and carrying the Apolipoprotein E (Apo E) ε4 allele (a genetic susceptibility risk factor for AD) were at higher risk for dementia than those with no diabetes and no ε4 allele, or either one alone. This interaction between diabetes and Apo E genotype has also been reported for cognitive outcomes (Kalmijn et al. 1996; Haan et al. 1999).

The vulnerability of the brain to diabetes was further supported in cross-sectional analyses of data from the ACCORD MIND participants. We found that variation in cognitive function among the diabetes patients was significantly associated with duration of diabetes as well as with baseline A1C levels. In particular, we found, after adjusting for age, sex, education and depression score, a 1% higher A1C was associated with a 1.43 (95% CI $-0.94$ to $-1.92$; $p < 0.0001$) lower DSST score, a 0.14 (95% CI $-0.06$ to $-0.22$; $p = 0.001$) lower MMSE score, and a 0.09 (95% CI $-0.01$ to $-0.17$; $p = 0.02$) lower memory score (Cukierman-Yaffe 2009).

## 4 Brain Structure

In these large studies, the pattern of cognitive impairment does not seem specific for T2D, which is consistent with clinical studies. However, they do suggest both neurodegenerative and vascular lesions contribute to cognitive impairments in persons with T2D. What is the anatomical basis for poorer performance in persons with diabetes compared to normoglycemic subjects? Consistent with the findings on clinical AD, we found in the HAAS autopsy study that those with diabetes and an Apo E ε4 allele, compared to those with neither, had an increased risk for cerebral amyloid angiopathy (CAA), neuritic plaques, and neurofibrillary tangles, all common markers in AD (Peila et al. 2002). The CAA risk was particularly high (6.6 (95% CI 1.5, 29.6), suggesting some synergism between vessel wall integrity and B-amyloid clearance. Further, we found that, compared to those with no diabetes, those with diabetes were at increased risk for hippocampal atrophy, which is frequently seen in AD (Kantarci and Jack 2003), as well as infarcts on MRI. In the AGES-RS cohort, persons with diabetes were more likely to have a single (OR = 1.60; 95% CI 1.27, 2.02) or multiple (OR = 2.51; 95% CI 1.79, 3.52) cerebral infarct(s).

If hippocampal volumes are smaller, and more infarcts are present in T2D, do persons with T2D also have more global, diffuse brain disease? Adjusting for demographic and cardiovascular risk factors, and expressed as the percentage of tissue volume to intracranial volume, compared to normoglycemic subjects, those with T2D had significantly lower total brain (72.2% vs 71.5%; $p < 0.001$) and white matter volume (and 25.7% vs 25.2%, $p < 0.001$) and lower gray matter volumes (45.2% vs 44.7%). Longer duration of T2D was associated with lower total (multi-variate adjusted p for trend < 0.001), gray (multi-variate adjusted p for trend = 0.002), and white matter (multi-variate adjusted p for trend < 0.001) volumes, but not more white matter lesions. Of note, in none of these analyses were there significant differences between those with diabetes and those with IGF or 1–6 yrs of diabetes (Saczynski et al. 2009).

Controlling for large and small vessel disease, loss of brain tissue suggests there may be microvascular damage occurring in the brain. To further address the question of microvascular disease in older persons with diabetes, we examined associations of cerebral microbleeds (CMBs) to retinopathy and diabetes. Microbleeds are small hemorrhages in the gray matter detected with the help of a T2* weighted gradient echo type echo planar MRI sequence. In AGES-Reykjavik, persons with T2D were more likely to have multiple CMBs than normoglycemic subjects (14% vs 4%, respectively; adjusted OR 1.58 (1.04–2.39)). A stronger marker for microvascular disease is the presence of both retinal and cerebral micro lesions. We found an increased risk for multiple CMBs in the presence of retinal microvascular disease (arteriol-venous nicking and microaneurysms/hemorrhages). Persons with T2D *and* retinal lesions had odds of 2.54 (1.43–4.50) of having CMBs compared to persons with only T2D and no retinal lesions (1.48 0.88–2.48; Qiu et al. 2008).

In summary, compared to non-diabetic individuals, the data suggest that those with diabetes have brain structural changes that reflect neuronal degeneration as well as vascular damage. Experimental data are needed to articulate mechanisms (i.e., Stranahan et al. 2008). From an epidemiologic perspective, longitudinal assessments of macro and micro structural changes will help us to better understand the trajectory and functional consequences of cerebral pathologic changes in T2D. A better understanding of the factors that modulate the association of diabetes with cognitive disorders will help to develop more targeted prevention strategies. For instance, modulation of risk may come from co-morbidities of diabetes, as has been reported in studies of the interaction of diabetes and hypertension on brain structure and function (Elias et al. 1997; Schmidt et al. 2004). The data also suggest that those with IGF do not have a significantly different risk for brain function and structure pathology than normoglycemic individuals. However, we do not know how long people have been in that metabolic state; data strongly suggest that the longer the duration of diabetes the higher the risk for brain pathology, and this may be similar for IGF. We also do not know whether subtypes of insulin and glucose dysregulation are differently related to cerebral disease. Additional areas of interest to further investigate include pathophysiology of diabetes and the brain, genetic contributions to the associations of diabetes and brain, and the clinical and functional consequences of diabetes.

**Acknowledgments** AGES-Reykjavik is supported by the National Institutes of Health contract N01-AG-12100, the National Institute on Aging Intramural Research Program, Hjartavernd (the Icelandic Heart Association), and the Althingi (the Icelandic Parliament). Components of the study were also supported by the National Eye Institute, the National Institute on Deafness and Other Communication Disorders, and the National Heart, Lung and Blood Institute. The HAAS has been supported by the Intramural Research Program of the NIH, the National Institute on Aging (grants U01 AG019349 and R01 AG0-17155 S1), and the National Heart, Lung, and Blood Institute (grant N01 HC05102).

## References

Action to Control Cardiovascular Risk in Diabetes Study Group, Gerstein HC, Miller ME, Byington RP, Goff DC Jr, Bigger JT, Buse JB, Cushman WC, Genuth S, Ismail-Beigi F, Grimm RH Jr, Probstfield JL, Simons-Morton DG, Friedewald WT (2008) Effects of intensive glucose lowering in type 2 diabetes. N Engl J Med 358:2545–2559

Baumbach GL (1994) Changes in the cerebral circulation in chronic hypertension. In: Bevan RD and Bevan JA (eds) The human brain circulation. Humana Press, Totowa, NJ pp. 421–431

Buse JB, Ginsberg HN, Bakris GL, Clark NG, Costa F, Eckel R, Fonseca V, Gerstein HC, Grundy S, Nesto RW, Pignone MP, Plutzky J, Porte D, Redberg R, Stitzel KF, Stone NJ (2007) Primary prevention of cardiovascular diseases in people with diabetes mellitus. Diabetes Care 30:162–172

Chaturvedi N (2007) The burden of diabetes and its complications: trends and implications for intervention. Diabetes Res Clin Pract. 76 Suppl 1:S3–12

Coker L, Schumaker SA (2003) Type 2 diabetes mellitus and cognition: an understudied issue in women's health. J Psychosom Res 54:129–139

Craft S, Watson GS (2004) Insulin and neurodegenerative disease: shared and specific mechanisms. Lancet Neurol 3:169–178

Cukierman-Yaffe T, Gerstein HC, Williamson JD, Lazar RM, Lovato L, Miller ME, Coker LH, Murray A, Sullivan MD, Marcovina SM, Launer LJ (2009) Action to ControlCardiovascular Risk in Diabetes-Memory in Diabetes (ACCORD-MIND) Investigators. Relationship between baseline glycemic control and cognitive function in individuals with type 2 diabetes and other cardiovascular risk factors: the action to control cardiovascular risk in diabetes-memory in diabetes (ACCORD-MIND) trial. Diabetes Care. 32:221–226

Diabetes Genetics Initiative of Broad Institute of Harvard and MIT, Lund University, and Novartis Institutes of BioMedical Research (2007) Genome-wide association analysis identifies loci for type 2 diabetes and triglyceride levels. Science 316:1331–1336

Elias PK, Elias MF, D'Agostino RB, Cupples LA, Wilson PW, Silbershatz H, Wolf PA (1997) NIDDM and blood pressure as risk factors for poor cognitive performance. The Framingham Study. Diabetes Care 20:1388–1395

Fernando MS, Ince PC, MRC Cognitive Function and Ageing Neuropthology Study Group (2004) Vascular pathologies and cognition in a population-based cohort of elderly people. J Neurol Sci 226:13–17

Gregg EW, Yaffe K, Cauley JA, Rolka DB, Blackwell TL, Narayan KM, Cummings SR (2000) Is diabetes associated with cognitive impairment and cognitive decline among older women? Study of Osteoporotic Fractures Research Group. Arch Intern Med 160:174–180

Gustafson D (2006) Adiposity indices and dementia. Lancet Neurol 5:713–720

Haan MN, Shemanski L, Jagust WJ, Manolio T, Kuller L (1999) The role of Apo E ε4 in modulating effects of other risk factors for cognitive decline in elderly persons, JAMA 281:40–46

Harris MI, Flegal KM, Cowie CC, Eberhardt MS, Goldstein DE, Little RR, Wiedmeyer HM, Byrd-Holt DD (1998) Prevalence of diabetes, impaired fasting glucose, and impaired glucose tolerance in U.S. adults. The Third National Health and Nutrition Examination Survey, 1988–1994. Diabetes Care 21:518–524

Harris TB, Launer LJ, Eiriksdottir G, Kjartansson O, Jonsson PV, Sigurdsson G, Thorgeirsson G, Aspelund T, Garcia ME, Cotch MF, Hoffman HJ, Gudnason V (2007). Age, Gene/Environment Susceptibility-Reykjavik Study: multidisciplinary applied phenomics. Am J Epidemiol 165:1076–1078

Havlik RJ, Foley DJ, Sayer B, Masaki K, White L, Launer LJ (2002) Variability in midlife systolic blood pressure is related to late-life brain white matter lesions: The Honolulu-Asia Aging Study. Stroke 33:26–30

Hoyer S (1998) Is sporadic Alzheimer's disease the brain type of non-insulin dependent diabetes mellitus? A challenging hypothesis. J Neural Transm 105:415–422

Irie F, Fitzpatrick AL, Lopez OL, Kuller LH, Peila R, Newman AB, Launer LJ (2008). Enhanced risk for Alzheimer disease in persons with type 2 diabetes and APOE epsilon4: the Cardiovascular Health Study Cognition Study. Arch Neurol 65:89–93

Kalmijn S, Feskens EJM, Launer LJ, Kromhout D (1996) Cerebrovascular disease, the apolipoprotein e4 allele, and cognitive decline in a community-based sample of elderly men. Stroke 27:2230–2235

Kalmijn S, Feskens EJM, Launer LJ, Stijen T, Kromhout D (1995) Glucose intolerance, hyperinsulemia, and cognitive function in a general population of elderly men. Diabetologica 38:1096–1102

Kantarci K, Jack CR (2003) Neuroimaging in Alzheimer's Disease: an evidence-based review. Neuroimaging Clin N Am 13:197–209

Klein JP, Waxman SG (2003) The brain in diabetes: molecular changes in neurons and their implications for end-organ damage. Lancet Neurol 2:548–554

Korf-Pelgrim E, White LR, Scheltens PH, Launer LJ (2006) Brain aging in very old men with type 2 diabetes: the Honolulu-Asia Aging Study. Diabetes Care 29:2268–74

Launer LJ, Ross GW, Petrovitch H, Masaki K, Foley D, White LR, Havlik RJ (2000) Midlife blood pressure and dementia: The Honolulu-Asia Aging Study. Neurobiol Aging 21:49–55

Leibson CL, Rocca WA, Hanson VA, Cha R, Kokmen E, O'Brien PC, Palumbo PJ (1997) Risk of dementia among persons with diabetes mellitus: a population-based cohort study. Am J Epidemiol 145:301–308

Lobo A, Launer LJ, Fratiglioni L, Andersen K, Di Carlo A, Breteler MMB, Copeland JRM, Dartigues J-F, Jagger C, Martinez-Lage J, Soininen H, Hofman A, for the Neurologic Diseases in the Elderly Research Group (2000) Prevalence of dementia and major subtypes in Europe: A collaborative study of population-based cohorts. Neurology 54:S4-S9

Lopez OL, Jagust WJ, DeKosky ST, JT, Fitzpatrick A, Dulberg C, Breitner J, Lyketsos C, Jones B, Kawas C, Carlson M, Kuller LH (2003) Prevalence and classification of mild cognitive impairment in the Cardiovascular Health Study Cognition Study: part 1. Arch Neurol. 60:1385–1389

Luchsinger JA, Tang MX, Stern Y, Shea S, Mayeux R (2001) Diabetes mellitus and risk of Alzheimer's disease and dementia with stroke in a multiethnic cohort. Am J Epidemiol 154:635–641

Marlowe L, Peila R, Benke KS, Hardy J, White LR, Launer LJ, Myers A (2006) Insulin-degrading enzyme haplotypes affect insulin levels but not dementia risk. Neurodegener Dis 3:320–326

Ott A, Stolk RP, van Harskamp F, Pols HA, Hofman A, Breteler MM (1999) Diabetes mellitus and the risk of dementia: The Rotterdam Study. Neurology 53:1937–1942

Peila R, Rodriguez BL, Launer LJ (2002) Type 2 diabetes, APOE gene, and the risk for dementia and related pathologies: The Honolulu-Asia Aging Study. Diabetes 2002: 51:1256–1262

Peila R, Rodriguez BL, White LR, Launer LJ (2004) Fasting insulin and incident dementia in an elderly population of Japanese-American men. Neurology 63:228–233

Petrovitch H, White LR, Izmirlian G, Ross GW, Havlik RJ, Markesbery W, Nelson J, Davis DG, Hardman J, Foley DJ, Launer LJ (2000) Midlife blood pressure and neuritic plaques, neurofibrillary tangles, and brain weight at death: the HAAS. Neurobiol Aging 21:57–62

Qiu C, Cotch MF, Sigurdsson S, Garcia M, Klein R, Jonasson F, Klein BE, Eiriksdottir G, Harris TB, van Buchem MA, Gudnason V, Launer LJ (2008) Retinal and cerebral microvascular signs and diabetes: The Age, Gene/Environment Susceptibility-Reykjavik Study. Diabetes 57:1645–50

Saczynski JS, Jónsdóttir MK, Garcia ME, Jonsson PV, Peila R, Eiriksdottir G, Olafsdottir E, Harris TB, Gudnason V, Launer LJ (2008) Cognitive impairment: an increasingly important complication of type 2 diabetes: the age, gene/environment susceptibility–Reykjavik study. Am J Epidemiol 168:1132–1139

Saczynski JS, Siggi S, Jonsson PV, Eiriksdottir G, Olafsdottir E, Kjartansson O, Harris TB, van Buchem M, Gudnason V, Launer LJ (2009) Glycemic status and regional brain volume and Infarcts: The Age Gene/Environment Susceptibility – Reykjavik Study. Diabetes Care 32:1608–1613

Sasaki N, Fukatsu R, Tsuzuki K, Hayashi Y, Yoshida T, Fujii N, Koike T, Wakayama I, Yanagihara R, Garruto R, Amano N, Makita Z (1998) Advanced glycation end products in Alzheimer's disease and other neurodegenerative diseases. Am J Pathol 153:1149–1155

Schmidt R, Launer LJ, Nilsson LG, Pajak A, Sans S, Berger K, Breteler MM, De Ridder M, Dufouil C, Fuhrer R, Giampaoli S, Hofman A (2004) Magnetic resonance imaging of the brain in diabetes: The Cardiovascular Determinants of Dementia (CASCADE) Study. Diabetes 53:687–692

Skyler Jay S, Bergenstal R, Bonow RO, Buse, J, Deedwania P, Gale EAM, Howard BV, Kirkman MS, Kosiborod MAC, Reaven P, Sherwin RS (2009) Intensive Glycemic Control and the Prevention of Cardiovascular Events: Implications of the ACCORD, ADVANCE, and VA Diabetes Trials: A Position Statement of the American Diabetes Association and a Scientific Statement of the American College of Cardiology Foundation and the American Heart Association. Circulation 119:351–357

Stranahan AM, Lee K, Pistell PJ, Nelson CM, Readal N, Miller MG, Spangler EL, Ingram DK, Mattson MP (2008) Diabetes impairs hippocampal function through glucocorticoid-mediated effects on new and mature neurons. Nature Neurosci 11:309–317

van Harten B, de Leeuw FE, Weinstein HC, Scheltens P, Biessels GJ (2006) Brain imaging in patients with diabetes: a systematic review. Diabetes Care 29:2539–2548

White L, Petrovitch H, Ross GW, Masaki KH, Abbott RD, Teng EL, Rodriguez BL, Blanchette PL, Havlik RJ, Wergowske G, Chiu D, Foley DJ, Murdaugh C, Curb JD. (1996) Prevalence of dementia in older Japanese-American men in Hawaii: The Honolulu-Asia Aging Study. JAMA 276:955–960

Williamson JD, Miller ME, Bryan RN, Lazar RM, Coker LH, Johnson J, Cukierman T, Horowitz KR, Murray A, Launer LJ; ACCORD Study Group (2007) The Action to Control Cardiovascular Risk in Diabetes Memory in Diabetes Study (ACCORD-MIND): rationale, design, and methods. Am J Cardiol 99:112i–122i

Zacchigna S, Lambrechts D, Carmeliet P (2008) Neurovascular signaling defects in neurodegeneration Nature Rev Neurosci. 9:169–181

# Cognition in Type 2 Diabetes: Brain Imaging Correlates and Vascular and Metabolic Risk Factors

**Geert Jan Biessels**

**Abstract** Patients with type 2 diabetes (DM2) manifest cognitive dysfunction relative to individuals without DM2. The domains of psychomotor efficiency, executive function, and learning and memory are most often affected, with the size of the effects varying from 0.3 to 1.0 standard deviation units in cross-sectional studies. These modest decrements in cognitive functioning appear to be due to a modest decrease in cognitive performance across the whole patient group, rather than a large deficit among a few individuals. Some individuals, however, do progress to more profound stages of cognitive impairment; longitudinal population based studies identify DM2 as a risk factor of aging-related cognitive decline and dementia.

Structural brain imaging studies in older adults with DM2 show that cerebral atrophy is relatively more pronounced and that lacunar infarcts are more common, relative to people without DM2. The association between DM2 and white matter hyperintensities has been a topic of some controversy, but analyses with sensitive volumetric techniques show a modest increase in lesion load.

Studies on risk factors for cognitive decrements and abnormalities on brain imaging in DM2 have mostly been performed in non-demented individuals. Thus far, the majority of these studies are cross-sectional. Currently available studies indicate that atherosclerotic disease, cerebrovascular or other, may be an important determinant of cognitive decrements in DM2. There may also be associations with worse glycemic control, microvascular complications, hyperinsulinemia and hypertension. Longitudinal observational studies, and ultimately intervention studies, are required to fully appreciate the relevance of these risk factors for cognitive dysfunction in DM2.

G.J. Biessels
Department of Neurology, G03.228, University Medical Center, PO Box 85500, 3508 GA Utrecht, The Netherlands
e-mail: g.j.biessels@umcutrecht.nl

S. Craft and Y. Christen (eds.), *Diabetes, Insulin and Alzheimer's Disease*,
Research and Perspectives in Alzheimer's Disease,
DOI 10.1007/978-3-642-04300-0_6, © Springer-Verlag Berlin Heidelberg 2010

# 1 Introduction

People with diabetes mellitus type 2 (DM2) are at increased risk of cognitive dysfunction. This review addresses the nature and severity of the cognitive decrements that occur in association with DM2, brain imaging correlates and metabolic and vascular risk factors.

A key message is that, to better understand risk factors for dementia and cognitive decline in DM2, the dynamics of these conditions need to be appreciated. Both the transition from normal cognition to dementia and the transition from normal glucose tolerance to DM2 are continuous, multifactorial processes. Early abnormalities in cognition do not necessarily lead to dementia and early abnormalities in glucose metabolism do not necessarily lead to DM2. Age affects cognition and modifies the levels of metabolic and vascular risk factors. Therefore, possible determinants of subtle cognitive decrements in DM2 or pre-DM stages in midlife are not necessarily the same factors that predispose individuals with DM2 to dementia later in life. The challenge will be to identify modifiable risk factors for cognitive problems that matter and to identify individuals at risk for accelerated cognitive decline when cognition is still largely intact in order to take preventive measures.

# 2 Defining Cognitive Dysfunction

DM2-related cognitive dysfunction cannot be regarded as a single condition and therefore cannot be described by a single definition. On one end of the spectrum, there is frank dementia and, on the other, subtle cognitive decrements. Both ends of this spectrum are bridged by intermediate stages of cognitive impairment.

Subtle cognitive decrements can be detected with sensitive neuropsychological tests. These subtle decrements may not even lead to complaints and generally have no significant impact on day-to-day functioning. In the general population, such decrements are more common in older individuals and may be considered to reflect "normal cognitive aging." While not all patients with subtle decrements or more pronounced impairments progress to dementia, dementia is always preceded by intermediate stages of cognitive impairment. In fact, the threshold where "predementia stages" become "dementia" is quite arbitrary, and it has been argued that to make progress in dementia care and research it is time to shift the focus from defining thresholds to considering cognitive impairment as a continuum (Hachinski 2008).

Dementia is not a clinical entity but rather a syndrome caused by various underlying diseases, each characterized by a specific constellation of signs and symptoms in combination with a presumed underlying pathology, the most common of which are Alzheimer's type pathology and vascular damage (van der Flier and Scheltens 2005). In fact, particularly in the oldest old, the majority of affected individuals have mixed pathology (Schneider et al. 2007).

## 3 Defining Diabetes

Current definitions of diabetes are based on cut-off values for blood glucose levels (i.e., fasting plasma glucose $\geq$7.0 mmol/l; American Diabetes Association 2005). The use of such cut-off values creates a dichotomy that is to some extent arbitrary, particularly for DM2. When it comes to potential cerebral complications of DM2, it is important to emphasize that DM2 is preceded by "pre-DM stages" that entail insulin resistance and abnormal glucose levels that fail to reach the cut-off values for a diagnosis of DM. In these pre-DM stages, glucose dysmetabolism is already clustered with vascular risk factors such as hypertension, dyslipidemia and obesity, also referred to as the metabolic syndrome. These risk factors may also affect the brain, and it is unlikely that adverse effects of DM2 and its preceding stages on the brain only start to evolve once glucose values cross the DM threshold. In this respect, it may be more appropriate to consider levels of glucose dysmetabolism and related risk factors as a continuum.

## 4 Cognition and Dementia in DM2 and Pre-diabetic Stages

A large number of studies have addressed cognition in non-demented individuals with DM2 (for reviews, see Allen et al. 2004; Awad et al. 2004; van den Berg et al. 2009). Studies on intermediate stages of glucose dysmetabolism or pre-DM stages are somewhat scarcer (Awad et al. 2004; van den Berg et al. 2009), but studies on vascular risk factors related to DM2, such as hypertension, are abundant (Review van den Berg et al. 2009).

Cross-sectional studies in non-demented individuals with DM2 report cognitive decrements, particularly on psychomotor efficiency, executive function, and learning and memory skills, relative to controls without DM (for systematic reviews, see Awad et al. 2004; Stewart and Liolitsa 1999). For those who are more than 65 years of age, effect sizes range from 0.4 to 1.0 standard deviation units (Awad et al. 2004; Stewart and Liolitsa 1999). Somewhat smaller effect sizes (<0.5) are found in relatively younger adults with DM2 (<60 years of age; Awad et al. 2004; Stewart and Liolitsa 1999). Cross-sectional psychometric studies in pre-DM stages, such as impaired glucose tolerance, generally report effect sizes relative to controls that are equal or slightly smaller than those observed in individuals with DM2 (for review, see van den Berg et al. 2009). Studies on vascular risk factors related to DM2, in particular hypertension and to a lesser extent dyslipidemia and obesity, also report cognitive decrements relative to individuals without these risk factors (van den Berg et al. 2009).

Our research group has now applied a standardized neuropsychological testing protocol in several study populations with different stages of (pre-)DM (Brands et al. 2007; Ruis et al. 2009; van den Berg et al. 2008). We observed modest cognitive decrements, in particular on information processing speed, executive functioning and memory, in individuals with the metabolic syndrome without DM2 (effect sizes ~0.3 relative to age, sex, and premorbid intelligence matched controls; van den Berg et al. 2008), in individuals with screening-detected DM2 of 2 years duration (effect sizes ~0.2; Ruis et al. 2009), or 5–10 years duration (effect sizes ~0.3), and in individuals with 10 years duration of DM2, diagnosed in regular care (effect sizes ~0.3; Brands et al. 2007). The overall picture that emerges from these studies is that the decrements are present in early (pre-)DM stages and show limited progression with increasing DM duration. We have recently confirmed this finding in a longitudinal study, with 4 years follow-up, in which we observed no accelerated decline in the well-controlled DM2 group relative to controls (manuscript in preparation). These observations are in line with other longitudinal studies on cognition in non-demented individuals with DM2. Although some of these studies do show accelerated cognitive decline in DM2 patients (Fontbonne et al. 2001; Gregg et al. 2000; Kanaya et al. 2004), the magnitude of the effect of DM2 on top of that of "normal aging" is at most 50% (Fontbonne et al. 2001; Gregg et al. 2000; Kanaya et al. 2004). Hence, the effect size of the decrement relative to controls observed in these studies increased only slightly over time.

This result does not imply, however, that cognitive decrements that are associated with DM2 always have a benign nature. In longitudinal studies of aging-related cognitive decline, patients with DM2 are clearly overrepresented in the subgroup of individuals with frank decline. The same studies that report modest effects of DM2 in the normal range of cognitive decline report increased odds for major cognitive decline in individuals with DM2 (Gregg et al. 2000; Kanaya et al. 2004). Moreover, in a systematic review of longitudinal population-based studies, we have shown that the incidence of dementia in DM patients was increased by 50% to 100% relative to non-DM individuals (Biessels et al. 2006). This increased risk involved both Alzheimer's disease (AD) and vascular dementia, with a ~50% to 100% increased risk of AD and a ~100% increased risk of vascular dementia (Biessels et al. 2006). Nevertheless, the exact relation between DM and these dementia subtypes is still uncertain. A key issue is the reliability of clinical diagnostic criteria for the underlying pathology (i.e., AD-type or vascular). While autopsy studies thus far have observed no association between DM and AD-type pathology (Alafuzoff et al. 2008; Arvanitakis et al. 2006a; Schnaider Beeri M. et al. 2005), they do show that the vascular lesion load is increased in older individuals with DM with or without dementia (Arvanitakis et al. 2006a; Sonnen et al. 2009).

There are still relatively few epidemiological studies on the risk of dementia associated with pre-DM stages. Still, impaired glucose tolerance (Curb et al. 1999; Xu et al. 2007), hyperinsulinemia (Luchsinger et al. 2004; Muller et al. 2007; Peila et al. 2004) and the metabolic syndrome (Raffaitin et al. 2009; Razay et al. 2007) have all been linked to increased dementia risk, although the results are not always consistent across studies. In addition, vascular risk factors related to DM2, including hypertension, dyslipidemia and obesity, also have all been related to an increased dementia risk (for review, see Kloppenborg et al. 2008).

# 5 Brain Imaging in DM2

Aging-related changes on brain imaging include focal and global atrophy and vascular lesions, in particular infarcts, white-matter hyperintensities (WMHs) and (micro)hemorrhages. Both atrophy and vascular lesions are related to accelerated cognitive decline and an increased dementia risk, although this relation is subject to considerable inter-individual variation.

The majority of brain imaging studies in DM2 are cross-sectional and predominantly involve individuals without frank cognitive impairments (for review, see van Harten et al. 2006). These studies consistently report modest degrees of global atrophy relative to controls (for reviews, see Jongen and Biessels 2008; van Harten et al. 2006). DM2 is also associated with reduced hippocampal and/or amygdalar volumes on MRI (den Heijer et al. 2003; Korf et al. 2006; Korf et al. 2007), but it is not yet clear whether atrophy of these temporal lobe structures is out of proportion to atrophy elsewhere in the brain. The prevalence and incidence of lacunar infarcts are also increased by a factor 1.5 to 2 (for systematic review, see van Harten et al. 2006). The relationship between DM2 and WMHs has long been subject to debate. Several large population-based studies did not observe a significant association between DM2 and WMHs (for systematic review, see van Harten et al. 2006), but these results may be affected by the use of relatively crude WMH rating scales. Recent case-control studies that applied a more refined WMH rating scale or volumetric measurements reported a modest increase in WMH severity in patients with DM2 (Jongen and Biessels 2008; van Harten et al. 2007), and there are now clear indications that DM2 is a risk factor for WMH progression (Gouw et al. 2008).

# 6 Risk Factors for Impaired Cognition and Brain Imaging Abnormalities in DM2

Possible risk factors for impaired cognition and brain imaging abnormalities in DM2 include demographic, socioeconomic or lifestyle factors, DM-specific factors (e.g., hyperglycemia), factors that are linked to DM but that are not specific to DM (e.g., hypertension, dyslipidemia, stroke, depression), factors related to DM treatment, and genetic factors. Because the brain may already be affected in pre-DM stages, factors associated with the pre-DM conditions should obviously be considered.

Unfortunately, there is still considerable uncertainty regarding the main risk factors for cognitive decrements in DM2, due to the fact that these factors have been addressed in a limited number of studies. These studies have mostly been cross-sectional and have addressed determinants of cognition and brain MRI findings predominantly in individuals without frank impairments of cognition. It is as yet unclear whether risk factors for relative cognitive decrements in this patient population are predictors for frank cognitive decline. In addition, levels of risk factors and their relation with cognition may change over time, possibly as a result of aging, although there are also indications that the dementia process itself is associated with changes in risk factor levels (for reviews, see Kloppenborg et al. 2008; Qiu et al. 2005).

Currently available studies show an association between elevated A1C or blood glucose levels and impaired cognition (Gallacher et al. 2005; Kanaya et al. 2004; Manschot et al. 2006; Reaven et al. 1990), DM duration (Bruce et al. 2008b; Elias et al. 1997), a history of macrovascular disease (Bruce et al. 2008b; Manschot et al. 2006; Umegaki et al. 2007), microalbuminuria (Bruce et al. 2008a), smoking (Arvanitakis et al. 2006b), hypertension (Elias et al. 1997), and hypoglycaemic episodes (Whitmer et al. 2009), although observations are not always consistent across studies.

## 7 Conclusion

DM2 and pre-DM stages are associated with dementia and pre-dementia stages. Imaging studies show association with vascular lesions and brain atrophy. A key message of this review is that the dynamics of the development of DM2 and those of dementia need to be appreciated to identify the determinants and to understand the mechanisms of DM2-associated cerebral complications.

To unravel the complex relation between DM2 and frank impairments of cognition, several questions still need to be answered. Is DM2 itself, or the context in which it develops, the driving factor for the increased dementia risk? Should the subtle cognitive decrements that are associated with DM2, but do not appear to progress rapidly over time, be regarded as pre-dementia stages? If not, what do studies on imaging correlates and risk factors for these subtle cognitive decrements tell us about determinants for frank cognitive decline in DM? Answering these questions is not trivial. DM2 and DM2-related vascular risk factors are extremely common and have a substantial impact on the overall risk of dementia at the population level. Better understanding of the nature of the relation between these factors and dementia is a key step in the development of preventive measures. The challenges will be to identify modifiable risk factors for clinically relevant cognitive deficits, to identify individuals at risk of accelerated cognitive decline when cognition is still largely intact, and to develop preventive measures. These goals will require large longitudinal observational studies and, ultimately, intervention studies. Fortunately, several such studies are well underway.

## References

Alafuzoff I, Aho L, Helisalmi S, Mannermaa A, Soininen H (2008) beta-Amyloid deposition in brains of subjects with diabetes. Neuropathol Appl Neurobiol 35:60–68

Allen KV, Frier BM, Strachan MWJ (2004) The relationship between type 2 diabetes and cognitive dysfunction: longitudinal studies and their methodological limitations. Eur J Pharmacol 490:169–175

American Diabetes Association (2005) Diagnosis and classification of diabetes mellitus. Diabetes Care 28:S37–S42
Arvanitakis Z, Schneider JA, Wilson RS, Wilson RS, Li Y, Arnold SE, Wang Z, Bennett DA (2006a) Diabetes is related to cerebral infarction but not to AD pathology in older persons. Neurology 67:1960–1965
Arvanitakis Z, Wilson RS, Li Y, Aggarwal NT, Bennett DA (2006b) Diabetes and function in different cognitive systems in older individuals without dementia. Diabetes Care 29:560–565
Awad N, Gagnon M, Messier C (2004) The relationship between impaired glucose tolerance, type 2 diabetes, and cognitive function. J Clin Exp Neuropsychol 26:1044–1080
Beeri MS, Silverman JM, Davis KL, Marin D, Grossman HZ, Schmeidler J, Purohit DP, Perl DP, Davidson M, Mohs RC, Haroutunian V (2005) Type 2 diabetes is negatively associated with Alzheimer's disease neuropathology. J Gerontol A Biol Sci Med Sci 60:471–475
Biessels GJ, Staekenborg S, Brunner E, Brayne C, Scheltens P (2006) Risk of dementia in diabetes mellitus: a systematic review. Lancet Neurol 5:64–74
Brands AMA, van den Berg E, Manschot SM, Biessels GJ, Kappelle LJ, De Haan EH, Kessels RP (2007) A detailed profile of cognitive dysfunction and its relation to psychological distress in patients with type 2 diabetes mellitus. J Int Neuropsychol Soc 13:288–297
Bruce DG, Davis WA, Casey GP, Starkstein SE, Clarnette RM, Almeida OP, Davis TM (2008a) Predictors of cognitive decline in older people with diabetes. Diabetologia 51:241–248
Bruce DG, Davis WA, Casey GP, Starkstein SE, Clarnette RM, Foster JK, Almeida OP, Davis TM (2008b) Predictors of cognitive impairment and dementia in older people with diabetes. Diabetologia 51:241–248
Curb JD, Rodriguez BL, Abbott RD, Petrovitch H, Ross GW, Masaki KH, Foley D, Blanchette PL, Harris T, Chen R, White LR (1999) Longitudinal association of vascular and Alzheimer's dementias, diabetes, and glucose tolerance. Neurology 52:971–975
den Heijer T, Vermeer SE, van Dijk EJ, Prins ND, Koudstaal PJ, Hofman A, Breteler MM (2003) Type 2 diabetes and atrophy of medial temporal lobe structures on brain MRI. Diabetologia 46:1604–1610
Elias PK, Elias MF, D'Agostino RB, Cupples LA, Wilson PW, Silbershatz H, Wolf PA (1997) NIDDM and blood pressure as risk factors for poor cognitive performance. The Framingham Study. Diabetes Care 20:1388–1395
Fontbonne A, Berr C, Ducimetiere P, Alperovitch A (2001) Changes in cognitive abilities over a 4-year period are unfavorably affected in elderly diabetic subjects: results of the epidemiology of vascular aging study. Diabetes Care 24:366–370
Gallacher JE, Pickering J, Elwood PC, Bayer AJ, Yarnell JW, Ben-Shlomo Y (2005) Glucoregulation has greater impact on cognitive performance than macro-vascular disease in men with type 2 diabetes: data from the Caerphilly study. Eur J Epidemiol 20:761–768
Gouw AA, van der Flier WM, Fazekas F, van Straaten EC, Pantoni L, Poggesi A, Inzitari D, Erkinjuntti T, Wahlund LO, Waldemar G, Schmidt R, Scheltens P, Barkhof F; LADIS Study Group. (2008) Progression of white matter hyperintensities and incidence of new lacunes over a 3-year period. The Leukoaraiosis and Disability Study. Stroke 39:1414–1420
Gregg EW, Yaffe K, Cauley JA, Rolka DB, Blackwell TL, Narayan KM, Cummings SR. (2000) Is diabetes associated with cognitive impairment and cognitive decline among older women? Study of Osteoporotic Fractures Research Group. Arch Intern Med 160:174–180
Hachinski V (2008) Shifts in thinking about dementia. JAMA 300:2172–2173
Jongen C, Biessels GJ (2008) Structural brain imaging in diabetes: A methodological perspective. Eur J Pharmacol 585:208–218
Kanaya AM, Barrett-Connor E, Gildengorin G, Yaffe K (2004) Change in cognitive function by glucose tolerance status in older adults: a 4-year prospective study of the Rancho Bernardo study cohort. Arch Intern Med 164:1327–1333
Kloppenborg RP, van den Berg E, Kappelle LJ, Biessels GJ (2008) Diabetes and other vascular risk factors for dementia: Which factor matters most? A systematic review. Eur J Pharmacol 585:97–108

Korf ES, White LR, Scheltens P, Launer LJ (2006) Brain aging in very old men with type 2 diabetes: the Honolulu-Asia Aging Study. Diabetes Care 29:2268–2274

Korf ES, van Straaten EC, de Leeuw FE, van der Flier WM, Barkhof F, Pantoni L, Basile AM, Inzitari D, Erkinjuntti T, Wahlund LO, Rostrup E, Schmidt R, Fazekas F, Scheltens P LADIS Study Group. (2007) Diabetes mellitus, hypertension and medial temporal lobe atrophy: the LADIS study. Diabet Med 24:166–171

Luchsinger JA, Tang MX, Shea S, Mayeux R (2004) Hyperinsulinemia and risk of Alzheimer disease. Neurology 63:1187–1192

Manschot SM, Brands AM, van der Grond J, Kessels RP, Algra A, Kappelle LJ, Biessels GJ (2006) Brain magnetic resonance imaging correlates of impaired cognition in patients with type 2 diabetes. Diabetes 55:1106–1113

Muller M, Tang MX, Schupf N, Manly JJ, Mayeux R, Luchsinger JA (2007) Metabolic syndrome and dementia risk in a multiethnic elderly cohort. Dement Geriatr Cogn Disord 24:185–192

Peila R, Rodriguez BL, White LR, Launer LJ (2004) Fasting insulin and incident dementia in an elderly population of Japanese-American men. Neurology 63:228–233

Qiu C, Winblad B, Fratiglioni L (2005) The age-dependent relation of blood pressure to cognitive function and dementia. Lancet Neurol 4:487–499

Raffaitin C, Gin H, Empana JP, Helmer C, Berr C, Tzourio C, Portet F, Dartigues JF, Alpérovitch A, Barberger-Gateau P (2009) Metabolic syndrome and risk for incident Alzheimer's disease or vascular dementia: the three-city study. Diabetes Care 32:169–174

Razay G, Vreugdenhil A, Wilcock G (2007) The metabolic syndrome and Alzheimer disease. Arch Neurol 64:93–96

Reaven GM, Thompson LW, Nahum D, Haskins E (1990) Relationship between hyperglycemia and cognitive function in older NIDDM patients. Diabetes Care 13:16–21

Ruis C, Biessels GJ, Gorter KJ, van den DM, Kappelle LJ, Rutten GE (2009) Cognition in the early stage of type 2 Diabetes Mellitus. Diabetes Care doi: 10.2337/dc08-2143

Schneider JA, Arvanitakis Z, Bang W, Bennett DA (2007) Mixed brain pathologies account for most dementia cases in community-dwelling older persons. Neurology 69:2197–2204

Sonnen JA, Larson EB, Brickell K, Crane PK, Woltjer R, Montine TJ, Craft S. (2009) Different patterns of cerebral injury in dementia with or without diabetes. Arch Neurol 66:315–322

Stewart R, Liolitsa D (1999) Type 2 diabetes mellitus, cognitive impairment and dementia. Diabet Med 16:93–112

Umegaki H, Iimuro S, Kaneko T, Araki A, Sakurai T, Ohashi Y, Iguchi A, Ito H (2008) Factors associated with lower mini mental state examination scores in elderly Japanese diabetes mellitus patients. Neurobiol Aging 29:1022–1026

van den Berg E, Dekker JM, Nijpels G, Kessels RP, Kappelle LJ, de Haan EH, Heine RJ, Stehouwer CD, Biessels GJ (2008) Cognitive functioning in elderly persons with type 2 diabetes and metabolic syndrome: the Hoorn study. Dement Geriatr.Cogn Disord 26:261–269

van den Berg E, Kloppenborg RP, Kessels RP, Kappelle LJ, Biessels GJ (2009) Type 2 diabetes mellitus, hypertension, dyslipidemia and obesity: A systematic comparison of their impact on cognition. Biochim Biophys Acta 1792:470–481

van der Flier WM, Scheltens P (2005) Epidemiology and risk factors of dementia. J Neurol Neurosurg Psychiat 76 (Suppl 5):v2–v7

van Harten B, de Leeuw FE, Weinstein HC, Scheltens P, Biessels GJ (2006) Brain imaging in patients with diabetes: a systematic review. Diabetes Care 29:2539–2548

van Harten B, Oosterman JM, Potter van Loon BJ, Scheltens P, Weinstein HC (2007) Brain lesions on MRI in elderly patients with type 2 diabetes mellitus. Eur Neurol 57:70–74

Whitmer RA, Karter AJ, Yaffe K, Quesenberry CP Jr, Selby JV (2009) Hypoglycemic episodes and risk of dementia in older patients with type 2 diabetes mellitus. JAMA 301:1565–1572

Xu W, Qiu C, Winblad B, Fratiglioni L (2007) The effect of borderline diabetes on the risk of dementia and Alzheimer's disease. Diabetes 56:211–216

# The Relationship Between the Continuum of Elevated Adiposity, Hyperinsulinemia, and Type 2 Diabetes and Late-onset Alzheimer's Disease: An Epidemiological Perspective

**José A. Luchsinger**

**Abstract** This manuscript provides a comprehensive review of the epidemiologic evidence linking the continuum of elevated adiposity, hyperinsulinemia, and type 2 diabetes (T2D) with late-onset Alzheimer's disease (LOAD). The mechanisms relating this continuum to LOAD may be vascular and non-vascular. Elevated adiposity in middle age is related to a higher risk of LOAD but the data on this association in old age are conflicting. Several studies have shown that hyperinsulinemia, a consequence of higher adiposity and insulin resistance, is also related to a higher risk of LOAD. Studies have consistently shown a relationship between T2D and higher LOAD risk. A large proportion of the world population may be at increased risk of LOAD, given the trends for increasing prevalence of overweight, obesity, hyperinsulinemia, and T2D. However, these associations may present a unique opportunity for prevention and treatment of LOAD. Several studies in the prevention and treatment of T2D are currently conducting or have planned cognition ancillary studies. In addition, clinical trials using insulin sensitizers in the treatment or prevention of LOAD are under way.

## 1 Introduction

Late-onset Alzheimer's disease (LOAD) is the most common form of dementia, accounting for between 70% and over 90% of all cases (Ritchie and Lovestone 2002), and its prevalence is expected to quadruple by the year 2047 in the United States (Brookmeyer et al. 1998). As much as 50% of the population aged 85 years and older, the fastest growing segment of the population, may have LOAD (Evans

J.A. Luchsinger
Department of Epidemiology, Joseph P. Mailman School of Public Health, Columbia University, New York, NY, USA
e-mail: jal94@columbia.edu

S. Craft and Y. Christen (eds.), *Diabetes, Insulin and Alzheimer's Disease*, Research and Perspectives in Alzheimer's Disease,
DOI 10.1007/978-3-642-04300-0_7, 

et al. 1989). The risk factors for LOAD can be classified as genetic and non-genetic. Three genes have been identified in familial early-onset AD: amyloid precursor protein (APP), presenilin 1, and presenilin 2 (Selkoe 1997). These genes affect less than 5% of AD cases, have full penetrance and expressivity, and usually affect persons in middle age (Cummings 2004). This review will address LOAD. Robust risk factors that have been identified for LOAD include older age, lower education, and the APOE-ε4 allele (Cummings 2004). Importantly, APOEε4 has been found to modulate the effect of other putative risk factors (Haan et al. 1999), such as type 2 diabetes (T2D) and hyperinsulinemia (Peila et al. 2002; Luchsinger et al. 2004). Current treatment options for LOAD only address symptoms, and no treatments are available that focus on delaying the actual disease process (Sano 2003). One of the currently accepted hypotheses of the pathogenesis of LOAD is that the main culprit is the accumulation of amyloid β in the brain, resulting in synapse disruption and neuronal destruction (Selkoe 1997, 2000).

Mild cognitive impairment (MCI) has been used to describe a transitional state between normal cognitive function and LOAD (Petersen et al. 1999; Luis et al. 2003) and has thus been targeted for interventions (Petersen et al. 2005). Individuals with MCI do not have dementia but have memory complaints without loss of function in their daily activities (Petersen et al. 1999). While general cognitive performance is well preserved, memory performance on standardized tests falls below expectations for age and education. Persons with MCI progress to LOAD at the rate of nearly 10% to 15% per year (Luis et al. 2003) compared to 1 to 2% in elderly persons with normal cognition (Petersen et al. 1999). MCI can be classified as amnestic and non-amnestic MCI. It is believed that amnestic MCI is an early stage of LOAD, whereas non-amnestic MCI, such as executive MCI, is less specific to LOAD (Luis et al. 2003). The prevalence of amnestic MCI varies between 3% and 20% depending on the criteria applied (Busse et al. 2003) and increases from about 1% in persons 60 years old to 25% at age 85 (Yesavage et al. 2002). LOAD can be studied in epidemiologic studies through the outcomes of memory impairment or decline, amnestic MCI, and Alzheimer's dementia, and this review will cover evidence examining these outcomes.

## 2 The Continuum of Adiposity, Hyperinsulinemia, and Type 2 Diabetes

There is a worrisome epidemic of obesity, insulin resistance and T2D in the world (Hill and Bessesen 2003). With the aging of the population and greater longevity, the long-term consequences of these conditions will have a significant public health impact in elderly populations. Adiposity refers to the amount of adipose (fat) tissue in the body (Reaven and Laws 1999). Adiposity is a continuum, and the normal or ideal threshold of adiposity is not clear. However, as adiposity increases, it is associated with higher risks of insulin resistance, T2D, hypertension, dyslipidemia, cardiovascular disease, degenerative joint disease, cancer, and respiratory diseases (Pi-Sunyer 2002; Poirier et al. 2006). Definitions of elevated levels of adiposity have been devised using existing measures and according to their relationship with adverse outcomes (Clinical guidelines... 1998). Adiposity is usually measured indirectly with anthropological measures (Mueller et al. 1991) such as the body mass index (BMI), defined as weight in kilograms divided by height in meters squared (k/m$^2$). BMI is strongly correlated with total body fat tissue and is a good indirect measure of adiposity (Pi-Sunyer 2002), although this correlation decreases in older age (Baumgartner et al. 1995). Another commonly used measure of adiposity is waist circumference. Waist circumference is meant to measure the accumulation of adipose tissue in the abdomen, the largest depot of adipose tissue, and it is, therefore, perhaps a more direct measure of adiposity than BMI (Mueller et al. 1991; Wahrenberg et al. 2005). Elevated waist circumference is also related to a higher risk of T2D, hypertension, dyslipidemia, and heart disease, and some studies have shown that it is a better predictor of adverse cardiovascular outcomes than BMI (Janssen et al. 2004). Some have advocated its use as the best measure of adiposity (Mueller et al. 1991). A commonly used cutoff to define elevated waist circumference is 102 cm for men and 88 cm for women (Janssen et al. 2004). Other, less frequently used anthropologic measures of adiposity include skinfolds and waist to hip ratio (Mueller et al. 1991). Overweight, obesity (Flegall et al. 2002) and elevated waist circumference (Ford et al. 2003) are increasing in adults in the United States. More worrisome, these trends are also observed in children and adolescents (Hedley et al. 2004). Two thirds of the United States population are overweight or obese (Hedley et al. 2004); 30% are obese, and the prevalence of obesity is higher in women than men.

Insulin sensitivity is the ability of insulin to dispose of a glucose load. Insulin resistance refers to the resistance of tissues that dispose of glucose to the actions of insulin. Insulin resistance results in an increase in insulin secretion in the pancreas to overcome that insulin resistance. Fasting insulin levels are used in epidemiological studies as indicators of the risk of T2D (Haffner et al. 1990; Lundgren et al. 1990; Charles et al. 1991; Lillioja et al. 1993). Fasting insulin is accepted as a measure of insulin resistance that is highly correlated with more complicated measures of insulin resistance such as the euglycemic clamp (Laakso 1993) and the homeostasis model assessment (Haffner et al. 1997).

Glucose intolerance and T2D are abnormal elevations of blood glucose that put people at risk for microvascular (nephropathy, neuropathy, retinopathy) and macrovascular disease (coronary artery disease, cerebrovascular disease, peripheral vascular disease; DeFronzo 2000). It is important to point out that this review addresses type 2 diabetes, not type 1 diabetes. The American Diabetes Association currently defines diabetes as a fasting glucose elevation $>$ 126 mg/dl, and glucose intolerance as an elevation of glucose between 110 and 126 mg/dl (Clark et al. 2000). It is difficult to establish an absolute threshold for the definition of glucose intolerance and diabetes. Previously, the definition of diabetes was a fasting glucose $>$ 140 mg/dl, and people currently defined as having diabetes were then considered non-diabetic (Luchsinger 2001). The fasting glucose threshold for the diagnosis of diabetes will change again and persons currently considered to have glucose intolerance will be considered to be diabetic. These changes underscore the caveats of using cutoffs to define conditions that have continuous (linear or non-linear) associations with disease: depending on the cutoff used, persons at risk may be classified as normal or abnormal (and vice versa). This is true for measures of adiposity, insulin resistance, and measures of glucose tolerance.

Adiposity, hyperinsulinemia, glucose intolerance, and T2D are often treated as separate constructs. However, they are related sequentially and often occur simultaneously, and understanding this relationship is fundamental in the study of the role of adiposity, insulin resistance, and T2D in LOAD. Keeping glucose at normal levels is achieved by balancing between the ability of peripheral tissues (muscle, adipose tissue, liver) to take glucose into cells and the pancreas' ability to secrete insulin, the hormone in charge of glucose tissue uptake (DeFronzo 2000). Thus, abnormal glucose levels are caused by a resistance of tissues to the action of insulin (insulin resistance) and by the pancreas' inability to secrete enough insulin at normal levels or higher than normal insulin levels (hyperinsulinemia) to overcome insulin resistance in tissues (Festa et al. 2006). Insulin resistance increases with age, and the organism maintains normal glucose levels as long as it can produce enough insulin (hyperinsulinemia). Some individuals are less capable than others of mounting sustained hyperinsulinemia and will develop glucose intolerance and T2D (Festa et al. 2006). Other individuals with insulin resistance will maintain normal glucose levels at the expense of hyperinsulinemia, but their pancreas will eventually burn out, i.e., will not be able to sustain hyperinsulinemia and will develop glucose intolerance and T2D (Festa et al. 2006). Others will continue having insulin resistance, may have or not have glucose intolerance, will not develop T2D but will have hyperinsulinemia and suffer its consequences. The most frequent modifiable determinant of insulin resistance and hyperinsulinemia is elevated adiposity (Reaven and Laws 1999; Reaven 2005), although adipose tissue is not the only factor. Insulin resistance can reside in other tissues, including muscle, liver, and the pancreas itself (Accili 2004). The susceptibility to adiposity, that is, the risk of developing insulin resistance and T2D in response to adiposity, varies by gender (Pi-Sunyer 2002) and particularly by ethnicity. For example, Chinese and southeast Asians are more susceptible than Europeans to developing insulin resistance with comparable increases of adiposity (Reaven and Laws 1999). The distribution of factors related to insulin resistance and the metabolic syndrome, including adiposity, is different in whites and blacks (Kraja et al. 2005). Thus, conventional ways to classify adiposity may not capture its relationship to adverse outcomes and this should be taken into account. High adiposity and hyperinsulinemia are both accompanied by dyslipidemia, hypertension, and inflammation (Reaven 2005), and these should also be taken into account.

An implication of the continuum described above is that, when an epidemiologic study finds a relationship between the components of this continuum and LOAD, we cannot be certain if we are looking at a surrogate marker of one of the other components (e.g., T2D is a marker of past adiposity or hyperinsulinemia, obesity is a marker of hyperinsulinemia) or if the important exposure is the one we are examining. The answer could be that there is an aggregate effect of all the components of the continuum. The metabolic syndrome, an increasingly popular term in clinical practice and research, and reported to be associated with a higher risk of cognitive decline (Yaffe et al. 2004) is a constellation of adiposity, hypertension, glucose intolerance, and dyslipidemia that is associated mainly with insulin resistance and hyperinsulinemia (Grundy et al. 2005; Luchsinger 2006). However, the definition of the metabolic syndrome is somewhat arbitrary, intended to capture the clustering of cardiovascular risk factors particularly in middle-aged populations, and its validity in elderly populations at risk for LOAD is not clear (Luchsinger 2006). The difficulty of arriving at precise metabolic syndrome criteria is reflected by the fact that, over the years, at least six different definitions have been developed that share several characteristics (Grundy et al. 2005).

## *2.1 Potential Mechanisms Relating the Continuum of Elevated Adiposity, Hyperinsulinemia, and Type 2 Diabetes with LOAD*

This continuum is related to cerebrovascular disease (Benson and Sacco 2000; Boden-Albala and Sacco 2000; Sacco et al. 2001; Sacco 2002; Suk et al. 2003; Boden-Albala et al. 2008). Elevated adiposity (Pi-Sunyer 2002), hyperinsulinemia, T2D (Sacco et al. 1997), and their clustering with other vascular risk factors (Grundy et al. 2005) are risk factors for stroke. In addition, insulin or related byproducts may affect the amyloid cascade (Craft and Watson 2004). Thus, we classify the mechanisms linking this continuum with LOAD as cerebrovascular and non-cerebrovascular.

## *2.2 Cerebrovascular Mechanisms. Brain Infarcts*

Strokes, ascertained by clinical history (Honig, Tang et al. 2003) or as brain infarcts on MRI (Vermeer et al. 2003), are related to a higher risk of dementia, including LOAD. The mechanisms for this association are not clear. However, pathology studies have demonstrated that the presence of amyloid plaques is lower in brains of persons with dementia who also have infarcts (Snowdon et al. 1997; Schneider et al. 2007), suggesting that the presence of infarcts is an insult that lowers the threshold of amyloid in the brain that is necessary to cause dementia. T2D has been shown in pathology studies to be related to infarcts but not LOAD pathology in persons with the clinical expression of LOAD (Arvanitakis et al. 2006). This observation suggests that the main mechanism linking hyperinsulinemia to LOAD clinical expression is the presence of infarcts, which lower the burden of amyloid necessary to cause memory decline and dementia.

## 2.3 White Matter Disease

White matter disease, ascertained as white matter hyperintensities (WHI) or leukoaraiosis on brain imaging, represents microvascular disease in the brain or demyelination. The nature of WHI is still a matter of controversy. WHI are thought to be ischemic in origin in the same way that infarcts are (Pantoni 2006) and have thus been proposed as surrogate markers of cerebrovascular disease (Pantoni 2006). However, recent evidence shows that WHI are common in LOAD and may be related to cerebral amyloid angiopathy (Alonzo et al. 1998; Haglund and Englund 2002; Gurol et al. 2006; Nakata-Kudo et al. 2006). Thus, some WHI may be due to amyloid disease and contribute to the development of LOAD. WHI are common correlates of cognitive impairment in T2D (Manschot et al. 2006), but it is unclear whether these WHI are markers of microvascular injury or represent a process related to amyloid deposition.

## 2.4 Hyperinsulinemia and Beta Amyloid Clearance

Hyperinsulinemia is a plausible risk factor for LOAD because 1) insulin can cross the blood-brain barrier (BBB; Park 2001) and peripheral insulin infusion in the elderly increases Aβ42 levels in the CSF (Watson, Peskind et al. 2003), a surrogate marker of Aβ clearance in the brain and an indirect marker of LOAD risk; 2) there are insulin receptors in the brain, including the hippocampus and entorhinal cortex (Frolich et al. 1998), structures affected early in LOAD (Small et al. 1999); 3) insulin-degrading enzyme (IDE) has been linked to clearance of Aβ in the brain, and insulin and Aβ are both competing substrates for IDE (Farris et al. 2003); and 4) insulin in the brain can increase the deposition of Aβ and Tau protein phosphorylation, which are central to the pathogenesis of LOAD (Park 2001).

The pathways relating insulin in the periphery with Aβ clearance in the brain are multiple and complex. We cannot discuss all due to space constraints. Craft et al. (2007) have reviewed how peripheral hyperinsulinemia affects amyloid beta clearance in the brain. A potential pathway is that peripheral hyperinsulinemia down regulates insulin uptake in the BBB due to saturation over physiologic levels (Banks et al. 1997), which may result in reduction of insulin levels in the brain, down-regulation of expression of IDE (Zhao et al. 2004) and reduction in IDE-mediated amyloid reduction (Farris et al. 2003). This complex observation has been used to support the use of rosiglitazone, an insulin sensitizer (Watson et al. 2005; Risner et al. 2006), and intranasal insulin (Reger et al. 2006) in the treatment of LOAD. However, there are potential mechanisms (affected by hyperinsulinemia) related to amyloid clearance in the periphery (not the brain) that may affect the amount of amyloid translocated to the brain from the periphery (Deane et al. 2004; Zlokovic 2008). These mechanisms include advanced products of glycosilation (AGE). AGE are most closely linked with glycemia and diabetes, as elevated glucose concentration promotes the Maillard reaction and AGE accrual. In a hyperglycemic environment, diabetic animal and human tissues contain increased AGE and upregulation of its receptor (RAGE, Basta et al. 2004; Goldin et al. 2006; Negrean et al. 2007). In fact, the most recognized AGE, hemoglobin $A_{1c}$, represents the standard-of-care for tracking glycemia. AGE contribute importantly to diabetic complications. AGE in the basement membranes of vessels promote vascular leakage (Vlassara et al. 1994). Glomerular AGE deposition is associated with diabetic nephropathy, and treatment with thiazolium compounds, thought to break AGE cross-links, reduces it (Thallas-Bonke et al. 2004; Peppa et al. 2006). AGE accumulation also contributes to complications such as retinal neovascularization (Stitt 2001; Schalkwijk et al. 2002) and diabetic neuropathy (Sullivan and Feldman 2005). Increased expression of RAGE is observed in LOAD (Yan et al. 1996; Schmidt et al. 2000; Lue et al. 2001). Similarly, vasculature with deposited Aβ from patients with LOAD also display increased RAGE antigen compared to age-matched controls (Yan et al. 1996; Deane et al. 2003). Expression of RAGE is enhanced in blood vessels near Aβ deposits in LOAD brain (Yan et al. 1996; Deane et al. 2003). Along with increased total amount of RAGE in AD brain, there is a shift of RAGE distribution from neuron to microvasculature (Donahue et al. 2006). The role of RAGE on the BBB was further revealed when mice infused with a radiolabeled Aβ showed RAGE-dependent brain capillary uptake and transport across the BBB of radiolabeled peptide. Further experiments demonstrated that Aβ(1-40) and Aβ(1-42) transport can be blocked by RAGE-specific IgG or sRAGE but are not affected by fucoidan (an inhibitor of Aβ binding to the type A macrophage scavenger receptor), antibodies to β1-integrin, the RHDS sequence of Aβ or nonimmune IgG. The BBB transport of Aβ was also completely blocked in mice deficient in the RAGE gene (Deane et al. 2003).

Lipoprotein-related proteins (LRP) are a family of lipoprotein receptors that affect lipid metabolism. LRP-1, found in the liver and other tissues, clears Aβ in the periphery (Donahue et al. 2006; Tamaki et al. 2007). LRP-1 is diminished in insulin resistance without affecting lipid levels (Tamaki et al. 2007). Thus, LRP-1 is a plausible mechanism linking hyperinsulinemia with Aβ and LOAD. Soluble LRP (sLRP) facilitates the clearance of Aβ by LRP-1 and occurs naturally (Deane et al. 2004; Sagare et al. 2007).

# 3 Summary of Prospective Epidemiological Studies Linking Adiposity, Hyperinsulinemia, and Diabetes to AD

## 3.1 *Adiposity*

Elevated BMI in middle age may be associated with higher dementia risk (Kivipelto et al. 2005; Whitmer et al. 2005). A recent study showed that central adiposity in middle age was related to a higher risk of dementia in older age (Whitmer et al. 2008). Higher BMI at ages 70, 75 and 79 years may also predict higher dementia risk (Gustafson et al. 2003). However, there have been reports of no association at mid-life (Stewart et al. 2005) and of lower BMI related to higher LOAD risk (Nourhashemi et al. 2003; Atti et al. 2008) at older ages. A study in northern New York City (Luchsinger et al. 2007) found that, in younger elderly (65 to 76 years of age), the association between BMI quartiles and LOAD resembles a U shaped-curve, whereas in the oldest old (> 76 years), higher BMI is related to a lower LOAD risk. This U-shaped association has been reported for the relationship between adiposity and cardiovascular mortality (Stevens et al. 1998) and underscores the difficulty of studying the effects of adiposity in older age (Stevens 2000). This study also found that higher waist circumference is related to higher LOAD risk in the younger elderly, but not in the oldest old. The Cardiovascular Health Study recently reported that elevated self-reported BMI at age 50 years was associated with a higher risk of dementia, whereas BMI at age 65 or older in the same individuals did not (Fitzpatrick et al. 2009).

This study underscores the importance of the period in life at which adiposity is ascertained in relation to dementia. The most important explanation for the paradox linking low weight in old age to dementia seems to be weight loss. The mechanisms for this are not entirely clear. They may include loss of olfaction (Devanand et al. 2000; Tabert et al. 2005), one of the earliest manifestations of LOAD, which may lead to decreased caloric intake, forgetfulness of meals (Gustafson 2008), and metabolic changes related to LOAD that are not well understood. In this regard, LOAD is accompanied by abnormalities in brain insulin signaling (Steen et al. 2005), which could affect appetite and food intake. It is not clear if weight loss is a consequence of LOAD, is a parallel process, or is related to potential causes of LOAD, such as insulin resistance (Wedick et al. 2001). The role of weight loss and nutritional supplementation in the progression of LOAD is beyond the scope of this review and can be found elsewhere (Reyes-Ortega et al. 1997; Guyonnet et al. 1998; Andrieuh et al. 2001; Reynish et al. 2001).

## 3.2 *Hyperinsulinemia*

Several cross-sectional studies show an association between hyperinsulinemia and an increased risk of LOAD (Razay and Wilcock 1994; Kuusisto et al. 1997; Stolk 1997). Two longitudinal studies, one in elderly Japanese Americans in Hawaii (Peila et al. 2004) and another in elderly black, Caribbean Hispanic, and Non-Hispanic whites in New York City (Luchsinger et al. 2004) found that the risk of incident LOAD was higher in persons with hyperinsulinemia. These studies also found that the risk of LOAD related to hyperinsulinemia was higher among persons with the APOE-ε4 allele. The Nurses' health study found that higher C-peptide levels, a measure of insulin secretion (Harris et al. 2002), and fasting insulin levels are related to cognitive decline in women (Okereke et al. 2005; Okereke et al. 2008; van Oijen et al. 2008). There is a paucity of prospective epidemiologic studies exploring the relationship between markers of hyperinsulinemia and LOAD, and more are needed.

### 3.3 Type 2 Diabetes

T2D has been related to a two-fold higher risk of developing MCI among postmenopausal women (Yaffe et al. 2004). A multiethnic study in elderly from New York City found that T2D was related to a higher risk of cognitive impairment-no dementia with stroke, although the effect on cognitive impairment-no dementia without stroke was not evident after adjusting for demographic variables and the presence of the Apo E-ε4 allele (Luchsinger et al. 2001). An Italian study showed a non-statistically significant increase in MCI with T2D in an elderly population (Solfrizzi et al. 2004), whereas a Canadian study found that T2D was related only to vascular cognitive impairment-no dementia (MacKnight et al. 2002). A study in New York City found that T2D was related to a higher risk of both amnestic and non-amnestic MCI, underlining the importance of T2D for both LOAD-related and vascular cognitive impairment (Luchsinger et al. 2007). A recent study in Olmstead county, Minnesota, found that the presence of T2D was not related to MCI risk, but longer T2D duration and treatment with insulin, a surrogate marker of T2D duration, were related to higher MCI risk (Roberts et al. 2008).

T2D has been found consistently to be related to vascular dementia (VD), but its relation to LOAD is less clear. A study of Japanese subjects aged 65 years and older found that T2D was related to a higher risk of both LOAD and VD (Yoshitake et al. 1995). A longitudinal study from the Netherlands in over 5,000 subjects aged 55 years and older without dementia at baseline found a higher risk of LOAD in persons with T2D (Ott et al. 1999). This association was stronger in subjects with T2D who reported insulin treatment. Another European study found that the risk of all cause-dementia was increased by T2D, but this relation was weaker with LOAD (Brayne et al. 1998). A study from Rochester, Minnesota, found a doubling of LOAD risk in relation to T2D (Leibson et al. 1997), similar to the study from the Netherlands. A study of Catholic nuns, priests, and brothers 55 years and older found that T2D was associated with a higher risk of LOAD (Arvanitakis et al. 2004). The Honolulu Asia Aging Study also found that T2D in old age was related to a higher risk of LOAD and AD pathology on autopsy, particularly in subjects with the APOE-ε4 allele (Peila et al. 2002). A study from Canada found that T2D had a weak, non-statistically significant relation to LOAD but was related to VD. A Swedish study found a similar, non-significant relation to LOAD and a significant relation to a higher risk of VD (Xu et al. 2004). A prospective study in over 1,000 subjects from New York City who were mostly African American and Caribbean-Hispanic, with a mean age of 75 years and without dementia at baseline, found a higher risk of LOAD in relation to T2D that was not statistically significant after adjustment for other variables, but T2D was significantly related to higher risk of a composite outcome of LOAD and cognitive impairment-no dementia (Luchsinger et al. 2001). The risk of LOAD was also increased in those treated with insulin, indicating a higher risk of LOAD in subjects with long-standing T2D. This study also found a stronger association between T2D and VD. A recent reanalysis of these data with longer follow-up showed that the risk of LOAD associated with T2D was stronger than previously reported, independent of other vascular conditions (hypertension, heart disease, stroke), and not explained by misclassification of VD cases as LOAD (Luchsinger et al. 2005). A study in Sweden found that T2D increased the risk of VD but not of LOAD (Xu et al. 2004) and that this risk was higher in the presence of hypertension and heart disease. The same study recently reported that borderline T2D was associated with a higher risk of LOAD in persons without the APOE-ε4 allele (Xu et al. 2007).

Few studies have examined whether T2D in middle age leads to the development of dementia in older age. One study in the United States (Whitmer et al. 2005) and another in Israel (Schnaider Beeri et al. 2004) found that T2D at midlife increased the risk of dementia in the elderly. A study in Japanese-Americans found no association between T2D in middle age and dementia (Curb et al. 1999).

The diagnosis of T2D is somewhat arbitrary and many cases go undetected. Few studies have examined the relationship between continuous measures of glycemia and dementia. One study in postmenopausal women found that the risk of MCI and dementia increased with each 1% elevation in glycosilated hemoglobin, a stable measure of glucose levels, even in women without T2D (Yaffe et al. 2006). Glycosilated hemoglobin in persons without T2D correlates with both glucose intolerance and insulin resistance, and this study underscores the continuous nature of the relationship between these constructs and higher dementia risk.

### *3.4 Metabolic Syndrome*

There is limited evidence of the association between the metabolic syndrome and dementia in the elderly. One study in 2,632 black and white elders found that the metabolic syndrome was associated with a higher risk of cognitive decline, particularly among those with high inflammatory markers (Yaffe et al. 2004). A cross-sectional study in Europeans found that LOAD prevalence was higher in persons with the metabolic syndrome (Vanhanen et al. 2006). In northern New York City, the metabolic syndrome was not related to LOAD risk, whereas T2D and hyperinsulinemia were (Muller et al. 2007). The discrepancy between these studies could be due to the fact that the study in New York City was conducted in an older population, ethnically diverse, and with a high prevalence of vascular risk factors (Luchsinger et al. 2005). In Japanese Americans, the metabolic syndrome in middle age was associated with VD but not LOAD (Kalmijn et al. 2000).

## 4 Implications for the Prevention and Treatment of AD

There is very strong evidence that adiposity, hyperinsulinemia, and T2D are related to LOAD. However, this evidence falls short of being considered as proof of causation until we understand the mechanisms. If the relation between these conditions and LOAD were to be causal, the public health implications are enormous. As explained before, two-thirds of the adult population of the United States is overweight or obese, and the short-term trend is for this to worsen. These trends are also being observed worldwide. Epidemiologic studies suggest that adiposity in middle age is important as a predictor of LOAD risk. With increasing life expectancy, we are likely to increasingly see the cognitive consequences of increased adiposity, hyperinsulinemia, and T2D in old age. We estimated that in New York City the presence of T2D or hyperinsulinemia in elderly people could account for 39% of cases of LOAD (Luchsinger et al. 2004). However, the other implication is that a large proportion of cases of LOAD could be preventable or treatable. The Finnish Diabetes Prevention Study (FDPS; Tuomilehto et al. 2001) demonstrated that T2D can be prevented with lifestyle interventions. The Diabetes Prevention Program (DPP) in the United States demonstrated that T2D can be prevented through lifestyle interventions or metformin (Diabetes Prevention Program Research Group 2002), and this effect was largely mediated by improvements in insulin sensitivity and reductions in insulin levels (The Diabetes Prevention Program Research 2005). Cognition ancillary studies are planned for both the FDPS and DPP. These cognition ancillary studies will be unique opportunities to answer whether the prevention of T2D through improvement in adiposity and insulin sensitivity is related to improvements in the risk of cognitive impairment. Rosiglitazone, an insulin sensitizer used in the treatment of T2D but also shown to be effective in prevention (DREAM investigators 2006) has shown preliminary promise (Watson et al. 2005) in the treatment of LOAD, particularly in persons without the APOE-ε4 allele (Risner et al. 2006) . Phase III trials of rosiglitazone are under way although concerns about its safety may limit this drug's usefulness (Nathan 2007; Nathan and Berkwits 2007). The potential mechanisms linking the continuum of adiposity, hyperinsulinemia, and T2D are multiple, overlapping, and highly correlated. Thus, it may be difficult to elucidate that a single mechanism is the culprit. Importantly, the potential therapeutic interventions mentioned above impact virtually all potential mechanisms, that is, weight loss through lifestyle interventions or medications alters adipokine activity, improves hyperinsulinemia, inflammation, glucose tolerance, blood pressure, lipids, and the risk of cerebrovascular disease. One of the pitfalls of the clinical diagnosis of LOAD is that it may detect a heterogeneous disease with overlapping vascular and amyloid pathologies in a continuum, which may explain the observation of relationships of adiposity, hyperinsulinemia, and T2D with LOAD. It is possible that these relationships are explained mostly by cases of dementia with cerebrovascular disease misclassified as LOAD or that they are cases of mixed dementia. However, the possibility of misclassification does not reduce the potential public health importance of these observations and the potential for prevention and treatment.

Another important question is whether intense treatment can decrease the risk of cognitive impairment and LOAD in persons with T2D. Moreover, it is important to answer whether treatment of T2D with drugs that increase insulin levels vs. insulin sensitizers affects cognitive impairment. Some of these questions will be answered by a cognition ancillary study in the Action to Control Cardiovascular Risk in

Diabetes (ACCORD) clinical trial (Williamson et al. 2007). Elucidating the mechanisms linking adiposity, hyperinsulinemia and T2D to LOAD will help identify specific targets for treatment, and more research is needed in this regard.

**Acknowledgments** Dr. Luchsinger's work in this review was supported by grants from the National Institute on Aging (AG07232), by the Alzheimer's Association (IIRG-05-15053), the American Diabetes Association (7-08-CR-41) and by the Florence and Herbert Irving Clinical Research Scholar's Award

## References

Accili D (2004) Lilly Lecture 2003: The struggle for mastery in insulin action: from triumvirate to republic. Diabetes 53:1633–1642

Alonzo NC, Hyman BT, Rebeck GW, Greenberg SM (1998) Progression of cerebral amyloid angiopathy: accumulation of amyloid-β40 in affected vessels. J Neuropathol Exp Neurol 57:353–359

Andrieu S, Reynish W, Nourhashemi F, Ousset PJ, Grandjean H, Grand A, Albarede JL, Vellas B (2001) Nutritional risk factors for institutional placement in Alzheimer's disease after one year follow-up. J Nutrition Health Aging 5:113–117

Arvanitakis Z, Schneider JA, Wilson RS, Li Y, Arnold SE, Wang Z, Bennett DA (2006) Diabetes is related to cerebral infarction but not to AD pathology in older persons. Neurology 67:1960–1965

Arvanitakis Z, Wilson RS, Bienias JL, Evans DA, Bennett DA (2004) Diabetes mellitus and risk of Alzheimer disease and decline in cognitive function. Arch Neurol 61:661–666

Atti AR, Palmer K, Volpato S, Winblad B, De Ronchi D, Fratiglioni L (2008) Late-life body mass index and dementia incidence: nine-year follow-up data from the Kungsholmen Project. J Am Geriatr Soc 56:111–116

Banks WA. Jaspan JB, Huang W, Kastin AJ (1997) Transport of insulin across the blood-brain barrier: saturability at euglycemic doses of insulin. Peptides 18:1423–1429

Basta G, Schmidt AM, De Caterina R (2004) Advanced glycation end products and vascular inflammation: implications for accelerated atherosclerosis in diabetes. Cardiovasc Res 63:582–592

Baumgartner RN, Heymsfield SB, Roche AF (1995) Human body composition and the epidemiology of chronic disease. Obes Res 3:73–95

Benson RT, Sacco RL (2000) Stroke prevention: hypertension diabetes tobacco and lipids. Neurol Clin 18:309–319

Boden-Albala B, Cammack S, Chong J, Wang C, Wright C, Rundek T, Elkind MS, Paik MC, Sacco RL (2008) Diabetes fasting glucose levels and risk of ischemic stroke and vascular events: findings from the Northern Manhattan Study (NOMAS). Diabetes Care 31:1132–1137

Boden-Albala B, Sacco RL (2000) Lifestyle factors and stroke risk: exercise alcohol diet obesity smoking drug use and stress. Curr Atheroscler Rep 2:160–166

Brayne C, Gill C, Huppert FA, Barkley C, Gehlhaar E, Girling DM, O'Connor DW, Paykel ES (1998) Vascular risks and incident dementia: results from a cohort study of the very old. Dement Geriatr Cogn Disord 9:175–180

Brookmeyer R, Gray S, Kawas C (1998) Projections of Alzheimer's disease in the United States and the public health impact of delaying disease onset. Am J Public Health 88:1337–1342

Busse A, Bischkopf J, Riedel-Heller SG, Angermeyer MC (2003) Mild cognitive impairment: prevalence and incidence according to different diagnostic criteria. Results of the Leipzig Longitudinal Study of the Aged (LEILA75+). Br J Psychiat 182:449–454

Charles MA, Fontbonne A, Thibult N, Warnet JM, Rosselin GE, Eschwege E (1991) Risk factors for NIDDM in white population. Paris prospective study. Diabetes 40:796–799
Clark Jr MJ, Sterrett JJ, Carson DS (2000) Diabetes guidelines: a summary and comparison of the recommendations of the American Diabetes Association Veterans Health Administration and American Association of Clinical Endocrinologists. ClinTherapeut 22:899–910
Clinical guidelines on the identification, evaluation and treatment of overweight and obesity in adults–the evidence report. National Institutes of Health (1998). Obesity Res 6 Suppl 2:51S–209S
Craft S (2007) Insulin resistance and Alzheimer's disease pathogenesis: potential mechanisms and implications for treatment. Curr Alzheimer Res 4:147–152
Craft S, Watson GS (2004) Insulin and neurodegenerative disease: shared and specific mechanisms. Lancet Neurol 3:169–178
Cummings JL (2004) Alzheimer's disease. N Engl J Med 351:56–67
Curb JD, Rodriguez BL, Abbott RD, Petrovitch H, Ross GW, Masaki KH, Foley D, Blanchette PL, Harris T, Chen R, White LR (1999) Longitudinal association of vascular and Alzheimer's dementias diabetes and glucose tolerance. Neurology 52:971–975
Deane R, Du Yan S, Submamaryan RK, LaRue B, Jovanovic S, Hogg E, Welch D, Manness L, Lin C, Yu J, Zhu H, Ghiso J, Frangione B, Stern A, Schmidt AM, Armstrong DL, Arnold B, Liliensiek B, Nawroth P, Hofman F, Kindy M, Stern D, Zlokovic B (2003) RAGE mediates amyloid-β peptide transport across the blood-brain barrier and accumulation in brain. Nature Med 9:907–913
Deane R, Wu Z, Sagare A, Davis J, Du Yan S, Hamm K, Xu F, Parisi M, LaRue B, Hu HW, Spijkers P, Guo H, Song X, Lenting PJ, Van Nostrand WE, Zlokovic BV (2004) LRP/amyloid β-peptide interaction mediates differential brain efflux of Aβ isoforms. Neuron 43:333–344
DeFronzo RA (2000) Pharmacologic therapy for type 2 diabetes mellitus. Ann InternMed 133:73–74
Devanand DP, Michaels-Marston KS, Liu X, Pelton GH, Padilla M, Marder K, Bell K, Stern Y, Mayeux R (2000) Olfactory deficits in patients with mild cognitive impairment predict Alzheimer's disease at follow-up. Am J Psychiat 157:1399–1405
Diabetes Prevention Program Research Group (2002) Reduction in the incidence of type 2 diabetes with lifestyle intervention or metformin. N Engl J Med 346:393–403
Donahue JE, Flaherty SL, Johanson CE, Duncan JA 3rd, Silverberg GD, Miller MC, Tavares R, Yang W, Wu Q, Sabo E, Hovanesian V, Stopa EG (2006) RAGE LRP-1 and amyloid-β protein in Alzheimer's disease. Acta Neuropathol (Berl) 112:405–415
DREAM investigators (2006) Effect of rosiglitazone on the frequency of diabetes in patients with impaired glucose tolerance or impaired fasting glucose: a randomised controlled trial. Lancet 368:1096–1105
Evans DA, Funkenstein HH, Albert MS, Scherr PA, Cook NR, Chown MJ, Hebert LE, Hennekens CH, Taylor JO (1989) Prevalence of Alzheimer's disease in a community population of older persons Higher than previously reported. JAMA 262:2551–2556
Farris W, Mansourian S, Chang Y, Lindsley L, Eckman EA, Frosch MP, Eckman CB, Tanzi RE, Selkoe DJ, Guenette S (2003) Insulin-degrading enzyme regulates the levels of insulin amyloid β-protein and the β-amyloid precursor protein intracellular domain in vivo Proc Natl Acad Sci USA 100:4162–4167
Festa A, Williams K, D'Agostino R Jr, Wagenknecht LE, Haffner SM (2006) The natural course of β-cell function in nondiabetic and diabetic individuals: the insulin resistance atherosclerosis tudy. Diabetes 55:1114–1120
Fitzpatrick AL, Kuller HL, Lopez OL, Diehr P, O'Meara ES, Longstreth WT Jr, Luchsinger JA (2009) Midlife and late-life obesity and the risk of dementia: cardiovascular health study. Arch Neurol 66:336–342
Flegal KM, Carroll MD, Ogden CL, Johnson CL (2002) Prevalence and trends in obesity among US adults 1999–2000 JAMA 288:1723–1727

Ford ES, Mokdad AH, Giles WH (2003) Trends in waist circumference among US adults. Obes Res 11:1223–1231

Frolich L, Blum-Degen D, Bernstein HG, Engelsberger S, Humrich J, Laufer S, Muschner D, Thalheimer A, Türk A, Hoyer S, Zöchling R, Boissl KW, Jellinger K, Riederer P (1998) Brain insulin and insulin receptors in aging and sporadic Alzheimer's disease. J Neural Transmiss 105:423–438

Goldin A. Beckman JA, Schmidt AM, Creager MA (2006) Advanced glycation end products: sparking the development of diabetic vascular injury. Circulation 114:597–605

Grundy SM, Cleeman JI, Daniels SR, Donato KA, Eckel RH, Franklin BA, Gordon DJ, Krauss RM, Savage PJ, Smith SC Jr, Spertus JA, Costa F; American Heart Association; National Heart, Lung, and Blood Institute (2005) Diagnosis and management of the metabolic syndrome: an American Heart Association/National Heart Lung and Blood Institute Scientific Statement. Circulation 112:2735–2752

Gurol ME, Irizarry MC, Smith EE, Raju S, Diaz-Arrastia R, Bottiglieri T, Rosand J, Growdon JH, Greenberg SM (2006) Plasma β-amyloid and white matter lesions in AD MCI and cerebral amyloid angiopathy. Neurology 66:23–29

Gustafson D (2008) A life course of adiposity and dementia. Eur J Pharmacol 585(1): 163–175

Gustafson D, Rothenberg E, Blennow K, Steen B, Skoog I (2003) An 18-year follow-up of overweight and risk of Alzheimer disease Arch Intern Med 163:1524–1528

Guyonnet S, Nourhashemi F, Ousset PJ, Micas M, Ghisolfi A, Vellas B, Albarede JL (1998) Factors associated with weight loss in Alzheimer's disease. J Nutr Health Aging 2:107–109

Haan MN, Shemanski L, Jagust WJ, Manolio TA, Kuller L (1999) The role of APOE epsilon4 in modulating effects of other risk factors for cognitive decline in elderly persons. JAMA 282:40–46

Haffner SM, Miettinen H, Stern MP (1997) The homeostasis model in the San Antonio Heart Study. Diabetes Care 20:1087–1092

Haffner SM, Stern MP, Mitchell BD, Hazuda HP, Patterson JK (1990) Incidence of type II diabetes in Mexican Americans predicted by fasting insulin and glucose levels obesity and body-fat distribution Diabetes 39:283–288

Haglund M, Englund E (2002) Cerebral amyloid angiopathy white matter lesions and Alzheimer encephalopathy – a histopathological assessment. Dement Geriatr Cogn Disord 14:161–166

Harris MI, Cowie CC, Gu K, Francis ME, Flegal K, Eberhardt MS (2002) Higher fasting insulin but lower fasting C-peptide levels in African Americans in the US population. Diabetes/Metab Res Rev 18:149–155

Hedley AA, Ogden CL, Johnson CL, Carroll MD, Curtin LR, Flegal KM (2004) Prevalence of overweight and obesity among US children adolescents and adults 1999–2002. JAMA 291:2847–2850

Hill JO, Bessesen D (2003) What to do about the metabolic syndrome? Arch Intern Med 163:395–397

Honig LS, Tang MX, Albert S, Costa R, Luchsinger J, Manly J, Stern Y, Mayeux R (2003) Stroke and the risk of Alzheimer disease Arch Neurol 60:1707–1712

Janssen I, Katzmarzyk PT, Ross R (2004) Waist circumference and not body mass index explains obesity-related health risk. Am J Clin Nutr 79:379–384

Kalmijn S, Foley D, White L, Burchfiel CM, Curb JD, Petrovitch H, Ross GW, Havlik RJ, Launer LJ (2000) Metabolic cardiovascular syndrome and risk of dementia in Japanese- American elderly men. The Honolulu-Asia aging study. Arterioscler Thromb Vasc Biol 20:2255–2260

Kivipelto M, Ngandu T, Fratiglioni L, Viitanen M, Kåreholt I, Winblad B, Helkala EL, Tuomilehto J, Soininen H, Nissinen A (2005) Obesity and vascular risk factors at midlife and the risk of dementia and Alzheimer disease. Arch Neurol 62:1556–1560

Kraja AT, Hunt SC, Pankow JS, Myers RH, Heiss G, Lewis CE, Rao D, Province MA (2005) An evaluation of the metabolic syndrome in the HyperGEN study. Nutr Metab 2:2

Kuusisto J, Koivisto K, Kervinen K, Mykkänen L, Helkala EL, Vanhanen M, Hänninen T, Pyörälä K, Kesäniemi YA, Riekkinen P, Laasko M (1997) Association between features of the insulin resistance syndrome and Alzheimer's disease independently of apolipoprotein E4 phenotype: cross sectional population based study. BMJ 315:1045–1049

Laakso M (1993) How good a marker is insulin level for insulin resistance? Am J Epidemiol 137:959–965

Leibson CL, Rocca WA, Hanson VA, Cha R, Kokmen E, O'Brien PC, Palumbo PJ (1997) Risk of dementia among persons with diabetes mellitus: a population- based cohort study. Am J Epidemiol 145:301–308

Lillioja S, Mott DM, Spraul M, Ferraro R, Foley JE, Ravussin E, Knowler WC, Bennett PH, Bogardus C (1993) Insulin resistance and insulin secretory dysfunction as precursors of non-insulin-dependent diabetes mellitus. Prospective studies of Pima Indians. N Engl J Med 329:1988–1992

Luchsinger JA (2001) Diabetes. In: Aguirre-Molina M, Molina CW, Zambrana R (eds.) Health issues in the Latino community. San Francisco: Jossey-Bass, 2001, 277–300

Luchsinger JA (2006) A work in progress: the metabolic syndrome. Sci Aging Knowledge Environ 2006: pe19

Luchsinger JA, Patel B, Tang MX, Schupf N, Mayeux R (2007) Measures of adiposity and dementia risk in elderly persons. Arch Neurol 64:392–398

Luchsinger JA, Reitz C, Honig LS, Tang MX, Shea S, Mayeux R (2005) Aggregation of vascular risk factors and risk of incident Alzheimer disease. Neurology 65:545–551

Luchsinger JA, Reitz C, Patel B, Tang MX, Manly JJ, Mayeux R (2007) Relation of diabetes to mild cognitive impairment. Arch Neurol 64:570–575

Luchsinger JA, Tang M-X, Shea S, Mayeux R (2004) Hyperinsulinemia and risk of Alzheimer disease. Neurology 63:1187–1192

Luchsinger JA, Tang MX, Stern Y, Shea S, Mayeux R (2001) Diabetes mellitus and risk of Alzheimer's disease and dementia with stroke in a multiethnic cohort Am J Epidemiol 154:635–641

Lue LF, Walker DG, Brachova L, Beach TG, Rogers J, Schmidt AM, Stern DM, Yan SD (2001) Involvement of microglial receptor for advanced glycation endproducts (RAGE) in Alzheimer's disease: identification of a cellular activation mechanism. Exp Neurol 171:29–45

Luis CA, Loewenstein DA, Acevedo A, Barker WW, Duara R (2003) Mild cognitive impairment: directions for future research. Neurology 61:438–444

Lundgren H, Bengtsson C, Blohmé G, Lapidus L, Waldenström J (1990) Fasting serum insulin concentration and early insulin response as risk determinants for developing diabetes. Diabetic Med 7:407–413

MacKnight C, Rockwood K, Awalt E, McDowell I (2002) Diabetes mellitus and the risk of dementia Alzheimer's disease and vascular cognitive impairment in the Canadian Study of Health and Aging. Dement Geriatr Cogn Disord 14:77–83

Manschot SM, Brands AMA, van der Grond J, Kessels RP, Algra A, Kappelle LJ, Biessels GJ Utrecht Diabetic Encephalopathy Study Group (2006) Brain magnetic resonance imaging correlates of impaired cognition in patients with type 2 diabetes Diabetes 55:1106–1113

Mueller WH, Wear ML, Hanis CL, Emerson JB, Barton SA, Hewett-Emmett D, Schull WJ (1991) Which measure of body fat distribution is best for epidemiologic research? Am J Epidemiol 133:858–869

Muller M, Tang MX, Schupf N, Manly JJ, Mayeux R, Luchsinger JA (2007) Metabolic syndrome and dementia risk in a multiethnic elderly cohort. Dement Geriatr Cogn Disord 24:185–192

Nakata-Kudo Y, Mizuno T, Yamada K, Shiga K, Yoshikawa K, Mori S, Nishimura T, Nakajima K, Nakagawa M (2006) Microbleeds in Alzheimer disease are more related to cerebral amyloid angiopathy than cerebrovascular disease. Dement Geriatr Cogn Disord 22:8–14

Nathan DM (2007) Rosiglitazone and cardiotoxicity – weighing the evidence N Engl J Med 357:64–66

Nathan DM, Berkwits M (2007) Trials that matter: rosiglitazone ramipril and the prevention of type 2 diabetes. Ann Intern Med 146:461–463

Negrean M, Stirban A et al. , Stratmann B, Gawlowski T, Horstmann T, Götting C, Kleesiek K, Mueller-Roesel M, Koschinsky T, Uribarri J, Vlassara H, Tschoepe D Effects of low- and high-advanced glycation endproduct meals on macro- and microvascular endothelial function and oxidative stress in patients with type 2 diabetes mellitus. Am J Clin Nutr 85:1236–1243

Nourhashémi F, Deschamps V, Larrieu S, Letenneur L, Dartigues JF, Barberger-Gateau P PAQUID study. Personnes Agées Quid (2003) Body mass index and incidence of dementia: the PAQUID study. Neurology 60:117–119

Okereke O, Hankinson SE, Hu FB, Grodstein F (2005) Plasma C peptide level and cognitive function among older women without diabetes mellitus. Arch Intern Med 165:1651–1656

Okereke OI, Pollak MN, Hu FB, Hankinson SE, Selkoe DJ, Grodstein F (2008) Plasma C-peptide levels and rates of cognitive decline in older community-dwelling women without diabetes Psychoneuroendocrinology 33:455–461

Ott A, Stolk RP, van Harskamp F, Pols HA, Hofman A, Breteler MM (1999) Diabetes mellitus and the risk of dementia: The Rotterdam Study Neurology 53:1937–1942

Pantoni L (2006) White matter ischemia: Time to begin integrating experimental and clinical data. Eur Neurol 56:71–73

Park CR (2001) Cognitive effects of insulin in the central nervous system. Neurosci Biobehav Rev 25:311–323

Peila R, Rodriguez BL, Launer LJ; Honolulu-Asia Aging Study Type 2 diabetes APOE gene and the risk for dementia and related pathologies: The Honolulu-Asia Aging Study. Diabetes 51:1256–1262

Peila R, Rodriguez BL, White LR, Launer LJ (2004) Fasting insulin and incident dementia in an elderly population of Japanese-American men Neurology 63:228–233

Peppa M, Brem H, Cai W, Zhang JG, Basgen J, Li Z, Vlassara H, Uribarri J (2006) Prevention and reversal of diabetic nephropathy in db/db mice treated with alagebrium (ALT-711) Am J Nephrol 26:430–436

Petersen RC, Smith GE, Waring SC, Ivnik RJ, Tangalos EG, Kokmen E (1999) Mild cognitive impairment: clinical characterization and outcome. Arch Neurol 56:303–308

Petersen RC, Thomas RG, Grundman M, Bennett D, Doody R, Ferris S, Galasko D, Jin S, Kaye J, Levey A, Pfeiffer E, Sano M, van Dyck CH, Thal LJ (2005) Vitamin E and Donepezil for the treatment of mild cognitive impairment. N Engl J Med: 352:2379–2388

Pi-Sunyer FX (2002) The obesity epidemic: pathophysiology and consequences of obesity. Obes Res 10:97S–104

Poirier P, Giles TD, Bray GA, Hong Y, Stern JS, Pi-Sunyer FX, Eckel RH (2006) Obesity and cardiovascular disease: pathophysiology evaluation and effect of weight loss: an update of the 1997 American Heart Association Scientific Statement on Obesity and Heart Disease from the Obesity Committee of the Council on Nutrition Physical Activity and Metabolism. Circulation 113:898–918

Razay G, Wilcock GK (1994) Hyperinsulinaemia and Alzheimer's disease. Age Ageing 23:396–399

Reaven G (2005) Insulin resistance type 2 diabetes mellitus and cardiovascular disease: the end of the beginning. Circulation 112:3030–3032

Reaven GM, Laws A (1999) Insulin resistance: the metabolic syndrome X. Totowa N J Humana Press

Reger MA, Watson GS, Green PS, Baker LD, Cholerton B, Fishel MA, Plymate SR, Cherrier MM, Schellenberg GD, Frey WH 2nd, Craft S (2006) Effects of intranasal insulin on cognition in memory-impaired older adults: modulation by APOE genotype. Neurobiol Aging 27:451–458

Reyes-Ortega G,Guyonnet S, Ousset PJ, Nourhashemi F, Vellas B, Albarède JL, De Glizezinski I, Riviere D, Fitten LJ (1997) Weight loss in Alzheimer's disease and resting energy expenditure (REE): a preliminary report. J Am Geriatr Soc 45:1414–1415

Reynish W, Andrieu S, Nourhashemi F, Vellas B (2001) Nutritional factors and Alzheimer's disease. J Gerontol A Biol Sci Med Sci 56:M675–680

Risner ME, Saunders AM, Altman JF, Ormandy GC, Craft S, Foley IM, Zvartau-Hind ME, Hosford DA, Roses AD; Rosiglitazone in Alzheimer's Disease Study Group (2006) Efficacy of rosiglitazone in a genetically defined population with mild-to-moderate Alzheimer's disease. Pharmacogenomics J 6:246–254

Ritchie K, Lovestone S (2002) The dementias. Lancet 360:1759–1766

Roberts RO, Geda YE, Knopman DS, Christianson TJ, Pankratz VS, Boeve BF, Vella A, Rocca WA, Petersen RC (2008) Association of duration and severity of diabetes mellitus with mild cognitive impairment. Arch Neurol 65:1066–1073

Sacco R L (2002) Reducing the risk of stroke in diabetes: what have we learned that is new? Diabetes Obes Metab 4 Suppl 1: S27–34

Sacco RL, Benjamin EJ, Broderick JP, Dyken M, Easton JD, Feinberg WM, Goldstein LB, Gorelick PB, Howard G, Kittner SJ, Manolio TA, Whisnant JP, Wolf PA (1997) American Heart Association Prevention Conference. IV. Prevention and rehabilitation of stroke risk factors. Stroke 28:1507–1517

Sacco RL, Benson RT, Kargman DE, Boden-Albala B, Tuck C, Lin IF, Cheng JF, Paik MC, Shea S, Berglund L (2001) High-density lipoprotein, cholesterol and ischemic stroke in the elderly: The Northern Manhattan Stroke Study. JAMA 285:2729–2735

Sagare A, Deane R, Bell RD, Johnson B, Hamm K, Pendu R, Marky A, Lenting PJ, Wu Z, Zarcone T, Goate A, Mayo K, Perlmutter D, Coma M, Zhong Z, Zlokovic BV (2007) Clearance of amyloid-β by circulating lipoprotein receptors. Nature Med 13:1029–1031

Sano M (2003) Noncholinergic treatment options for Alzheimer's disease. J Clin Psychiatr 64 (Suppl 9): 23–28

Schalkwijk CG, Lieuw-a-Fa M, van Hinsbergh VW, Stehouwer CD (2002) Pathophysiological role of Amadori-glycated proteins in diabetic microangiopathy. Semin Vasc Med 2:191–197

Schmidt AM, Yan SD, Yan SF, Stern DM (2000) The biology of the receptor for advanced glycation end products and its ligands. Biochim Biophys Acta 1498:99–111

Schnaider Beeri M, Goldbourt U, Silverman JM, Noy S, Schmeidler J, Ravona-Springer R, Sverdlick A, Davidson M (2004) Diabetes mellitus in midlife and the risk of dementia three decades later Neurology 63:1902–1907

Schneider JA, Arvanitakis Z, Bang W, Bennett DA (2007) Mixed brain pathologies account for most dementia cases in community-dwelling older persons. Neurology 69:2197–2204

Selkoe DJ (1997) Alzheimer's disease: genotypes, phenotypes and treatments. Science 275:630–631

Selkoe DJ (2000) The origins of Alzheimer disease: a is for amyloid. JAMA 283:1615–1617

Small SA, Perera GM, DeLaPaz R, Mayeux R, Stern Y (1999) Differential regional dysfunction of the hippocampal formation among elderly with memory decline and Alzheimer's disease. Ann Neurol 45:466–472

Snowdon DA, Greiner LH, Mortimer JA, Riley KP, Greiner PA, Markesbery WR (1997) Brain infarction and the clinical expression of Alzheimer disease. The Nun Study. JAMA 277:813–817

Solfrizzi V. Panza F, Colacicco AM, D'Introno A, Capurso C, Torres F, Grigoletto F, Maggi S, Del Parigi A, Reiman EM, Caselli RJ, Scafato E, Farchi G, Capurso A; Italian Longitudinal Study on Aging Working Group (2004) Vascular risk factors incidence of MCI and rates of progression to dementia. Neurology 63:1882–1891

Steen E, Terry BM, Rivera EJ, Cannon JL, Neely TR, Tavares R, Xu XJ, Wands JR, de la Monte SM (2005) Impaired insulin and insulin-like growth factor expression and signaling mechanisms in Alzheimer's disease–is this type 3 diabetes? J Alzheimers Dis 7:63–80

Stevens J (2000) Impact of age on associations between weight and mortality. Nutr Rev 58:129–137

Stevens J, Cai J, Pamuk ER, Williamson DF, Thun MJ, Wood JL (1998) The effect of age on the association between body-mass index and mortality. N Engl J Med 338:1–7

Stewart R, Masaki K, Xue QL, Peila R, Petrovitch H, White LR, Launer LJ (2005) A 32-year prospective study of change in body weight and incident dementia: The Honolulu-Asia Aging Study. Arch Neurol 62:55–60

Stitt AW (2001) Advanced glycation: an important pathological event in diabetic and age related ocular disease. Br J Ophthalmol 85:746–753

Stolk RP. Breteler MM, Ott A, Pols HA, Lamberts SW, Grobbee DE, Hofman A (1997) Insulin and cognitive function in an elderly population. The Rotterdam Study. Diabetes Care 20:792–795

Suk SH, Sacco RL, Boden-Albala B, Cheun JF, Pittman JG, Elkind MS, Paik MC; Northern Manhattan Stroke Study (2003) Abdominal obesity and risk of ischemic stroke: the Northern Manhattan Stroke Study Stroke 34:1586–1592

Sullivan KA, Feldman EL (2005) New developments in diabetic neuropathy. Curr Opin Neurol 18:586–590

Tabert M, Liu X Doty RL, Serby M, Zamora D, Pelton GH, Marder K, Albers MW, Stern Y, Devanand DP (2005) A 10-item smell identification scale related to risk for Alzheimer's disease. Ann Neurol 58:155–160

Tamaki C, Ohtsuki S, Terasaki T (2007) Insulin facilitates the hepatic clearance of plasma amyloid β-peptide (1-40) by intracellular translocation of low-density lipoprotein receptor-related protein 1 (LRP-1) to the plasma membrane in hepatocytes. Mol Pharmacol 72:850–855

Thallas-Bonke V, Lindschau C, Rizkalla B, Bach LA, Boner G, Meier M, Haller H, Cooper ME, Forbes JM (2004) Attenuation of extracellular matrix accumulation in diabetic nephropathy by the advanced glycation end product cross-link breaker ALT-711 via a protein kinase C-α-dependent pathway. Diabetes 53:2921–2930

The Diabetes Prevention Program Research G (2005) Role of insulin secretion and sensitivity in the evolution of type 2 diabetes in the Diabetes Prevention Program: effects of lifestyle intervention and metformin. Diabetes 54:2404–2414

Tuomilehto J, Lindstrom JJ, Eriksson JG, Valle TT, Hämäläinen H, Ilanne-Parikka P, Keinänen-Kiukaanniemi S, Laakso M, Louheranta A, Rastas M, Salminen V, Uusitupa M; Finnish Diabetes Prevention Study Group (2001) Prevention of type 2 diabetes mellitus by changes in lifestyle among subjects with impaired glucose tolerance. N Engl J Med 344:1343–1350

van Oijen M, Okereke OI, Kang JH, Pollak MN, Hu FB, Hankinson SE, Grodstein F (2008) Fasting insulin levels and cognitive decline in older women without diabetes. Neuroepidemiology 30:174–179

Vanhanen M, Koivisto K, Moilanen L, Helkala EL, Hänninen T, Soininen H, Kervinen K, Kesäniemi YA, Laakso M, Kuusisto J (2006) Association of metabolic syndrome with Alzheimer disease: A population-based study. Neurology 67:843–847

Vermeer SE, Prins ND, den Heijer T, Hofman A, Koudstaal PJ, Breteler MM (2003) Silent brain infarcts and the risk of dementia and cognitive decline. N Engl J Med 348:1215–1222

Vlassara H, Striker LJ, Teichberg S, Fuh H, Li YM, Steffes M (1994) Advanced glycation end products induce glomerular sclerosis and albuminuria in normal rats. Proc Natl Acad Sci USA 91:11704–11708

Wahrenberg H, Hertel K, Leijonhufvud BM, Persson LG, Toft E, Arner P (2005) Use of waist circumference to predict insulin resistance: retrospective study. BMJ 330:1363–1364

Watson GS, Cholerton BA, Reger MA, Baker LD, Plymate SR, Asthana S, Fishel MA, Kulstad JJ, Green PS, Cook DG, Kahn SE, Keeling ML, Craft S (2005) Preserved cognition in patients with early Alzheimer disease and amnestic mild cognitive impairment during treatment with rosiglitazone: a preliminary study. Am J Geriatr Psychiat 13:950–958

Watson GS, Peskind ER, Asthana S, Purganan K, Wait C, Chapman D, Schwartz MW, Plymate S, Craft S (2003) Insulin increases CSF Aβ42 levels in normal older adults. Neurology 60:1899–1903

Wedick NM, Mayer-Davis EJ, Wingard DL, Addy CL, Barrett-Connor E (2001) Insulin resistance precedes weight loss in adults without diabetes : The Rancho Bernardo Study. Am J Epidemiol 153:1199–1205

Whitmer RA, Gunderson EP, Barrett-Connor E, Quesenberry CP Jr, Yaffe K (2005) Obesity in middle age and future risk of dementia: a 27 year longitudinal population based study. BMJ 330:1360

Whitmer RA, Gustafson DR, Barrett-Connor E, Haan MN, Gunderson EP, Yaffe K (2008) Central obesity and increased risk of dementia more than three decades later. Neurology 71:1057–1064
Whitmer RA, Sidney S, Selby J, Johnston SC, Yaffe K (2005) Midlife cardiovascular risk factors and risk of dementia in late life. Neurology 64:277–281
Williamson JD, Miller ME, Bryan RN, Lazar RM, Coker LH, Johnson J, Cukierman T, Horowitz KR, Murray A, Launer LJ, ACCORD Study Group (2007) The action to control cardiovascular risk in diabetes memory in diabetes study (ACCORD-MIND): rationale design and methods. Am J Cardiol 99:112i–122i
Xu W, Qiu C, Winblad B, Fratiglioni L (2007) The effect of borderline diabetes on the risk of dementia and Alzheimer's disease. Diabetes 56:211–216
Xu WL, Qiu CX, Wahlin A, Winblad B, Fratiglioni L (2004) Diabetes mellitus and risk of dementia in the Kungsholmen project: a 6-year follow-up study. Neurology 63:1181–1186
Yaffe K, Blackwell T, Kanaya AM, Davidowitz N, Barrett-Connor E, Krueger K (2004) Diabetes impaired fasting glucose and development of cognitive impairment in older women. Neurology 63:658–663
Yaffe K, Blackwell T, Whitmer RA, Krueger K, Barrett Connor E (2006) Glycosylated hemoglobin level and development of mild cognitive impairment or dementia in older women. J Nutr Health Aging 10:293–295
Yaffe K, Kanaya A, Lindquist K, Simonsick EM, Harris T, Shorr RI, Tylavsky FA, Newman AB (2004) The metabolic syndrome inflammation and risk of cognitive decline. JAMA 292: 2237–2242
Yan SD, Chen X, Fu J, Chen M, Zhu H, Roher A, Slattery T, Zhao L, Nagashima M, Morser J, Migheli A, Nawroth P, Stern D, Schmidt AM (1996) RAGE and amyloid-β peptide neurotoxicity in Alzheimer's disease. Nature 382:685–691
Yesavage JA, O'Hara R, Kraemer H, Noda A, Taylor JL, Ferris S, Gély-Nargeot MC, Rosen A, Friedman L, Sheikh J, Derouesné C (2002) Modeling the prevalence and incidence of Alzheimer's disease and mild cognitive impairment. J Psychiatr Res 36:281–286
Yoshitake T, Kiyohara Y, Kato I, Ohmura T, Iwamoto H, Nakayama K, Ohmori S, Nomiyama K, Kawano H, Ueda K (1995) Incidence and risk factors of vascular dementia and Alzheimer's disease in a defined elderly Japanese population: the Hisayama Study. Neurology 45:1161–1168
Zhao L, Teter B, Morihara T, Lim GP, Ambegaokar SS, Ubeda OJ, Frautschy SA, Cole GM (2004) Insulin-degrading enzyme as a downstream target of insulin receptor signaling cascade: implications for Alzheimer's disease intervention. J Neurosci 24:11120–11126
Zlokovic BV (2008) The blood-brain barrier in health and chronic neurodegenerative disorders. Neuron 57:178–201

Whitmer RA, Gustafson DR, Barrett-Connor E, Haan MN, Gunderson EP, Yaffe K (2008) Central obesity and increased risk of dementia more than three decades later. Neurology 71:1057–1064

Whitmer RA, Sidney S, Selby J, Johnston SC, Yaffe K (2005) Midlife cardiovascular risk factors and risk of dementia in late life. Neurology 64:277–281

Williamson JD, Miller ME, Bryan RN, Lazar RM, Coker LH, Johnson J, Cukierman T, Horowitz KR, Murray A, Launer LJ, ACCORD Study Group (2007) The action to control cardiovascular risk in diabetes memory in diabetes study (ACCORD-MIND): rationale, design, and methods. Am J Cardiol 99:112i–122i

Xu W, Qiu C, Winblad B, Fratiglioni L (2007) The effect of borderline diabetes on the risk of dementia and Alzheimer's disease. Diabetes 56:211–216

Xu WL, Qiu CX, Wahlin A, Winblad B, Fratiglioni L (2004) Diabetes mellitus and risk of dementia in the Kungsholmen project: a 6-year follow-up study. Neurology 63:1181–1186

Yaffe K, Blackwell T, Kanaya AM, Davidowitz N, Barrett-Connor E, Krueger K (2004) Diabetes, impaired fasting glucose, and development of cognitive impairment in older women. Neurology 63:658–663

Yaffe K, Blackwell T, Whitmer RA, Krueger K, Barrett Connor E (2006) Glycosylated hemoglobin level and development of mild cognitive impairment or dementia in older women. J Nutr Health Aging 10:293–295

Yaffe K, Kanaya A, Lindquist K, Simonsick EM, Harris T, Shorr RI, Tylavsky FA, Newman AB (2004) The metabolic syndrome, inflammation, and risk of cognitive decline. JAMA 292:2237–2242

Yan SD, Chen X, Fu J, Chen M, Zhu H, Roher A, Slattery T, Zhao L, Nagashima M, Morser J, Migheli A, Nawroth P, Stern D, Schmidt AM (1996) RAGE and amyloid-beta peptide neurotoxicity in Alzheimer's disease. Nature 382:685–691

Yesavage JA, O'Hara R, Kraemer H, Noda A, Taylor JL, Ferris S, Gely-Nargeot MC, Rosen A, Friedman L, Sheikh J, Derouesne C (2002) Modeling the prevalence and incidence of Alzheimer's disease and mild cognitive impairment. J Psychiatr Res 36:281–286

Yoshitake T, Kiyohara Y, Kato I, Ohmura T, Iwamoto H, Nakayama K, Ohmori S, Nomiyama K, Kawano H, Ueda K, Sueishi K, Tsuneyoshi M, Fujishima M (1995) Incidence and risk factors of vascular dementia and Alzheimer's disease in a defined elderly Japanese population: the Hisayama Study. Neurology 45:1161–1168

Zhao L, Teter B, Morihara T, Lim GP, Ambegaokar SS, Ubeda OJ, Frautschy SA, Cole GM (2004) Insulin-degrading enzyme as a downstream target of insulin receptor signaling cascade: implications for Alzheimer's disease intervention. J Neurosci 24:11120–11126

Zlokovic BV (2008) The blood-brain barrier in health and chronic neurodegenerative disorders. Neuron 57:178–201

# The Role of Insulin Dysregulation in Aging and Alzheimer's Disease

**Suzanne Craft**

**Abstract** Our knowledge of the multifaceted role of insulin in the central nervous system has expanded rapidly in recent years. It is now apparent that perturbation of this role by insulin resistance and hyperinsulinemia can increase the risk for aging-related neurodegenerative disorders such as Alzheimer's disease (AD) through a number of mechanistic pathways, delineated in elegant in vitro and animal studies. We have investigated these pathways in humans using models of insulin resistance and hyperinsulinemia. We will present results suggesting that insulin-associated effects on Aβ regulation and inflammation contribute to AD pathophysiology. This premise raises the possibility that treatments aimed at improving insulin resistance will benefit patients with AD. We will review data testing this possibility in pilot therapeutic trials using insulin sensitizers and intranasal insulin.

## 1 Insulin and the Brain

The peripheral effects of insulin, a hormone secreted by pancreatic β-cells, have been well-characterized. Recent evidence demonstrates that insulin is also active in the CNS. Although controversy exists as to whether insulin is synthesized in the adult brain, insulin is readily transported into the CNS across the blood brain barrier (BBB) by a saturable, receptor-mediated process (Banks et al. 1997b; Baskin et al. 1987; Baura et al. 1993). Raising peripheral insulin levels acutely elevates brain and CSF insulin levels, whereas prolonged peripheral hyperinsulinemia down-regulates BBB insulin receptors and reduces insulin transport into the brain (Schwartz et al. 1990; Wallum et al. 1987). Insulin receptors are located in the synapses of both astrocytes and neurons (Abbott et al. 1999). Although insulin and

S. Craft
GRECC S-182, VAPSHCS, Department of Psychiatry and Behavioral Sciences, University of Washington School of Medicine, 1660 South Columbian Way, Seattle, WA 98108, USA
e-mail: scraft@u.washington.edu

S. Craft and Y. Christen (eds.), *Diabetes, Insulin and Alzheimer's Disease*,
Reserach and Perspectives in Alzheimer's Disease,
DOI 10.1007/978-3-642-04300-0_8, © Springer-Verlag Berlin Heidelberg 2010

insulin receptors are abundant in the brain, they are selectively distributed, with high concentrations in the olfactory bulb, cerebral cortex, hippocampus, hypothalamus, amygdala, and septum (Havrankova et al. 1978a, b; Unger et al. 1991).

The localization of insulin receptors in hippocampus and medial temporal cortex is consistent with evidence that insulin influences memory. In rats, acute intracerebroventricular insulin administration enhanced memory on a passive-avoidance task (Unger et al. 1991). In humans, acute intravenous (IV) insulin administration while maintaining euglycemia reliably enhances story recall (Craft et al. 1996, 1999, 2003; Kern et al. 2001). Conversely, learning may also influence insulin receptor expression and function. For example, training rodents on a spatial memory task increased insulin receptor expression in the dentate gyrus and hippocampal CA1 field (Zhao et al. 1999). Thus, the act of learning is accompanied by changes in insulin signaling molecules in the hippocampus. Collectively, these studies suggest that insulin may contribute to normal memory functioning.

There are several mechanisms through which insulin may affect memory. One mechanism may be through effects on cerebral energy metabolism. Although insulin does not appear to influence glucose transport into brain, it may have more selective effects on cerebral glucose metabolism. Bingham et al. (2002) recently demonstrated an increase in cerebral glucose metabolism that was particularly pronounced in the cortex following administration of a low dose of insulin. The basis for regional insulin effects on glucose metabolism may be due to the distribution of glucose transporter isoforms (GLUTs) (Reagan et al. 2001; Schulingkamp et al. 2000). The insulin-sensitive GLUTs 4 and 8 are selectively distributed in the brain, and insulin increases brain GLUT4 expression. Notably, substantial co-localization exists for insulin-containing neurons, insulin receptors and GLUTs 4 and 8 (Apelt et al. 1999; Schulingkamp et al. 2000). These overlapping distributions are consistent with insulin-stimulated glucose uptake in selective brain regions, including medial temporal lobe structures that support learning and memory.

Other insulin-related mechanisms that are not directly related to modulation of glucose uptake have also been implicated in normal hippocampal functioning (Zhao and Alkon 2001). For example, insulin may modulate components of the long-term potentiation cascade, such as the cell membrane expression of NMDA receptors (Skeberdis et al. 2001), which affect the likelihood of LTP induction. Insulin also modulates CNS levels of neurotransmitters, such as acetylcholine and norepinephrine, that are known to influence cognitive function (Figlewicz et al. 1993; Kopf and Baratti 1999). Thus, insulin affects numerous mechanisms relating to neuronal activity and cognitive function supported by such activity.

Paradoxically, chronic hyperinsulinemia and insulin resistance, or reduced insulin effectiveness, may exert a negative influence on memory. For example, type 2 diabetes mellitus (T2DM) has been associated with impaired learning in animal and human studies (Greenwood and Winocur 2001). Furthermore, impaired verbal memory has been observed in individuals with chronic hyperinsulinemia in the absence of hyperglycemia (Vanhanen et al. 1998). Additionally, impaired glucose tolerance has been associated with reduced hippocampal volume and memory impairment (Convit et al. 2003). Taken together, these findings are consistent with the notion that acute and chronic hyperinsulinemia have opposing effects on the neural substrates of memory.

# 2 Insulin Resistance and AD

The association between AD and syndromes related to peripheral hyperinsulinemia and insulin resistance has been largely supported in epidemiological work, although some inconsistency has been noted. Special attention may be warranted for population-based studies that directly assessed diabetes, hyperinsulinemia and insulin resistance with oral glucose tolerance testing, and used neuroimaging to facilitate differential diagnosis of AD and vascular dementia. In general, such methodologically sound studies have shown an increased risk of both AD and vascular dementia associated with diabetes and insulin resistance. Two prospective, population-based cohort studies reported such an association (Ott et al. 1999; Peila et al. 2002). In the prospective, community-based Rotterdam study, Ott et al. (1999) found that T2DM significantly increased the risk for dementia and AD. Similar findings were reported by Leibson et al. (1997), who reported that hyperinsulinemia increases risk for dementia overall and that AD risk is raised independently from vascular or other dementias. Mayeux and colleagues reported that hyperinsulinemia was a risk factor for AD and general memory decline in a sample of 683 older adults (Luchsinger et al. 2004). In other cross-sectional and longitudinal population-based studies, fasting plasma insulin elevations were associated with AD and with faster memory decline for both diabetic and non-diabetic subjects, suggesting that insulin resistance independent of diabetes may increase disease risk (Kurochkin and Goto 1994).

## *2.1 Insulin Abnormalities and AD Pathology*

The ways in which insulin abnormalities may contribute to the symptoms and pathogenesis of AD have been examined in a variety of experimental models. Hoyer and colleagues were the first group to suggest that desensitization of the neuronal insulin receptor plays a role in AD (Hoyer 2002). More recently, animal and in vitro studies have documented relationships between insulin and mechanisms with clear pathogenic implications for AD. In vitro, insulin modulates levels of the Aβ peptide, the aggregation of which is a fundamental neuropathological hallmark of AD. For example, insulin promotes release of intracellular Aβ in neuronal cultures, affecting both its short (Aβ40) and long (Aβ42) forms and accelerating their trafficking from the Golgi and trans-Golgi network to the plasma membrane (Gasparini et al. 2001). Thus, low brain insulin may reduce the release of Aβ from intracellular to extracellular compartments. Interestingly, Aβ also regulates brain insulin signaling. Soluble Aβ binds to the insulin receptor and disrupts its signaling capacity and LTP induction in mouse hippocampal slice preparations (Townsend et al. 2007). These effects could be prevented by exposing tissue to insulin prior to Aβ exposure.

A growing understanding of the importance of impaired Aβ clearance as opposed to increased Aβ production in late-onset AD has created intense focus on mechanisms regulating Aβ degradation. Insulin may modulate Aβ degradation by regulating expression of insulin-degrading enzyme (IDE), a metalloprotease that catabolizes insulin (Zhao et al. 2004). IDE is highly expressed in brain as well as in liver, kidney, and muscle (Authier et al. 1996) and may play a critical role in Aβ clearance in brain (Kurochkin and Goto 1994; McDermott and Gibson 1997; Qiu et al. 1998). IDE has also been implicated in the intracellular degradation of Aβ (Sudoh et al. 2002). Furthermore, decreased IDE activity, levels, and mRNA have been observed in AD brain tissue, and IDE knockout mice have reduced degradation of Aβ and insulin in brain (Cook et al. 2003; Farris et al. 2003; Perez et al. 2000). Thus, low CNS insulin may reduce IDE levels in brain and thereby impair Aβ clearance. Conversely, excessively high insulin levels may act as a competitive substrate for IDE and inhibit its degradation of Aβ.

A recently studied mechanism through which insulin and Aβ may interact to modulate AD pathology is synaptotoxic effects. Loss of synapses is the earliest structural defect observed in AD. Soluble oligomeric species of Aβ are synaptotoxic, and insulin prevents binding of Aβ to synapses, thereby preserving synaptic integrity (De Felice et al. 2009). Insulin also reduced oligomer formation, which may have additional protective effects; a functional consequence of these effects appears to be protection against Aβ-induced disruption of long-term potentiation integrity, the process of synaptic remodeling believed to underlie memory formation (Lee et al. 2009).

Chronic peripheral hyperinsulinemia may thus lower brain insulin levels and interfere with peripheral Aβ clearance. Chronic peripheral hyperinsulinemia has been associated with a pattern in which brain insulin levels are initially higher, then decrease as transport of insulin into the brain is down-regulated (Banks et al. 1997a). We have reported that patients with AD showed lower cerebral spinal fluid (CSF) insulin levels, higher plasma insulin levels, and reduced CSF-to-plasma insulin ratios compared to healthy controls (Craft et al. 1998). In this study, 14 normal older adults and 25 patients with AD received a paired blood draw and lumbar puncture (LP) following a 10-hour fast. Subjects with moderate AD had higher plasma insulin levels ($p < 0.05$) but lower CSF insulin levels ($p < 0.02$) relative to controls with equivalent body mass indices and fasting plasma glucose. As a consequence, CSF-to-plasma ratios were lowest for subjects with moderate AD (Fig. 1; $p < 0.05$). Overall, CSF insulin levels were lowest among AD subjects with the highest plasma insulin levels. These results are consistent with animal models showing that chronic hyperinsulinemia reduces CNS insulin uptake This possibility was further explored in a second study, in which we examined the effects of raising plasma insulin to levels associated with insulin-resistant states on CSF insulin. The finding that AD patients show lower CSF insulin following intravenous insulin infusion than do normal older adults would support the notion of reduced brain insulin uptake in AD. Sixteen normal older adults and seven patients with

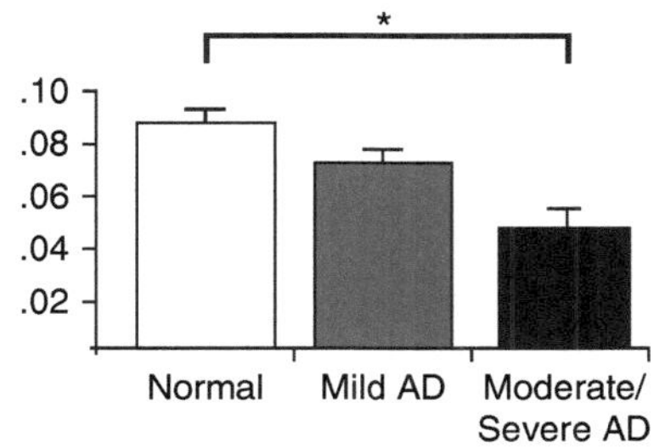


**Fig. 1** CSF-to-plasma insulin ratios for normal adults, adults with mild AD, and adults with moderate to severe AD (from Craft et al. 1998)

early AD received a saline and an insulin infusion on separate days in counterbalanced order. In the insulin condition, an insulin dose of 1.0 $mU{\cdot}kg^{-1}{\cdot}min^{-1}$ was used, resulting in plasma levels of about 80 μU/ml. Such levels are commonly observed post-prandially in adults with insulin resistance. Dextrose (20%) was infused as needed to keep plasma glucose at fasting levels of 90 mg/dl. In the control condition, only saline was infused and insulin remained at a fasting baseline (~10μU/ml). After 90 min of infusion, plasma was obtained and CSF was collected. Insulin infusion resulted in an increase in CSF insulin levels for normal adults ($p = 0.02$), presumably reflecting transport of insulin into the CNS, as has been well described in animal models. CSF insulin was not reliably increased for AD patients, a result consistent with reduced brain uptake.

In previous studies, we have modeled the effects of hyperinsulinemia associated with insulin resistance on Aβ42 and inflammation (Fishel et al. 2005). Healthy older adults ($n = 16$, mean age 69) received infusions of saline or insulin (1.0 $mU{\cdot}kg^{-1}{\cdot}min^{-1}$) with variable dextrose to maintain euglycemia on separate days in counterbalanced order. After 90 min of infusion, CSF was collected and frozen immediately. All participants received both infusions and two LPs, allowing them to serve as their own controls. Insulin administration increased plasma insulin to levels typical of post-prandial insulin in insulin-resistant adults (85 μU/ml). CSF insulin also increased ($p = 0.02$), presumably reflecting increased transport into the CNS. As predicted, CSF Aβ42 increased in an age-dependent manner (Fig. 2; $p = 0.02$), with the greatest effects noted for adults older than 70 years of age. Notably, greater insulin-induced increases in CSF insulin levels predicted greater increases in CSF Aβ42 levels ($r = 0.76$, $p = 0.05$). CSF inflammatory cytokines and F2-isoProstane (IL-1α, IL-6, TNFα, F2-isoP) all increased in response to insulin (all p values $< 0.01$). Higher insulin-induced F2-isoprostane levels predicted greater increases in CSF Aβ42 ($p = 0.009$).

Together, chronic peripheral hyperinsulinemia and associated reductions in brain insulin uptake may affect insulin's role in several pathophysiologic processes in AD, including Aβ regulation and inflammation. High plasma insulin levels may interfere with degradation of Aβ transported out of the brain, thereby obstructing a peripheral Aβ-clearing "sink." Concomitantly, low brain insulin levels reduce release of Aβ from intracellular compartments into extracellular compartments where clearance is believed to occur. Thus, for some patients with AD, high

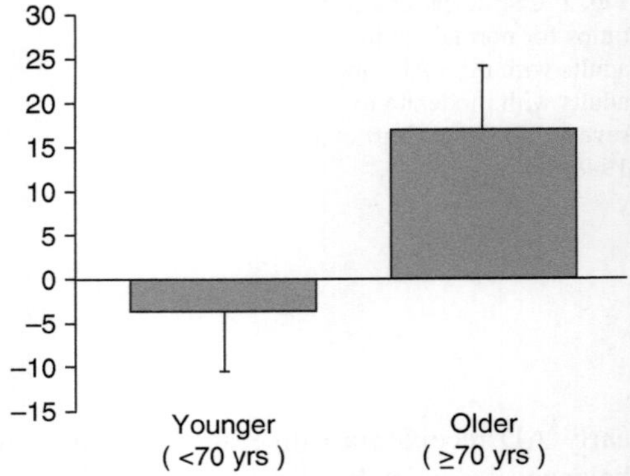


**Fig. 2** Percent change in CSF Aβ42 levels with insulin infusion relative to saline infusion for normal adults <70 or ≥70 years of age

peripheral insulin levels and low brain insulin levels would result in reduced clearance of Aβ both in brain and in the periphery.

Support for the validity of this model is provided by a recent study that induced insulin resistance in the T2576 mouse model of AD with a high fat diet. Diet manipulation resulted in a metabolic profile of high peripheral insulin and low brain insulin and IDE levels compared to Tg2576 mice fed a normal diet (Ho et al. 2004). Diet-induced insulin resistance caused two-fold increases in Aβ40 and 42 and, earlier, larger Aβ deposits compared to non-insulin-resistant Tg2576 mice. Furthermore, insulin-resistant mice had impaired learning on a water maze test. In another model of insulin resistance, APP/PS1 mice were given sucrose-sweetened beverages and also demonstrated increased brain Aβ deposition and reduced Morris water maze learning (Cao et al. 2007). Together these results suggest that insulin resistance can precipitate the neuropathological and behavioral features of AD and that raising brain insulin levels may reduce neuropathological changes related to AD. Recent research has also focused intensively on the contribution of vascular dysfunction to AD pathogenesis, as well as to vascular cognitive impairment.

## 2.2 *Insulin Resistance and Vascular Dysfunction*

Insulin resistance has many negative effects on vascular function, often related to dyslipidemia. High triglycerides and low density lipoprotein impair endothelial integrity, in part via inflammation. Impaired suppression of adipose tissue lipolysis and mild post-prandial hyperglycemia favor free fatty acid (FFA) utilization and oxidation, increasing levels of reactive oxygen species and cytokines. Additionally, insulin directly affects vasoreactivity and hemodynamic functions, such as capillary recruitment, vasodilation and regional blood flow. Hemodynamic and metabolic effects working in concert enhance energy substrate delivery (Cersosimo and DeFronzo 2006). Insulin normally increases NO-mediated vasodilation and regulates vasoconstriction via endothelin-1. Conversely, insulin resistance decreases NO and increases endothelin-1 activity, favoring vasoconstriction and reducing capillary recruitment. In turn, endothelial dysfunction reduces insulin transport, ultimately reducing capillary recruitment and microvascular blood flow. This process exacerbates glucose and lipid abnormalities and establishes a negative feedback loop between progressive endothelial dysfunction and increasing insulin resistance (Cersosimo and DeFronzo 2006). In the brain, vasoconstriction and reduced capillary recruitment may interfere with functions of the neurovascular unit, the coordinated interaction of astrocyte, neuron, and endothelium that couples neural activity with increased blood flow. Glucose may thus be unavailable to neurons. Reduced capillary recruitment by insulin results in a 50% decrease in glucose extractional fraction (Cersosimo and DeFronzo 2006).

Recent neuropathological studies have illustrated the complex interrelationship between markers of vascular dysfunction and AD pathology. An interesting pattern has emerged, in which patients with treated diabetes demonstrate a reduced amyloid load compared to non-diabetics with similar levels of dementia (Beeri et al. 2008). In a recent study, we reported that patients with treated diabetes and dementia had Aβ plaque loads similar to non-demented cases and instead had increased microvascular infarcts and IL-6 (Sonnen et al. 2009). In contrast, untreated diabetics with dementia had plaque loads that were similar to non-diabetic dementia cases. One intriguing interpretation of these results is that diabetic treatment affected amyloid load but not degree of dementia. If true, this finding would raise questions about the role of amyloid plaques in dementia symptoms. This interpretation must be considered to be speculative, however, given the small number of cases and the fact that treated diabetics typically have more severe diabetes. Similarly, the finding of increased microvascular injury in treated diabetics with dementia, but not in similarly affected treated diabetics without dementia, raises the question of which diabetic factors are associated with microvascular infarcts? Clearly not all treated diabetics developed dementia, only those cases with concomitant microvascular infarcts. The nature of this microvascular injury is also an important consideration. Given their small volume, it is unlikely that microvascular infarcts directly cause dementia but rather serve as a marker for more extensive microvascular dysfunction. These findings illustrate the importance of careful assessment of treatment and metabolic status in future neuropathologic studies.

## 3 Therapeutic Implications of Insulin Resistance as a Pathogenetic Mechanism in AD

### 3.1 *Pharmacologic Insulin Sensitization*

Given the relationship between insulin resistance and memory impairment, therapeutic strategies aimed at treating early T2DM may also benefit those MCI patients who are at risk for developing AD. PPAR-γ agonists have been shown to improve insulin sensitivity by decreasing circulating insulin and increasing insulin-mediated glucose uptake, with minimal risk of hypoglycemia (Olefsky 2000). In addition to improving insulin sensitivity, several investigators have reported that PPAR-γ activity may reduce both Aβ accumulation and inflammatory reactants and protect against neurotoxicity (Combs et al. 2000; Delerive et al. 2001; Paik et al. 2000). Activation of PPAR-γ receptors has been shown to regulate inflammatory responses and apoptosis (Corton et al. 2000; Escher and Wahli 2000). PPAR-γ agonists inhibit Aβ-stimulated secretion of pro-inflammatory products, arrest the evaluation of activated macrophages, and inhibit the expression of cyclooxygenase-2 (Combs et al. 2000). Decreased oxidative stress and reduced inflammation have been demonstrated with PPAR-γ treatment in both in vitro and in vivo models (Hirsch et al. 2003; Schmidt et al. 2004). PPAR-γ agonists are thus attractive candidates for the treatment of insulin resistance and inflammation associated with early cognitive decline.

Rosiglitazone, a compound that binds with high affinity to PPAR-γ (Lehmann et al. 1995), is approved by the FDA for use as an antidiabetic agent and acts to reduce hyperglycemia by increasing insulin sensitivity. Reduced insulin resistance is likely to produce indirect CNS effects as a result of decreased circulating insulin, which would presumably help to restore an appropriate balance of peripheral and central insulin levels and enhance peripheral Aβ clearance. Rosiglitazone, when administered to insulin-resistant Tg2576 mice, both normalizes the insulin response and ameliorates the associated impaired stress response in these animals (Pedersen and Flynn 2004). Animal models suggest that treatment with rosiglitazone significantly reduces inflammation in response to acute injury and also improves endothelium-dependent vasodilation and coronary arteriole function (Bagi et al. 2004; Cuzzocrea et al. 2004; Tao et al. 2003).

We recently completed a parallel group, double-blind, placebo-controlled trial to test the hypothesis that treatment with the PPAR-γ agonist, rosiglitazone, which both improves insulin sensitivity and reduces inflammation, would produce beneficial cognitive effects for patients with amnestic MCI and early AD. Thirty subjects [21 diagnosed with AD per NINCDS-ADRDA criteria, 9 diagnosed with amnestic MCI per the Petersen criteria (Petersen et al. 1999)] received a single daily oral dose of 4 mg of rosiglitazone ($n = 20$) or matched placebo ($n = 10$) for six months. Delayed memory on the Buschke list-learning task was preserved over the six-month trial for the rosiglitazone-treated group, whereas the placebo-assigned group showed the expected decline in memory performance (Fig. 3A). Differences between the two treatment groups were observed after four months ($p = 0.04$) and six months ($p = 0.001$). The degree of memory preservation was related to treatment response, as indexed by fasting plasma insulin levels at six months. For the rosiglitazone-treated group, lower insulin levels, indicative of a positive treatment response, were associated with a greater increase in delayed list recall from baseline (Spearman $r = -0.48$, $p = 0.04$). Performance on the Stroop Color-Word Interference Test was substantially improved for rosiglitazone-treated participants. These subjects showed fewer errors on the Stroop interference test than did placebo-assigned subjects after six months of treatment (Fig. 3B; $p = 0.03$). A similar relationship was observed between six-month treatment response and changes in verbal fluency for categories, a sensitive measure of semantic search that is impaired at the earliest stages of AD (Spearman $r = -0{\cdot}52$, $p = 0{\cdot}0184$).

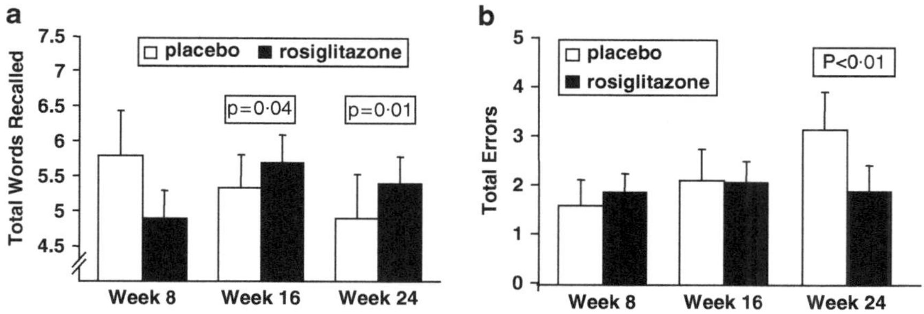


**Fig. 3** (**a**) Number of items recalled after a delay on Buschke Selective Reminding Test and (**b**) number of errors on Stroop Color Word Interference Test for participants with AD/aMCI after 8, 16, or 24 weeks of receiving placebo or 4 mg/d rosiglitazone (from Watson et al. 2005)

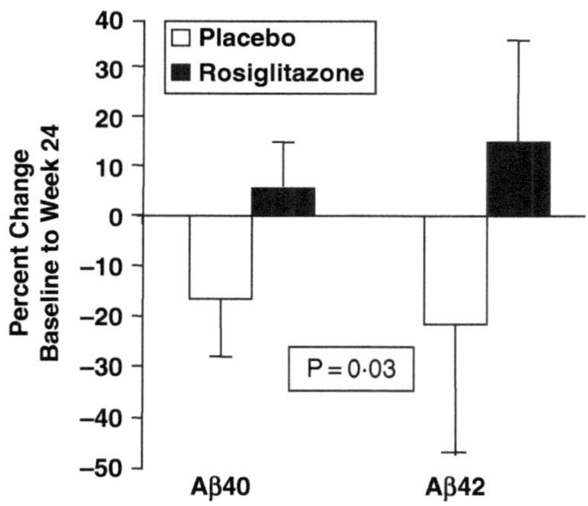


**Fig. 4** Percent change from baseline in plasma Aβ40 and Aβ42 levels for participants with AD/aMCI after 24 weeks of receiving placebo or 4 mg/d rosiglitazone (from Watson et al. 2005)

Both plasma Aβ40 and Aβ42 levels declined over the six-month treatment period and Aβ42 (Fig. 4). This pattern of decreased Aβ42 is in keeping with results reported in a recent, large longitudinal study of early AD patients (Mayeux et al. 2003). In contrast, Aβ levels remained stable in the rosiglitazone-treated group. The observed effects of rosiglitazone on cognitive measures and plasma Aβ suggest that treating insulin resistance can ameliorate the progressive cognitive decline associated with early AD. In light of these findings, PPAR-γ agonists may constitute a potential novel class of therapeutic agents for these patients.

## 4 Intranasal Insulin

Multiple studies have demonstrated that IV insulin administration (while maintaining euglycemia) increases CNS insulin and reliably improves cognition (Craft et al. 1999, 2003; Park et al. 2000). However, chronic peripheral insulin administration is not a viable therapeutic option, due to risks associated with hypoglycemia. In addition, it is likely that such an approach would exacerbate peripheral hyperinsulinemia, with possible negative effects on Aβ clearance. Any long-term treatment strategy for normalizing CNS insulin levels in persons with AD must avoid significantly increasing insulin in the periphery. Such an approach is possible with an intranasal administration technique.

## 4.1 Intranasal Pathways to the CNS

The nasal cavity is unique in that olfactory sensory neurons are directly exposed to the external environment in the upper nasal cavity while their axons extend through the cribriform plate to the olfactory bulb. Following intranasal administration, drugs can be directly transported to the CNS, bypassing the periphery. Several extraneuronal and intraneuronal pathways from the nasal cavity to the CNS are possible. The extraneuronal pathways appear to rely on bulk flow transport through perineural channels to the brain or CSF. In recent studies, labeled intranasal insulin or a closely related peptide, insulin-like growth factor-I (IGF-I), was administered to rodents (Francis et al. 2008, Thorne et al. 2004). Within 30 minutes, IGF-I signal was detected along olfactory and trigeminal channels, with robust signal evident in hippocampus, amygdale and cortex. An additional extracellular pathway was identified with quick access to the CSF after absorption into the submucosa along the olfactory nerve and cribriform plate (Born et al. 2002; Frey 2002; Thorne et al. 2004). These extracellular pathways provide direct access to the CNS within minutes of intranasal administration. Additionally, an intraneuronal pathway delivers drugs to the CNS hours or days later. Anterograde axoplasmic transport within olfactory sensory neurons has been demonstrated.

## 4.2 Intranasal Drug Administration

Although use of intranasal pathways for drug delivery is considered novel, the existence of such pathways has been known for many years. Viruses and micro-organisms (Bodian and Howe 1941; Faber 1938; Fairbrother and Hurst 1930), metals (Czerniawska 1970; Gopinath et al. 1978), dyes (Clark and Gros 1929; Faber 1937), amino acids (Weiss and Holland 1967) and proteins (Kristensson and Olsson 1971; Shipley 1985, Thorne et al. 1995) have been shown for decades to enter the CNS via nasal routes. There are a variety of drugs that either cannot permeate the BBB to reach targets in the CNS or that penetrate the BBB but can have harmful effects in the periphery. Substances with lower molecular weights are more likely to be transported to the CNS along intranasal pathways.

A common approach to determining whether a specific drug can be transported to the CNS is to measure the drug in CSF or brain following intranasal administration while controlling or adjusting for drug transport to the periphery. Such research demonstrates that significant numbers of therapeutic compounds are successfully delivered to the CNS following intranasal administration, including insulin (Born et al. 2002), neurotrophic factors (Thorne and Frey 2001), antibiotics (Sakane et al. 1991), antivirals (Seki et al. 1994), adrenergics (Anand Kumar et al. 1976), antineoplastics (Wang et al. 2003, 2004), estrogen and progesterone (Anand Kumar et al. 1974, 1982), vasopressin (Pietrowsky et al. 1996a), cholecystokinin (Pietrowsky et al. 1996b), corticotropin-releasing hormone (Born et al. 2002), DNA plasmids (Oh et al. 2001) and cocaine (Chow et al. 1999).

## 4.3 Intranasal Insulin Effects in the CNS

Kern and colleagues (Born et al. 2002) administered 40 IU of insulin intranasally in young, healthy adults. CSF and blood were sampled every 10 to 20 minutes for 80 minutes following administration. Insulin treatment resulted in increased CSF insulin levels within 10 minutes of administration compared to placebo, with peak levels noted within 30 minutes. CSF insulin levels had not returned to baseline by the end of the 80-minute study. Blood glucose and insulin levels did not change, demonstrating that the effects in CSF were not due to transport from the nasal cavity to systemic circulation. This finding is consistent with extensive literature that demonstrates insulin's poor transport from the nasal cavity into blood (Illum 2002). Although elevated CSF insulin levels do not conclusively demonstrate that brain insulin levels are similarly elevated, animal studies have shown significant drug uptake to hippocampus and cortex. Francis et al. (2008) showed that intranasal insulin reversed the effects of diabetes in a murine model, reducing brain atrophy, increasing markers of synaptic function, increasing insulin receptors and phosphorylation, reversing diabetes-related reductions in choline acetyltranferase levels, reducing neuronal NFkB activation, and increasing activation of Akt, CREB, and GSK3β. These multifaceted effects were accompanied by a striking preservation of memory, as measured by the Morris water maze and radial arm tasks (Francis et al. 2008).

Functional and cognitive studies of the acute and chronic effects of intranasal administration also support insulin's transport to the CNS. Sixty minutes of intranasal insulin treatment (20 IU every 15 minutes) induced changes in auditory-evoked brain potentials (AEPs) compared to placebo (Kern et al. 1999). We have also demonstrated that intranasal insulin acutely improves verbal memory in memory-impaired persons without affecting plasma insulin or glucose levels (Reger et al. manuscript in preparation). Thirteen memory-impaired [MI; six with early AD and seven with amnestic MCI (aMCI)] and 35 normal adults received saline and four doses of intranasally administered regular insulin (10, 20, 40, or 60 IU insulin) on separate mornings. Percent change in memory (story recall) relative

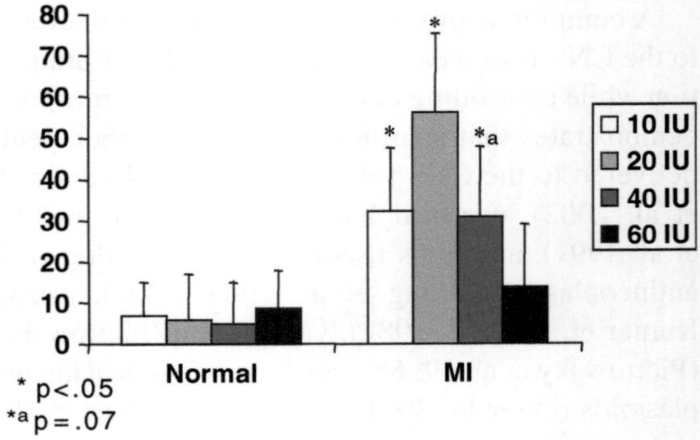


**Fig. 5** Percent change compared to saline in story recall after 10, 20, 40, and 60 IU of intranasal insulin for normal adults and memory-impaired (MI) adults

to the placebo condition was calculated for each of the four insulin doses. Story recall was enhanced following 10 ($p = .04$), 20 ($p = .005$) and 40 ($p = .07$) IU of intranasal insulin for the MI group (Fig. 5). Normal adults showed smaller but consistent enhancement in story recall that only approached statistical significance at the highest dose.

With respect to effects of chronic intranasal insulin administration, several studies reported that two months of daily insulin administration (4 × 40 IU/day) significantly improved verbal memory and enhanced mood in young, healthy adults (Benedict et al. 2004, 2008). In a recent pilot clinical trial. we examined the effects of short-term daily intranasal insulin administration in 25 adults with AD ($n = 11$) or aMCI ($n = 14$), who were randomly assigned to receive insulin (20 IU bid; $n = 13$) or placebo ($n = 12$). Treatment groups were comparable in terms of age, education, body mass index, or dementia severity as assessed by the Mattis Dementia Rating Scale. Saline or insulin was administered with a ViaNase$^{TM}$ Electronic Atomizer (Bothell, Washington). The ViaNase device has been specifically designed to deliver drugs to the olfactory region to maximize drug transport to the CNS. This device releases a metered insulin (20 IU) or saline dose into a chamber covering the subject's nose, which then is inhaled by breathing evenly over a 2 minute period. This method allows administration of smaller particle sizes to increase drug deposition in the upper nasal cavity without transporting the drug to the lungs (Djupesland et al. 2004). Intranasal treatment was administered twice daily, immediately after breakfast and dinner, for 21 days under the supervision of the caregiver for patients with AD.

The primary outcome measure was the amount of story recall retained over the delay (calculated as immediate recall/delayed recall). Secondary measures included a frontal-executive test of selective attention and response inhibition (Stroop Color-Word Interference Task). Blood was collected at baseline and after 21 days of treatment from fasting participants for analysis of plasma Aβ. Relative to their baseline performance, insulin-treated subjects had greater memory savings at day 21 than did placebo-assigned subjects (Fig. 6A). Insulin treatment also resulted in faster average performance on the selective attention test (Fig. 6B). Insulin-treated

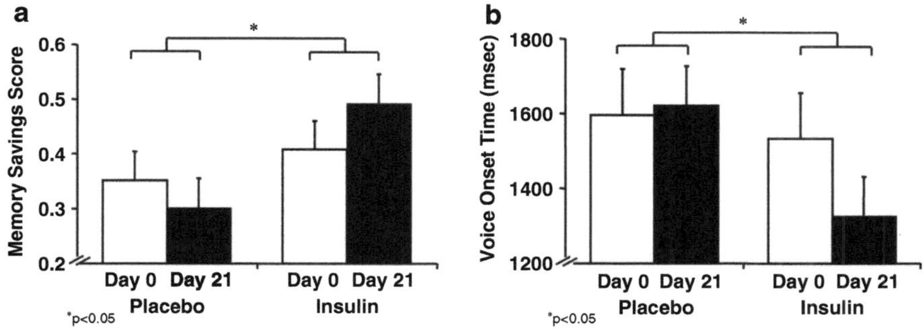


**Fig. 6** (**a**) Memory savings scores for story recall and (**b**) voice onset time for discordant trials on the Stroop Color-Word Interference Test after 21 days of placebo or intranasal insulin for adults with aMCI or AD (from Reger et al. 2008)

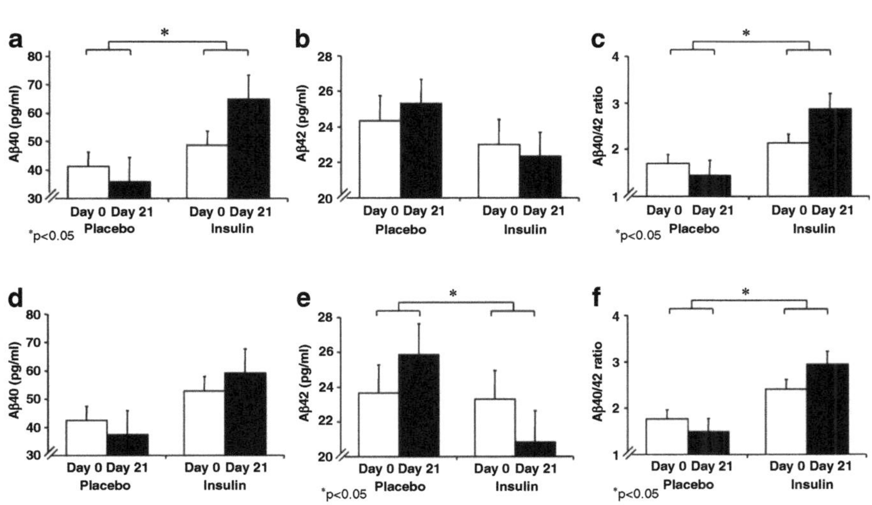


**Fig. 7** Fasting plasma (**a**) Aβ40, (**b**) Aβ42, and (**c**) Aβ40/42 ratio and post-prandial (**d**) Aβ40, (**e**) Aβ42, and (**f**) Aβ40/42 ratios at baseline and Day 21 for placebo-assigned and intranasal insulin-treated participants (from Reger et al. 2008)

adults' fasting Aβ40/42 ratios increased over the 21-day period compared to placebo-assigned participants (Fig. 7; p = 0.0207), reflecting increased Aβ40 levels relative to Aβ42. These results provide the first evidence of cognitive improvement following daily intranasal peptide administration for patients with early AD and support brain insulin signaling as a promising target in the search for new therapeutic avenues in AD. An important question is whether longer-term administration would provide additional benefit and a comparably favorable safety profile. We are currently investigating this question in an ongoing trial.

# 5 Conclusions

Insulin contributes to multiple functions in the central nervous system, and disruption of its effects by insulin resistance and hyperinsulinemia can increase the risk for aging-related neurodegenerative disorders such as AD through a number of mechanistic pathways. Human studies have shown that insulin modulates markers of AD pathology, such as CSF Aβ42 levels and inflammatory cytokines. These effects suggest that treatments aimed at improving insulin resistance will benefit patients with AD. Preliminary data from pilot studies of insulin sensitizers, exercise, and intranasal insulin support this possibility and provide a strong foundation for expanded studies in these areas.

**Aknowledgments** This work was supported by the Department of Veterans Affairs, the National Institute on Aging (RO1 AG10880), the Institute for the Study of Aging, and GlaxoSmithKline.

# References

Abbott MA, Wells DG, Fallon JR (1999) The insulin receptor tyrosine kinase substrate p58/53 and the insulin receptor are components of CNS synapses. J Neurosci 19:7300–7308

Anand Kumar TC, David GFX, Umberkoman B, Saini KD (1974) Uptake of radioactivity by body fluids and tissues in rhesus monkeys after intravenous injection or intranasal spray of tritium-labeled estradiol and progesterone. Curr Sci 43:435–439

Anand Kumar TC, David GFX, Kumar K, Umberkoman B, Krishnamoorthy MS (1976) A new approach to fertility regulation by interfering with neuroendocrine pathways. In: Anand Kumar TC (ed) Proceedings of International Symposium on Neuroendocrine Regulation of Fertility, Basel, Karger

Anand Kumar TC, David GF, Sankaranarayanan A, Puri V, Sundram KR (1982) Pharmacokinetics of progesterone after its administration to ovariectomized rhesus monkeys by injection, infusion, or nasal spraying. Proc Natl Acad Sci USA 79:4185–4189

Apelt J, Mehlhorn G, Schliebs R (1999) Insulin-sensitive GLUT4 glucose transporters are colocalized with GLUT3-expressing cells and demonstrate a chemically distinct neuron-specific localization in rat brain. J Neurosci Res 57:693–705

Authier F, Posner BI, Bergeron JJ (1996) Insulin-degrading enzyme. Clin Invest Med 19:149–160

Bagi Z, Koller A, Kaley G (2004) PPARgamma activation, by reducing oxidative stress, increases NO bioavailability in coronary arterioles of mice with Type 2 diabetes. Am J Physiol Heart Circ Physiol 286:H742–748

Banks WA, Jaspan JB, Kastin AJ (1997a) Selective, physiological transport of insulin across the blood-brain barrier: novel demonstration by species-specific radioimmunoassays. Peptides 18:1257–1262

Banks WA, Jaspan JB, Huang W, Kastin AJ (1997b) Transport of insulin across the blood-brain barrier: saturability at euglycemic doses of insulin. Peptides 18:1423–1429

Baskin DG, Figlewicz DP, Woods SC, Porte D Jr., Dorsa DM (1987) Insulin in the brain. Annu Rev Physiol 49:335–347

Baura GD, Foster DM, Porte D Jr., Kahn SE, Bergman RN, Cobelli C, Schwartz MW (1993) Saturable transport of insulin from plasma into the central nervous system of dogs in vivo. A mechanism for regulated insulin delivery to the brain. J Clin Invest 92:1824–1830

Beeri MS, Schmeidler J, Silverman JM, Gandy S, Wysocki M, Hannigan CM, Purohit DP, Lesser G, Grossman HT, Haroutunian V (2008) Insulin in combination with other diabetes medication is associated with less Alzheimer neuropathology. Neurology 71:750–757

Benedict C, Hallschmid M, Hatke A, Schultes B, Fehm HL, Born J, Kern W (2004) Intranasal insulin improves memory in humans. Psychoneuroendocrinology 29:1326–1334

Benedict C, Kern W, Schultes B, Born J, Hallschmid M (2008) Differential sensitivity of men and women to anorexigenic and memory-improving effects of intranasal insulin. J Clin Endocrinol Metab 93:1339–1344

Bingham EM, Hopkins D, Smith D, Pernet A, Hallett W, Reed L, Marsden PK, Amiel SA (2002) The role of insulin in human brain glucose metabolism: an 18fluoro-deoxyglucose positron emission tomography study. Diabetes 51:3384–3390

Bodian D, Howe HA (1941) Experimental studies on intraneuronal spread of poliomyelitis virus. Bull Johns Hopkins Hosp 68:248–267

Born J, Lange T, Kern W, McGregor GP, Bickel U, Fehm HL (2002) Sniffing neuropeptides: a transnasal approach to the human brain. Nature Neurosci 5:514–516

Cersosimo E, DeFronzo RA (2006) Insulin resistance and endothelial dysfunction: the road map to cardiovascular diseases. Diabetes Metab Res Rev 22:423–436

Chow HS, Chen Z, Matsuura GT (1999) Direct transport of cocaine from the nasal cavity to the brain following intranasal cocaine administration in rats. J Pharm Sci 88:754–758

Clark WE, Gros L (1929) Anatomical investigation into the routes by which infections may pass from the nasal cavities into the brain. Ministry of Health Report on Puiblic Health and Medical Subjects, vol 54. London: H.M. Stationary Office

Combs CK, Johnson DE, Karlo JC, Cannady SB, Landreth GE (2000) Inflammatory mechanisms in Alzheimer's disease: inhibition of β-amyloid-stimulated proinflammatory responses and neurotoxicity by PPARγ agonists. J Neurosci 20:558–567

Convit A, Wolf OT, Tarshish C, de Leon MJ (2003) Reduced glucose tolerance is associated with poor memory performance and hippocampal atrophy among normal elderly. Proc Natl Acad Sci USA 100:2019–2022

Cook DG, Leverenz JB, McMillan PJ, Kulstad JJ, Ericksen S, Roth RA, Schellenberg GD, Jin LW, Kovacina KS, Craft S (2003) Reduced hippocampal insulin-degrading enzyme in late-onset Alzheimer's disease is associated with the apolipoprotein E-epsilon4 allele. Am J Pathol 162:313–319

Corton JC, Anderson SP, Stauber A (2000) Central role of peroxisome proliferator-activated receptors in the actions of peroxisome proliferators. Annu Rev Pharmacol Toxicol 40:491–518

Craft S, Newcomer J, Kanne S, Dagogo-Jack S, Cryer P, Sheline Y, Luby J, Dagogo-Jack A, Alderson A (1996) Memory improvement following induced hyperinsulinemia in Alzheimer's disease. Neurobiol Aging 17:123–130

Craft S, Peskind E, Schwartz MW, Schellenberg GD, Raskind M, Porte D (1998) Cerebrospinal fluid and plasma insulin levels in Alzheimer's disease - Relationship to severity of dementia and apolipoprotein E genotype. Neurology 50:164–168

Craft S, Asthana S, Newcomer JW, Wilkinson CW, Matos IT, Baker LD, Cherrier M, Lofgreen C, Latendresse S, Petrova A, Plymate S, Raskind M, Grimwood K, Veith RC (1999) Enhancement of memory in Alzheimer disease with insulin and somatostatin, but not glucose. Arch Gen Psychiat 56:1135–1140

Craft S, Asthana S, Cook DG, Baker LD, Cherrier M, Purganan K, Wait C, Petrova A, Latendresse S, Watson GS, Newcomer JW, Schellenberg GD, Krohn AJ (2003) Insulin dose-response effects on memory and plasma amyloid precursor protein in Alzheimer's disease: interactions with apolipoprotein E genotype. Psychoneuroendocrinology 28:809–822

Cuzzocrea S, Pisano B, Dugo L, Ianaro A, Maffia P, Patel NS, Di Paola R, Ialenti A, Genovese T, Chatterjee PK, Di Rosa M, Caputi AP, Thiemermann C (2004) Rosiglitazone, a ligand of the peroxisome proliferator-activated receptor-gamma, reduces acute inflammation. Eur J Pharmacol 483:79–93

Czerniawska A (1970) Experimental investigations on the penetration of 198Au from nasal mucous membrane into cerebrospinal fluid. Acta Otolaryngol 70:58–61

De Felice FG, Vieira MN, Bomfim TR, Decker H, Velasco PT, Lambert MP, Viola KL, Zhao WQ, Ferreira ST, Klein WL (2009) Protection of synapses against Alzheimer's-linked toxins: insulin signaling prevents the pathogenic binding of Aβ oligomers. Proc Natl Acad Sci USA 106:1971–1976

Delerive P, Fruchart JC, Staels B (2001) Peroxisome proliferator-activated receptors in inflammation control. J Endocrinol 169:453–459

Djupesland PG, Skretting A, Winderen M, Holand T (2004) Bi-directional nasal delivery of aerosols can prevent lung deposition. J Aerosol Med 17:249–259

Escher P, Wahli W (2000) Peroxisome proliferator-activated receptors: insight into multiple cellular functions. Mutat Res 448:121–138

Faber HK (1938) The early lesions of poliomyelitis after intranasal inoculation. J Pediat 13:10–37

Faber WM (1937) The nasal mucosa and the subarachnoid space. Am J Anat 62:121–148

Fairbrother RW, Hurst EW (1930) The pathogenesis of, and propagation of the virus in experimental poliomyeltis. J Path Bact 33:17–45

Farris W, Mansourian S, Chang Y, Lindsley L, Eckman EA, Frosch MP, Eckman CB, Tanzi RE, Selkoe DJ, Guenette S (2003) Insulin-degrading enzyme regulates the levels of insulin, amyloid beta-protein, and the β-amyloid precursor protein intracellular domain in vivo. Proc Natl Acad Sci USA 100:4162–4167

Figlewicz DP, Szot P, Israel PA, Payne C, Dorsa DM (1993) Insulin reduces norepinephrine transporter mRNA in vivo in rat locus coeruleus. Brain Res 602:161–164

Fishel MA, Watson GS, Montine TJ, Wang Q, Green PS, Kulstad JJ, Cook DG, Peskind ER, Baker LD, Goldgaber D, Nie W, Asthana S, Plymate SR, Schwartz MW, Craft S (2005) Hyperinsulinemia provokes synchronous increases in central inflammation and beta-amyloid in normal adults. Arch Neurol 62:1539–1544

Francis GJ, Martinez JA, Liu WQ, Xu K, Ayer A, Fine J, Tuor UI, Glazner G, Hanson LR, Frey WH 2nd, Toth C (2008) Intranasal insulin prevents cognitive decline, cerebral atrophy and white matter changes in murine type I diabetic encephalopathy. Brain 131:3311–3334

Frey WH 2nd (2002) Intranasal delivery: bypassing the blood-brain barrier to deliver therapeutic agents to the brain and spinal cord. Drug Deliv Technol 2:46–49

Gasparini L, Gouras GK, Wang R, Gross RS, Beal MF, Greengard P, Xu H (2001) Stimulation of beta-amyloid precursor protein trafficking by insulin reduces intraneuronal beta-amyloid and requires mitogen-activated protein kinase signaling. J Neurosci 21:2561–2570

Gopinath PG, Gopinath G, Kumar TCA (1978) Target site of intranasally sprayed substances and their transport across the nasal mucosa: A new insight into the intranasal route of drug delivery. Curr Ther Res 23:596–607

Greenwood CE, Winocur G (2001) Glucose treatment reduces memory deficits in young adult rats fed high-fat diets. Neurobiol Learn Mem 75:179–189

Havrankova J, Roth J, Brownstein M (1978a) Insulin receptors are widely distributed in the central nervous system of the rat. Nature 272:827–829

Havrankova J, Schmechel D, Roth J, Brownstein M (1978b) Identification of insulin in rat brain. Proc Natl Acad Sci USA 75:5737–5741

Hirsch EC, Breidert T, Rousselet E, Hunot S, Hartmann A, Michel PP (2003) The role of glial reaction and inflammation in Parkinson's disease. Ann NY Acad Sci 991:214–228

Ho L, Qin W, Pompl PN, Xiang Z, Wang J, Zhao Z, Peng Y, Cambareri G, Rocher A, Mobbs CV, Hof PR, Pasinetti GM (2004) Diet-induced insulin resistance promotes amyloidosis in a transgenic mouse model of Alzheimer's disease. Faseb J 18:902–904

Hoyer S (2002) The aging brain. Changes in the neuronal insulin/insulin receptor signal transduction cascade trigger late-onset sporadic Alzheimer disease (SAD). A mini-review. J Neural Transm 109:991–1002

Illum L (2002) Nasal drug delivery: new developments and strategies. Drug Discov Today 7:1184–1189

Kern W, Born J, Schreiber H, Fehm HL (1999) Central nervous system effects of intranasally administered insulin during euglycemia in men. Diabetes 48:557–563

Kern W, Peters A, Fruehwald-Schultes B, Deininger E, Born J, Fehm HL (2001) Improving influence of insulin on cognitive functions in humans. Neuroendocrinology 74:270–280

Kopf SR, Baratti CM (1999) Effects of posttraining administration of insulin on retention of a habituation response in mice: participation of a central cholinergic mechanism. Neurobiol Learn Mem 71:50–61

Kristensson K, Olsson Y (1971) Uptake of exogenous proteins in mouse olfactory cells. Acta Neuropathol 19:145–154

Kurochkin IV, Goto S (1994) Alzheimer's beta-amyloid peptide specifically interacts with and is degraded by insulin degrading enzyme. FEBS Lett 345:33–37

Lee CC, Kuo YM, Huang CC, Hsu KS (2009) Insulin rescues amyloid beta-induced impairment of hippocampal long-term potentiation. Neurobiol Aging 30:377–387

Lehmann JM, Moore LB, Smith-Oliver TA, Wilkison WO, Willson TM, Kliewer SA (1995) An antidiabetic thiazolidinedione is a high affinity ligand for peroxisome proliferator-activated receptor gamma (PPAR $\gamma$). J Biol Chem 270:12953–12956

Leibson CL, Rocca WA, Hanson VA, Cha R, Kokmen E, O'Brien PC, Palumbo PJ (1997) The risk of dementia among persons with diabetes mellitus: a population-based cohort study. Ann NY Acad Sci 826:422–427

Luchsinger JA, Tang MX, Shea S, Mayeux R (2004) Hyperinsulinemia and risk of Alzheimer disease. Neurology 63:1187–1192

Mayeux R, Honig LS, Tang MX, Manly J, Stern Y, Schupf N, Mehta PD (2003) Plasma Aβ40 and Aβ42 and Alzheimer's disease: relation to age, mortality, and risk. Neurology 61:1185–1190

McDermott JR, Gibson AM (1997) Degradation of Alzheimer's beta-amyloid protein by human and rat brain peptidases: involvement of insulin-degrading enzyme. Neurochem Res 22:49–56

Oh YK, Kim JP, Hwang TS, Ko JJ, Kim JM, Yang JS, Kim CK (2001) Nasal absorption and biodistribution of plasmid DNA: an alternative route of DNA vaccine delivery. Vaccine 19:4519–4525

Olefsky JM (2000) Treatment of insulin resistance with peroxisome proliferator-activated receptor gamma agonists. J Clin Invest 106:467–472

Ott A, Stolk RP, van Harskamp F, Pols HA, Hofman A, Breteler MM (1999) Diabetes mellitus and the risk of dementia: The Rotterdam Study. Neurology 53:1937–1942

Paik JH, Ju JH, Lee JY, Boudreau MD, Hwang DH (2000) Two opposing effects of non-steroidal anti-inflammatory drugs on the expression of the inducible cyclooxygenase. Mediation through different signaling pathways. J Biol Chem 275:28173–28179

Park CR, Seeley RJ, Craft S, Woods SC (2000) Intracerebroventricular insulin enhances memory in a passive-avoidance task. Physiol Behav 68:509–514

Pedersen WA, Flynn ER (2004) Insulin resistance contributes to aberrant stress responses in the Tg2576 mouse model of Alzheimer's disease. Neurobiol Dis 17:500–506

Peila R, Rodriguez BL, Launer LJ (2002) Type 2 diabetes, APOE gene, and the risk for dementia and related pathologies: The Honolulu-Asia Aging Study. Diabetes 51:1256–1262

Perez A, Morelli L, Cresto JC, Castano EM (2000) Degradation of soluble amyloid β-peptides 1-40, 1-42, and the Dutch variant 1-40Q by insulin degrading enzyme from Alzheimer disease and control brains. Neurocheml Res 25:247–255

Petersen RC, Smith GE, Waring SC, Ivnik RJ, Tangalos EG, Kokmen E (1999) Mild cognitive impairment: clinical characterization and outcome. Arch Neurol 56:303–308

Pietrowsky R, Struben C, Molle M, Fehm HL, Born J (1996a) Brain potential changes after intranasal vs. intravenous administration of vasopressin: evidence for a direct nose-brain pathway for peptide effects in humans. Biol Psychiat 39:332–340

Pietrowsky R, Thiemann A, Kern W, Fehm HL, Born J (1996b) A nose-brain pathway for psychotropic peptides: evidence from a brain evoked potential study with cholecystokinin. Psychoneuroendocrinology 21:559–572

Qiu W, Walsh D, Ye Z, Vekrellis K, Zhang J, Podlisny M, Rosner M, Safavi A, Hersh L, Selkoe D (1998) Insulin-degrading enzyme regulates extracellular levels of amyloid beta-protein by degradation. J Biol Chem 273:32730–32738

Reagan LP, Gorovits N, Hoskin EK, Alves SE, Katz EB, Grillo CA, Piroli GG, McEwen BS, Charron MJ (2001) Localization and regulation of GLUTx1 glucose transporter in the hippocampus of streptozotocin diabetic rats. Proc Natl Acad Sci USA 98:2820–2825

Sakane T, Akizuki M, Yoshida M, Yamashita S, Nadai T, Hashida M, Sezaki H (1991) Transport of cephalexin to the cerebrospinal fluid directly from the nasal cavity. J Pharm Pharmacol 43:449–451

Schmidt S, Moric E, Schmidt M, Sastre M, Feinstein DL, Heneka MT (2004) Anti-inflammatory and antiproliferative actions of PPAR-gamma agonists on T lymphocytes derived from MS patients. J Leukoc Biol 75:478–485

Schulingkamp RJ, Pagano TC, Hung D, Raffa RB (2000) Insulin receptors and insulin action in the brain: review and clinical implications. Neurosci Biobehav Rev 24:855–872

Schwartz MW, Figlewicz DF, Kahn SE, Baskin DG, Greenwood MR, Porte D, Jr. (1990) Insulin binding to brain capillaries is reduced in genetically obese, hyperinsulinemic Zucker rats. Peptides 11:467–472

Seki T, Sato N, Hasegawa T, Kawaguchi T, Juni K (1994) Nasal absorption of zidovudine and its transport to cerebrospinal fluid in rats. Biol Pharm Bull 17:1135–1137

Shipley MT (1985) Transport of molecules from nose to brain: transneuronal anterograde and retrograde labeling in the rat olfactory system by wheat germ agglutinin-horseradish peroxidase applied to the nasal epithelium. Brain Res Bull 15:129–142

Skeberdis VA, Lan J, Zheng X, Zukin RS, Bennett MV (2001) Insulin promotes rapid delivery of N-methyl-D- aspartate receptors to the cell surface by exocytosis. Proc Natl Acad Sci USA 98:3561–3566

Sonnen JA, Larson EB, Brickell K, Crane PK, Woltjer R, Montine TJ, Craft S (2009) Different patterns of cerebral injury in dementia with or without diabetes. Arch Neurol 66:315–322

Sudoh S, Frosch MP, Wolf BA (2002) Differential effects of proteases involved in intracellular degradation of amyloid β-protein between detergent-soluble and -insoluble pools in CHO-695 cells. Biochemistry 41:1091–1099

Tao L, Liu HR, Gao E, Teng ZP, Lopez BL, Christopher TA, Ma XL, Batinic-Haberle I, Willette RN, Ohlstein EH, Yue TL (2003) Antioxidative, antinitrative, and vasculoprotective effects of a peroxisome proliferator-activated receptor-gamma agonist in hypercholesterolemia. Circulation 108:2805–2811

Thorne RG, Frey WH 2nd (2001) Delivery of neurotrophic factors to the central nervous system: pharmacokinetic considerations. Clin Pharmacokinet 40:907–946

Thorne RG, Emory CR, Ala TA, Frey WH 2nd (1995) Quantitative analysis of the olfactory pathway for drug delivery to the brain. Brain Res 692:278–282

Thorne RG, Pronk GJ, Padmanabhan V, Frey WH 2nd (2004) Delivery of insulin-like growth factor-I to the rat brain and spinal cord along olfactory and trigeminal pathways following intranasal administration. Neuroscience 127:481–496

Townsend M, Mehta T, Selkoe DJ (2007) Soluble Abeta inhibits specific signal transduction cascades common to the insulin receptor pathway. J Biol Chem 282:33305–33312

Unger JW, Livingston JN, Moss AM (1991) Insulin receptors in the central nervous system: localization, signalling mechanisms and functional aspects. Prog Neurobiol 36:343–362

Vanhanen M, Koivisto K, Kuusisto J, Mykkanen L, Helkala EL, Hanninen T, Riekkinen P Sr., Soininen H, Laakso M (1998) Cognitive function in an elderly population with persistent impaired glucose tolerance. Diabetes Care 21:398–402

Wallum BJ, Taborsky GJ Jr,, Porte D Jr, Figlewicz DP, Jacobson L, Beard JC, Ward WK, Dorsa D (1987) Cerebrospinal fluid insulin levels increase during intravenous insulin infusions in man. J Clin Endocrinol Metab 64:190–194

Wang F, Jiang X, Lu W (2003) Profiles of methotrexate in blood and CSF following intranasal and intravenous administration to rats. Int J Pharm 263:1–7

Wang F, Jiang X, Lu W (2004) Intranasal delivery of methotrexate to the brain in rats bypassing the blood-brain barrier. Drug Deliv Technol 4:48–55

Weiss P, Holland Y (1967) Neuronal dynamics and axonal flow, II. The olfactory nerve as a model test object. Proc Natl Acad Sci USA 57:258–264

Zhao L, Teter B, Morihara T, Lim GP, Ambegaokar SS, Ubeda OJ, Frautschy SA, Cole GM (2004) Insulin-degrading enzyme as a downstream target of insulin receptor signaling cascade: implications for Alzheimer's disease intervention. J Neurosci 24:11120–11126

Zhao W, Chen H, Xu H, Moore E, Meiri N, Quon MJ, Alkon DL (1999) Brain insulin receptors and spatial memory. Correlated changes in gene expression, tyrosine phosphorylation, and signaling molecules in the hippocampus of water maze trained rats. J Biol Chem 274:34893–34902

Zhao WQ, Alkon DL (2001) Role of insulin and insulin receptor in learning and memory. Mol Cell Endocrinol 177:125–134

# Is Alzheimer's a Disorder of Ageing and Why Don't Mice get it? The Centrality of Insulin Signalling to Alzheimer's Disease Pathology

**Simon Lovestone and Richard Killick**

**Abstract** The amyloid cascade hypothesis has had considerable importance in driving forward the molecular understanding of Alzheimer's disease (AD) pathology. One component of that cascade might be glycogen synthase kinase-3 (GSK3), a kinase that appears to be activated by Aβ and in turn phosphorylates tau. GSK3 is also inhibited by insulin signalling and insulin resistance, and diabetes is a major risk factor for AD. We hypothesise, as others have done, that insulin signalling is central to the pathological process, with evidence that both genetic and environmental risk factors for AD involve the insulin pathway. We also postulate that transgenic mice provide only a partial model for AD, as insulin signalling acts as a protective factor against Aβ toxicity; also, the well-established relationship between insulin signalling and longevity might explain why the single most important risk factor for AD is age.

## 1 Introduction

Alzheimer's disease (AD) is a common and costly disorder. With an estimated 700,000 people with dementia in the UK costing over £15 billion per annum, and with numbers set to rise inexorably as the population ages, AD can truly be said to be an unsustainable problem in urgent need of therapeutics (Albanese et al. 2007; Lowin et al. 2001). Few other disorders are so common or costly – the costs of AD are greater than all cancers and heart disease combined (Lowin et al. 2001) – and it is almost surely only because these costs are spread over health and social services, and even more importantly shared by families and individuals, that there is not an even greater sense of urgency to find a disease-modifying therapy than there is.

S. Lovestone (✉)
Institute of Psychiatry, King's College London, De Crespigny Park, London, SE5 8AF, UK

S. Craft and Y. Christen (eds.), *Diabetes, Insulin and Alzheimer's Disease*, Research and Perspectives in Alzheimer's Disease,
DOI 10.1007/978-3-642-04300-0_9, 

However, AD is not the only common and costly condition; diabetes is as common and set to become something of an epidemic, as not only the age but also the weight and waist measurement of the population rise. In contrast to AD, however, diabetes has a relatively sophisticated and growing set of diagnostics and therapeutics.

In this review we discuss the overlapping risk factors and pathologies of diabetes, or more accurately insulin and insulin-like growth factor (IGF) signalling (IIS) and AD, and the prospects therefore of shared therapeutics. Our understanding of AD pathogenesis is based almost entirely on extrapolation from autosomal dominant, early onset forms of the condition (familial AD; FAD). The discovery of mutations in APP (β-amyloid precursor protein) and PS1, and the overwhelming evidence that these mutations when modelled in cells and in animals increase either the total amount of β-amyloid (Aβ) or the relatively pathogenic form of Aβ42, gave rise to the amyloid cascade hypothesis (Hardy 2006; Hardy and Higgins 1992). This almost universally accepted outline of pathogenic processes in AD was reinforced by the discovery of mutations in MAPT, the gene encoding tau, causing frontotemporal dementia (FTD; Heutink 2000). As FTD is a disorder of tau-pathology and not amyloid pathology, and FAD is a disorder of both, this discovery seemed to be the final proof needed that the order of pathological processes postulated by the amyloid cascade hypothesis – amyloid before tau – is correct. However, although this hypothesis has been immensely powerful and has driven the research agenda for the past two decades, and correspondingly predicates almost all therapeutics programmes, caution is warranted. Is late onset AD the same as early onset FAD? Observations in both rodent models and in man suggest that whilst there is much to learn from rare forms of genetically determined AD, there are some differences. In rodents, one of the most striking observations, to which we return below, is that mice are astonishingly resistant to the full AD pathology; there are few examples of an intact amyloid cascade in mice and an overwhelming body of evidence that such a cascade is not induced as it appears to be in man. In man, although the core pathological lesions of AD, plaques and tangles, are present and apparently identical in both FAD and late onset AD, it is an awkward fact that these pathological lesions are also present in people without the symptoms of AD. Furthermore, in people with clinically apparent dementia, the range of other pathological processes in the brain at post mortem, including most obviously vascular damage and inflammation, is very extensive indeed (MRC CFAS 2001).

These observations are not in themselves enough to seriously challenge the amyloid cascade hypothesis – although they do add to a sense of nervousness as amyloid-based therapeutics appear to fail in clinical trials – but they do suggest that alternative processes either contribute to, or are even causative in, the vast majority of cases of AD. Increasingly, the amyloid cascade is being nuanced with suggestions that primary aetiopathological events might act at multiple points on the pathological cascade and might affect, independently and simultaneously, both plaque- and tangle-related pathologies (Mudher and Lovestone 2002; Small and Duff 2008). In this context, the term "plaque and tangle pathologies" is shorthand for the molecular pathology associated with these macromolecular lesions. For both lesions, this pathology seems likely to be oligomeric forms of the protein that aggregate to form the lesion, but although the evidence for this hypothesis is growing, it remains to be unambiguously proven (Brunden et al. 2008; Haass and Selkoe 2007).

In this chapter we suggest that one primary aetiopathological event that might increase both plaque- and tangle-related pathologies – the dual hypothesis – is altered IIS. Specifically, we suggest that defective IIS, manifested as insulin resistance, will result in altered regulation of a kinase, glycogen synthase kinase-3 (GSK3), that has effects not only on plaque- and tangle-related pathologies but also on the primary and early clinical symptom of long-term memory formation and plasticity. We review the evidence for a dysregualtion of GSK3 as a risk factor for AD and also its role in longevity, suggesting that one reason why rodent models of AD are so limited is the differences between mice and man in IIS and considering the effects of altered IIS in cellular and animal models. Our own findings suggest a complicated effect of altered IIS in mice, with paradoxical effects on the two principal pathological processes in AD. Together these findings suggest that therapeutic approaches seeking to modulate IIS might be a rich source of possible disease-modifying therapies in AD but that such approaches might have complex and sometimes unexpected effects.

## 2 The Role of GSK-3 in AD Pathology

The neurofibrillary tangle observed by Alois Alzheimer is one stage of a pathological process that appears to begin with the accumulation of highly phosphorylated tau in neuronal cell bodies and end with the extracellular, insoluble ghost or tombstone tangle. Careful studies attempting to correlate pathological lesions with symptoms have tended to suggest that the tangle is more closely related to severity of cognitive impairment than plaques (Nagy et al. 1995). Such studies are, however, almost certainly fundamentally flawed, both because of the inevitable delay between clinical assessment and pathological examination and also, perhaps more importantly, because tangle pathologies are relatively insoluble and hence more persistent whereas plaque pathologies have been shown in animal models to have a surprisingly rapid turnover. As a consequence, tangle pathologies will have a more linear accumulation in relation to length of disease and hence severity by virtue of biophysical characteristics alone. In fact, the closest association with cognitive impairment is almost certainly neither the plaque nor the tangle but the loss of synapses (Terry et al. 1991). Despite this caveat, it makes sense that there is a relationship between tangle-related pathologies and severity. Tau is a predominantly axonal protein that plays a key role in neuronal function (Avila et al. 2004). One of a number of microtubule-associated proteins, tau binds to microtubules and facilitates axonal transport, a process that is essential to neuronal stability and function as the synapse possesses relatively little protein translational apparatus. Tau binding to microtubules is regulated by phosphorylation and axonal transport is disrupted by both alteration in the expression of tau and by tau phosphorylation.

The mutations in the gene encoding tau also alter the normal properties of tau, with some directly affecting its ability to bind microtubules and some affecting tau phosphorylation (Dayanandan et al. 1999). In the tauopathies (AD, FTD and other disorders with tau pathology), there is evidence for all three abnormalities in tau in at least some cases – altered overall expression or expression of some isoforms of tau (Ingelsson et al. 2007; Myers et al. 2007), altered intracellular expression with an accumulation in the cell body together with aggregation and increased phosphorylation (Avila 2006).

Despite this consensus about the nature of the pathological events and the generalities of the amyloid cascade, the precise sequence of pathological events, and defining which are critical in causing symptoms, is disputed. The accumulation of tau in the cell body and the formation of intracellular tangles are both likely to result in cell toxicity, and it is obvious that neuronal loss will result in the symptoms of neurodegeneration. Neuropathological studies have shown that the earliest sign of tau-related pathology is the accumulation of highly phosphorylated tau in neuronal cell bodies and elsewhere, and these early pathological observations precede clinical symptoms (Braak and Braak 1991; Braak et al. 1994). However, whether the phosphorylation of tau and its altered intracellular distribution result in neuronal dysfunction in man in vivo is unclear, although animal models suggest that this explanation is plausible (Mudher et al. 2004; Hernandez et al. 2002). Most evidence does point to phosphorylation being an early event, and a likely sequence of events is that an increase in phosphorylation results in loss of normal microtubule binding function of tau, leading to a redistribution of tau and abnormal intracellular accumulation in the cell body and in the unbound fraction in axons (Avila 2006; Ferrer et al. 2005). In the intracellular environment, this is likely to result in aggregation of tau and the formation of oligomeric species and then fibrils in the form of paired helical filaments that form the core of the tangle. Either oligomeric or fibrillised tau may have a toxic gain of function that would complement the loss of normal function of highly phosphorylated tau, and together they may result in neuronal dysfunction and toxicity leading to neuronal loss.

This sequence of events places phosphorylation as an early and causative, but not necessarily essential, step in the process leading to tau-related pathology. It is possible that some aspects of this process are failed attempts to restore function; an increase in phosphorylation would compensate for an increase in expression, for example, by reducing the amount of tau binding to microtubules. Equally, in the case of autosomal dominant FTD, the primary event might be an increase in tau aggregation caused by a relative failure to bind microtubules or by a mutation-induced propensity to self-aggregation. Nonetheless, even in these instances, phosphorylation of tau is an invariable aspect of pathology and a likely contributor to the pathological process.

Given this primacy of tau phosphorylation in the pathological cascade, it becomes important to understand its regulation (Lovestone and Reynolds 1997). Tau is phosphorylated at more than 30 epitopes, most serines preceding a proline but also at tyrosines and threonines (Hanger et al. 2007). Both kinases and phosphatases have been extensively investigated, with an emerging consensus that no single enzyme is responsible for either normal or aberrant regulation but that the two key participants, although it must be stressed not the only ones, are GSK3 and protein phosphatase 2a (PP2a). The latter has been consistently shown to be the predominant phosphatase in cellular and animal models (Trojanowski and Lee 1995; Martin et al. 2008; Wang et al. 2007), and indeed PP2a is inhibited in response to increases in GSK3 activity, suggesting a positive feedback loop in relation to tau phosphorylation (Liu et al. 2008). GSK3 is in fact one of many tau kinases, any of which might play a role in disease pathogenesis in vivo. However, the weight of evidence for a primary role for GSK3 is compelling. GSK3 was first identified as a tau kinase in vitro, at the time one of many such candidates (Hanger et al. 1992; Mandelkow et al. 1992). Subsequently, we showed that GSK3, but not other proline-directed protein kinases, was a tau-kinase in cellular models (Lovestone et al. 1994). At this time, and independently, a tau kinase activity in bovine brain known as tau protein kinase 1 was identified as GSK3 (Ishiguro et al. 1993). Subsequently, we showed that in cellular models the phosphorylation of tau by GSK3 altered the normal microtubule-binding properties of tau (Lovestone et al. 1996), and these three pieces of evidence – that GSK3 was a tau kinase in cellular systems, that phosphorylation of tau by GSK3 altered its properties and that the tau protein kinase activity from mammalian brain was GSK3 – together placed GSK3 centre stage as a predominant tau kinase. Other subsequently identified tau kinases, such as CDK5, may play a role in phosphorylation of tau (Mazanetz and Fischer 2007) and in fact may act co-operatively with GSK3 as priming kinases (Sengupta et al. 1997), although other evidence suggests an antagonist effect, as CDK5 inhibits GSK3 (Plattner et al. 2006).

In line with the idea that there might be dual, or more, routes to disease, GSK3 is also implicated in APP metabolism and Aβ formation. GSK3 phosphorylates APP and thereby alters its maturation (Aplin et al. 1996, 1997). In cellular models and animal models, GSK3 inhibitors appear to reduce amyloid generation or pathology (Phiel et al. 2003; Ryder et al. 2003; Su et al. 2004). Most tellingly, in a screen of kinases that regulate membrane trafficking, using inhibitors and knock-down, GSK3 was identified as a key regulator both of trafficking processes in general and, more specifically, of trafficking resulting in altered secretion of Aβ (Adachi et al. 2009). In animal models, the mood-stabilising drugs valproate and lithium have both been reported to have effects on amyloid pathology. Both may be inhibitors of GSK3. Lithium is unambiguously a GSK3 inhibitor, although it has other targets as well; valproic acid is less obviously so. The outcomes on amyloid pathology are mixed, however; valproate reportedly reduced plaque pathology (Qing et al. 2008) but lithium reduced APP phosphorylation and decreased Aβ production in one model (Rockenstein et al. 2007) but had no effect in another (Caccamo et al. 2007). These mouse models are different; all are driven by some combination of overexpression of human genes possessing autosomal dominant mutations, but clearly the degree and location of these drivers differ, which may account for the variability in findings. The model above, reporting no effects of a GSK3 inhibitor, is driven by three different genes, all with disease-causing mutations, and overcoming such a comprehensive pathogenic load may be difficult.

As well as being the predominant tau kinase and having a possible role in the generation of Aβ, GSK3 may be the mediator of Aβ toxicity. Reproducing Aβ toxicity in vitro has not been straightforward as it is highly dependent on experimental design and, in particular, on the state of oligomerisation of Aβ. Nonetheless, in neuronal cultures, Aβ induces toxicity and tau phosphorylation, which is accompanied by an increase in GSK3 activity (Takashima et al. 1998). Further evidence for a role of GSK3 inhibition came from the findings that direct or indirect inhibition of GSK3 protects neurons from the adverse effects of Aβ (Alvarez et al. 1999; Inestrosa et al. 2002). In cell culture, muscarinic agonists also prevent Aβ-induced toxicity (Farias et al. 2004), and we demonstrated that they also reduced tau phosphorylation via inhibition of GSK3 (Forlenza et al. 2000), suggesting that this too is further evidence that GSK3 is a mediator of Aβ neurotoxicity. Much the same has been found inanimal models; for example, GSK3 inhibition prevents the Aβ induction of caspase activity in rabbits (Ghribi et al. 2003) and rescues Aβ-induced behavioural abnormalities in rats (De Ferrari et al. 2003).

If GSK3 is an APP and a tau kinase and transmits the Aβ toxicity, then increasing GSK3 activity in animal models should recapitulate some of the pathological processes of AD. The first animal model with overexpression of human GSK3 in the brain expressed very small amounts of protein, but nonetheless tau phosphorylation was increased (Brownlees et al. 1997). Other groups used conditional overexpression to increase transgene expression and found correspondingly increased pathological effects, including considerable amounts of tau phosphorylation, intraneuronal redistribution of tau to the somatodentridic compartment as in human AD and neuronal toxicity with an accompanying gliosis (Lucas et al. 2001). These mice have learning and memory deficits but, interestingly, no tangles were observed (Hernandez et al. 2002), suggesting that the neurotoxicity and behavioural effects are pre-tangle in nature. Crossing mice overexpressing GSK3 with mice possessing human, mutated tau results in much-enhanced pathology, including tangles, where either founder mouse did not have any, or increased severity or earlier onset of pathology (Engel et al. 2005; Muyllaert et al. 2006; Terwel et al. 2008).

These findings are matched in Drosophila, where overexpression of tau results in a phenotype (the nature of which is driver-, and hence expression-pattern, dependent; Ferber 2001; Williams et al. 2000) that is then exacerbated by dysregulation or overexpression of GSK3 (Jackson et al. 2002; Mudher et al. 2004). Indeed, we showed that the expression of tau in the absence of alteration of GSK3 results in a phenotype that is itself GSK3-dependent. Thus overexpression of tau in motor neurons results in both larval and adult motor phenotypes and a disruption in axonal transport that then disrupts the neurophysiology of the motor end plate (Chee et al. 2005; Mudher et al. 2004). Whilst this phenotype is exacerbated by GSK3, treatment of larvae overexpressing tau in the absence of overexpression of GSK3 with either lithium or specific GSK3 inhibitors abolished the phenotype, demonstrating that the effects of tau on axonal transport are at least GSK3-dependent (Mudher et al. 2004). In mice, also, some of the tau-induced phenotypes, including tau phosphorylation and aggregation as well as in some models the full neurodegenerative phenotype, are attenuated by GSK3 inhibition (Nakashima et al. 2005; Noble et al. 2005; Engel et al. 2006a, b). Even non-tau-dependent phenotypes of ageing, for example, apoptosis in a murine model of senescence, are also reduced by GSK3 inhibitors, notably lithium (Tajes et al. 2008).

Finally, we explored the normal role of GSK3 in neuronal plasticity by following the activity of GSK3 using its phosphorylation at Ser9 as a measure of relative inhibition (Hooper et al. 2007a). We found that induction of long-term potentiation (LTP) in the dentate was followed by an increase in Ser9 in the ipsilateral but not the control, contralateral, hippocampus, suggesting a relative inhibition of GSK3 following LTP. We speculated that GSK3 inhibition might be a necessary process to propagate LTP, and we tested this hypothesis in mice conditionally overexpressing GSK3. These mice, as discussed above, have no neurofibrillary tangles and, in young mice, even before any neurodegeneration, we observed an almost complete absence of LTP. When the mice were treated with lithium, LTP was restored completely (Hooper et al. 2007a). In complementary studies, Peineau et al. (2007) showed that GSK3 activity was increased following the induction of long term depression (LTD) and, conversely, that GSK3 inhibition blocked LTD induction. These and similar studies (Zhu et al. 2007) suggest that the inhibition of GSK3 is necessary for the plasticity events that most likely underlie the consolidation of memory and hence the core symptoms of AD.

These data provide powerful but still circumstantial evidence for a role for GSK3 in AD. The demonstration that GSK3 increases tau phosphorylation, results in loss of normal tau function, probably increases Aβ generation and certainly transmits the Aβ toxicity in cellular models first suggested that GSK3 is critical to AD pathogenesis. The findings from mouse and fly models that GSK3 induces neurodegeneration, enhances tau pathology and is critical for the central phenotype of the human disease all complement these findings. But in order to truly demonstrate a role for GSK3 in AD, it would be necessary first to show an alteration of GSK3 activity in the disease state and then, in the ultimate proof, demonstrate a reversal with GSK3 inhibitors. In fact, there is a relative paucity of data examining the activity of GSK3 in man in AD. Pei et al. (1997, 1999) have shown that GSK3 protein is increased, and active but not inactive forms of GSK3 colocalise with pretangle pathology in AD, and we showed that in white cells GSK3 protein and activity are increased in AD, even in the early stages, suggesting a systemic effect (Hye et al. 2005). However, neither set of data is sufficient to prove that GSK3 is actually altered in disease.

As lithium is an inhibitor of GSK3, then people treated with lithium would be predicted to be at less risk of AD if the "GSK3 hypothesis of AD" is correct (Hooper et al. 2008). However, such studies are fraught with confounds that make interpretation difficult. For one, depression, a common therapeutic indication for lithium is itself a risk factor or even prodrome for AD (Schweitzer et al. 2002; Jorm 2000); for another, older people with long-term lithium prescriptions are at increased risk of adverse effects and so clinicians tend to remove the lithium at precisely the time it might start to have beneficial effects in relation to AD. Nonetheless, in both large population-based studies (Kessing et al. 2008) and in small case-control studies (Nunes et al. 2007; Terao et al. 2006), lithium appears to reduce the risk of dementia in most but not all studies (Dunn et al. 2005). These findings support the possibility of GSK3 as a target for therapy and possibly even for the use of lithium in AD (Macdonald et al. 2008). However, GSK3, as a kinase at the centre of a number of critical signalling pathways, is a difficult target and, until more direct evidence for alterations in GSK3 signalling in AD are found, doubts will remain regarding its role in the disease.

## 3 Genetic and Environmental Risk Factors Converge on Insulin Signalling

Although it is widely assumed that there are both genetic and environmental influences on AD, surprisingly few of either are unequivocally known. Late-onset AD has a high inheritable component (Liddell et al. 2001) and, although one of the most highly reproduced and robust findings in the genetics of complex disease, the effect of APOE, the one unequivocal risk factor, in influencing disease is relatively modest. Variation in the GSK3 gene itself may influence risk of disease (Kwok et al. 2008), but only in an epistatic model with variation in the MAPT gene encoding the protein tau. However, there is evidence that genes on pathways that regulate GSK3 activity may have effects on disease risk. Case control and family-based studies converge on variation in LRP6 as a risk factor for AD (De Ferrari et al. 2007). LRP6, arrow in *Drosophila*, is part of the wnt receptor, one of the main negative GSK3 regulatory pathways. The other main regulatory pathway resulting in an inhibition of GSK3 is insulin signalling. We first looked for variation in two key genes on the pathway – the p85$\alpha$ subunit of PI3kinase and regulatory subunit 3 of protein phosphatase 1 (PPP1R3) – that had previously been implicated not only in insulin signalling but in risk of diabetes and insulin resistance (Liolitsa et al. 2002). We found modest association with disease and then went on to examine over 150 single nucleotide polymorphisms that had previously been examined as loci for quantitative traits related to insulin signalling (Barroso et al. 2003). In a large, two-stage study, we found evidence for association with SOS2, PCK1 and PPAR$\gamma$ and AD, which was an interesting finding not least because PCK1 was also identified as a risk gene in a data-driven, genome wide association study (Grupe et al. 2007) and was subsequently replicated (Feulner et al. 2009) in a study of the top hits on the AlzGene meta-analysis (Bertram et al. 2005). Other genes that are currently in the top ten AlzGene hits include GAB2, or GRB2-associated binding protein 2, part of the MAPK and EGF signaling pathways that interacts with insulin receptor substrates (KEGG ko04910; insulin pathway), and angiotensin-converting enzyme, long implicated in diabetes.

These putative susceptibility genes have not been fully replicated and it is likely that the ongoing genome-wide association studies will find other genes. Nonetheless, genetic variants on the IIS pathways do show some promise as susceptibility factors for AD. The epidemiological evidence for IIS as a risk factor for AD is, however, overwhelming, and systematic reviews of longitudinal, prospective studies show unequivocal evidence for association (Biessels et al. 2006; Stewart and Liolitsa 1999). Diabetes increases risk of vascular dementia, not surprisingly, because diabetes is one of the elements of the clinical history that prompts a diagnosis of vascular dementia. This finding is likely to lead to an overestimation of the association. However, an association remains between diabetes, and even pre-diabetes, and AD, even when vascular dementia is excluded as far as possible. In our own studies (Velayudhan et al; B J Psych In press), diabetes is the single most powerful influence on progression from MCI to AD, adding weight to this, one of very few, undisputed risk factors for disease.

## 4 Is AD a Disorder of Ageing?

Both genetic and environmental data converge, therefore, to suggest that a relative failure of insulin regulation of GSK3 leads to an increase risk of AD. This convergence prompts the question as to whether AD is a disorder of ageing or, more accurately, longevity. AD affects predominantly older people but this might be due to a number of possible factors. The question most commonly posed is: "Is AD a disorder of ageing or an age-related disorder?" That is, is AD simply a disorder of the ageing body, i.e., a disorder of failing systems brought on by the accumulation of a lifetime of environmental damage? Or is AD a disorder that occurs most commonly in people of a given age, and like most disorders, it just so happens that the age range of risk is at the end of life? If AD is a disorder that arises as a consequence of a long life and without any further aetiopathological consequence, then it would be expected that the incidence rates of AD would continue to rise though very late life. If, on the other hand, AD is an age-related disorder, then one might expect the supremely old to show declining incidence as they live beyond the age of risk. Epidemiological studies have not, to date, completely resolved this issue, as the numbers reaching very late life remain too small. A meta analysis of such studies found that, although the rate of increase in incidence in the very elderly slows, the actual incidence itself does not fall (Gao et al. 1998). On balance then, the weight of evidence tends to suggest that AD is a disorder of ageing rather than an age-related disorder; however, there is an alternative possibility – that AD is in fact not a disorder of ageing but of longevity itself.

Although intriguing, these two commonly posed questions do not take into account one of the most striking lines of enquiry in recent years in the field of ageing and longevity. As discovered first in C. elegans and replicated in Drosophila and then in mammals, the evidence is now overwhelming that IIS regulates longevity (Giannakou and Partridge 2007; McElwee et al. 2007). In flies, the *age-1* long-lived mutant carries a mutation in the PI3Kinase gene, the same gene we find to be a risk factor for AD (Liolitsa et al. 2002; Friedman and Johnson 1988). In C. elegans, the long-lived DAF-2 mutant exerts its effects via the forkhead transcription factor, DAF-16, the mammalian homologue of which is phosphorylated as a consequence of Aβ-induced toxicity and is altered in conjunction with changes in pathology in calorie-restricted mice with Aβ (Qin et al. 2008; Smith et al. 2005; Ogg et al. 1997). Mice too can have their life spans extended very considerably by disruption in IIS (Vijg and Campisi 2008; Bartke 2008). Timing and location matter, and studies in fly and worm have shown that the longevity-inducing effects of IIS disruption can be effected in adult stages and that these effects are not simply due to extension of the developmental period. Furthermore, the tissues in which the effect is most important are brain and fat, although for the brain the direction of effect is not entirely clear, with some studies in line with the idea that decreased IIS in neurons is sufficient for longevity induction, other studies suggesting it isn't even a necessary organ and yet others, for example in mice lacking IRS2 in brain, suggesting that loss of IIS may protect the brain from otherwise age-related hyperinsulinaemia (reviewed in Piper et al. 2008). Piper et al. (2008) conclude that appropriate levels of insulin signalling in the brain can be both protective for neurons and life span-increasing and, furthermore, that a reduction in systemic IIS, which is controlled by the inverterbrate brain, can also be life span-enhancing.

Other mechanisms, perhaps parallel or in some cases downstream of IIS, are increasingly being found that alter ageing and life span. One such is the p53 tumour suppressor, which not only protects from cancer but has an apparently independent effect on aging (Feng et al. 2008; Matheu et al. 2008; Rodier et al. 2007). P53 is not only implicated in longevity but is also altered in AD. Thus, P53 is increased in both brain and peripheral tissues in AD and, in cell models, induces tau phosphorylation (Hooper et al. 2007b; Kitamura et al. 1997; Lanni et al. 2007, 2008; Ves da et al. 2002). As another example, there is some evidence that the IIS effects on longevity may be mediated by autophagy (McPhee and Baehrecke 2009; Cuervo 2008), and this system is also altered in AD, as autophagy is probably necessary for the clearance of the protein aggregates that are characteristic of the condition (Sarkar et al. 2009). Finally, the sirtuin family of histone deacetylases alter longevity (Kwon and Ott 2008; li-Youcef et al. 2007), protect against AD processes in model systems (Anekonda and Reddy 2006) and may be genetic susceptibility factors for AD (Helisalmi et al. 2008).

These findings are remarkably consistent; a set of pathways, principally IIS but including also p53 signalling, autophagy and the sirtuins, all alter longevity and all also have evidence of either a genetic association with AD, are altered in AD, or have clear effects on AD pathology in model systems. In some cases, in particular IIS, all three interactions between longevity and AD have been found. What then does this mean? The most parsimonious explanation would be that AD is a disorder of ageing and, therefore, it is to be expected that there will be similarities between ageing and AD. However, the direction of effect does not allow such a straightforward explanation of the data without further consideration. Taking IIS alone: a defective IIS in worms, mice and mammals enhances life span but, as we review the epidemiological studies above, in man a defective IIS in insulin resistance and in diabetes *increases* the risk of AD. From a pathway perspective, defective IIS would be predicted to, and indeed in cellular models does, result in an increase in GSK3 activity, which would be expected to increase AD pathological events, in line with the epidemiology but opposite to what is predicted by the simple formula that "increased life span = less pathological processes."

We propose an alternative hypothesis that explains all the available data reviewed above. We suggest that longevity itself is in a pact with the devil of disease; that is the price paid for longevity is, in this case, AD, not through the uninteresting and general mechanisms of just living longer but through the very specific molecular pathway of longevity. In other words, the molecular processes that have evolved in some organisms to permit longevity specifically induce AD processes as well. The two are inextricably, for now, linked. We suggest that man is already a long-lived organism whose life span is considerably greater than his, or more accurately her, reproductive life span, in marked contrast to many if not most other organisms. At some point in our evolution we acquired a relatively defective IIS, a process that allows us to live longer but brings in its wake a relative loss of inhibitory regulation of GSK3 and a consequent increase in pathogenic processes leading to AD. This explains, we suggest, why mice do not get AD, despite the best efforts of multiple laboratories modeling AD processes in transgenic animals.

## 5 Why Don't Mice get AD?

A large number of animal models, especially transgenic mice, have been generated that recapitulate some of the pathogenic lesions of AD. These models have been extensively reviewed (see, for example, Woodruff-Pak 2008; Howlett and Richardson 2009; Duff 1998; Bornemann and Staufenbiel 2000). The first, and still the most extensively investigated, mice are those carrying human APP genes with mutations known to cause early-onset familial forms of the condition. These mice have amyloid deposits in the form of plaques and other lesions and in many cases these lesions are increased in number and occur with an earlier onset if the mouse is also carrying a PS1 gene, also with the mutations causing autosomal dominant forms of the condition. These models, extensively reviewed elsewhere, have been immensely informative and useful in AD research. Through these models we now understand much about the processing of Aβ and the relative importance of oligomeric versus fibrillised Aβ, of intracellular versus extracellular Aβ and even of the specific occurrence of certain species of Aβ, which may be the most important correlate of cognitive impairment and other behavioural deficits in these various models. It is these models that have permitted the development of a wide range of amyloid-based therapeutics, including the active and passive immunisation that is still probably the most widely anticipated in AD disease-modifying trials, the development of the inhibitors of BACE1 and gamma-secretase that are in clinical trials, and the various approaches to prevent Aβ aggregation, at least one of which went as far as multiple phase III trials.

These models have been of undoubted importance and are widely, almost universally, used. It would be difficult to exaggerate their role in the developing story of understanding the molecular pathogenesis of AD. However, there is one outstandingly obvious and dismayingly important fact about these animals: they do not have AD. Specifically, they have no neurofibrillary tangles, no substantive tau phosphorylation – just minor amounts in close proximity to the plaque (Howlett and Richardson 2009), and little detectable neuronal loss or toxicity. Life span is not hugely affected. This single observation should have raised huge concerns, and yet it has been repeated time and again in many laboratories in every single mouse model of APP over expression and Aβ generation. We emphasise this point as it is surely the most significant observation of the past few decades of exploration of the molecular pathogenesis of AD. Mice do not get AD; they do not seem to have an amyloid cascade.

It is not that mice cannot get tangles or tau pathology. Multiple perturbations in mouse demonstrate that they do, under the right conditions. Overexpression of normal human tau induces some tau aggregation pathology (Gotz et al. 2000) and overexpression of human tau-carrying, disease-causing mutations results in widespread tangle formation, neuronal loss, corresponding behavioural deficits and other characteristics of the tauopathies (Lewis et al. 2000; Gotz et al. 2001a; Tanemura et al. 2001). It might be thought that this then is a characteristic of human tau, but that also is not the case. As discussed above, inducing increased GSK3 activity results in widespread phosphorylation of endogenous mouse tau and substantial neuronal loss and significant neurodegeneration. In summary all of the aspects of tau-related pathology that occur in AD in man can also be made to occur in mice; just not in response to amyloid.

Mice with both amyloid plaques and tau aggregation in tangles can be generated, most notably in the triple transgenic line carrying the full house of AD and FTD mutations in APP, PS1 and tau transgenes (Oddo et al. 2003). However, as each gene overexpressed by itself induces its own pathology, this is not in itself surprising and is not evidence for an intact amyloid cascade. Some evidence has been suggested that in this mouse, and in other mice harbouring both mutated APP or mutated PS1 and MAPT transgenes and in mice with tau transgenes and Aβ intracerebral injections, the extent of the pathology is increased as a consequence of the presence of the Aβ (Gotz et al. 2001b; Lewis et al. 2001; Boutajangout et al. 2002). We would suggest that this evidence is, however, not striking. Where there is an enhancement of tau pathology, it is relatively modest and, given that multiple non-specific, extra-cerebral events can increase tau phosphorylation, including for example body temperature (Planel et al. 2007), these attempts to demonstrate that Aβ induces tau pathology are not entirely convincing. In cellular models, the reverse is true. Here Aβ induces neuronal toxicity rapidly and comprehensively. There is a difficulty in ensuring that the effects of Aβ are consistent, and the effects are highly sensitive to the preparation of Aβ used. Many studies use non-physiological levels of Aβ; nonetheless, there can be little doubt that in vitro Aβ induces tau phosphorylation and is neurotoxic. The amyloid cascade hypothesis itself is safe, but the finding of no effect of Aβ in vivo on endogenous, and a small and questionable effect of Aβ on mutated, overexpressed human tau already forming tangle pathology is remarkable.

Why mice do not get AD is a question that has not received as much attention as it ought to have done, given the primacy of these models and the amyloid cascade in developing novel therapies. In a seminal paper, Stein and Johnson (2002) asked this question using expression analysis in brain of the most widely used model: the Tg2576 line overexpressing human APP carrying the double Swedish mutation. Comparing these animals to littermates with transcript arrays, Stein and Johnson speculated that gene expression differences between model and wild type animals would reflect a protective mechanism that would prevent the toxic effects of Aβ generation. They found a remarkable increase in IGF2, accompanied by substantial upregulation of other IIS genes and the amyloid-binding chaperone transthyretin (Stein and Johnson 2002; Stein et al. 2004).

This finding suggests that the reason mice don't get AD or, to be more accurate, the reason the effects of Aβ are so attenuated in mice is that mice upregulate their IIS as a consequence of Aβ generation. This observation is in line with another widely made and oft-ignored observation that mice carrying APP transgenes, including Tg2576, have some elements of insulin resistance including an increase in peripheral insulin and altered glucose tolerance (Pedersen and Flynn 2004). Putting these two somewhat provocative statements together, we hypothesise that the reason mice do not get AD is the same reason that they do not live long (relative to say, reproductive age). Compared to man, we would suggest that mice have a dynamic and effective IIS, a system that prevents longevity but at the same time protects them from Aβ-induced neurodegeneration. Man, on the other hand, is a long-lived animal with an IIS that is relatively impaired, and it is this impairment that lets us live three or four times our effective reproductive age and that makes us vulnerable to AD and, of course, diabetes.

## 6 Insulin Signalling and Cellular and Animal Models of AD

In the sections above we have made a case for GSK3 as a kinase involved in multiple aspects of AD pathology and, most importantly, as transmitting the Aβ effects to tau and hence neuronal toxicity. The evidence for alteration of GSK3 in man is not strong, but genetic and environmental evidence converge to suggest that dysregulation of GSK3 is a risk factor for AD. We note the overwhelming evidence that IIS, one of the two main pathways regulating GSK3 activity, is the main mediator of longevity determination in invertebrates and in rodents, and we speculate that differences in regulation of GSK3 through IIS between mice and man might explain why mice seem so remarkably resistant to the toxic effects of Aβ. These diverse data suggest that insulin and IGF signalling should regulate Aβ production and tau phosphorylation in cellular and animal models. Multiple lines of evidence demonstrate that this is the case, but in vivo somewhat contradictory findings also suggest that the effects of IIS on AD pathology are some what complex.

In neurons in culture, insulin and insulin growth factors reduce tau phosphorylation (Hong and Lee 1997) and protect neurons from exogenous Aβ (Dore et al. 1997) and from the effects of APP-carrying FAD mutations (Niikura et al. 2001). The effects of IIS on tau phosphorylation are temporal and bimodal: a rapid increase followed by a substantial and more sustained decrease (Lesort et al. 1999; Lesort and Johnson 2000). Insulin promotes secretion of the non-amyloidogenic products of APP (Solano et al. 2000) and, in man, early studies showed that manipulation of insulin through ingestion of glucose altered APP levels in plasma (Boyt et al. 2000). These effects are not specific to IIS. We demonstrated much the same results in the activation of wnt signalling, the other mechanism that inhibited GSK3 and also reduced tau phosphorylation and increased non-amyloidogenic metabolism of APP (Mudher et al. 2001). Insulin signalling, as well as wnt signalling, affect APP processing, Aβ generation and tau phosphorylation in animal models, too Peripheral IGF1 reduces the amyloid burden of the ageing rat and, conversely, low levels of central IGF1 increase amyloid load in mice (Carro et al. 2002, 2005, 2006). Inducing insulin resistance through high fat diet or by sucrose-sweetened water increases Aβ load (Ho et al. 2004; Cao et al. 2007), and depleting insulin using streptozotocin increases tau phosphorylation (Clodfelder-Miller et al. 2006; Planel et al. 2007).

These are apparently consistent findings: in both cellular and animal models, insulin resistance increases both aspects of AD pathology. Some studies fail to reproduce some aspects of these experimental models (Lanz et al. 2007) and others note that the mechanisms are complex and sometimes unexpected, such as the effects of hypothermia on tau phosphorylation induced by changes in IIS (Planel et al. 2004, 2007). Harder to explain are the apparently completely contradictory findings from some studies suggesting that IIS has opposite effects from insulin resistance, being not disease inducing but protective. Thus impaired IIS has also been shown to reduce Aβ aggregation and toxicity in a C. elegans model with AD-associated pathology in muscle (Cohen et al. 2006) and, in a different model, it attenuated Aβ-induced paralysis via an autophagy-mediated mechanism (Florez-McClure et al. 2007). Thus a paradox exits; impaired IIS promotes longevity and yet increases AD-relevant pathologies in some models and offers some protection in others (Cohen and Dillin 2008).

It is difficult to resolve this paradox using the models that currently exist. Diet-induced insulin resistance has complex effects, including changes in immunity, liver function and, as discussed above, core body temperature. Streptozotocin induces IIS failure by destroying beta cells of the pancreas, and the relationship between the failure in circulating insulin and central IIS in brain is not at all understood. In an attempt to address these issues, we decided to generate mice with both amyloid pathology and a clearly defined genetic lesion resulting in insulin resistance. We used mice lacking IRS2 that exhibit insulin resistance with high levels of circulating insulin (Withers et al. 1998). Previously, these mice had been shown to have modestly increased tau phosphorylation, with cytoplasmic deposition of phosphorylated tau but no tangle pathology, in line with a role for disrupted IRS2 signalling in tau pathology (Schubert et al. 2003). We crossed these mice with the Tg2576 line expressing human APP with the Swedish mutations and hypothesised that, if the ability to increase IIS is the reason why mice fail to demonstrate the amyloid cascade, then in the context of depletion of IRS2 the amyloid deposition in these mice should induce increased tau phosphorylation and, correspondingly, enhance the known behavioural deficit (Killick et al. 2009).

At 12 months, mice lacking IRS2 had modestly increased tau phosphorylation at some but not all epitopes (Schubert et al. 2003), In line with our hypothesis, mice expressing both APPsw and lacking IRS2 had substantially increased tau phosphorylation, again at most but not all epitopes. Surprisingly, this increase in pathology was due to a substantial decrease in PP2a and not to GSK3, the activity of which was in fact reduced and not increased as predicted. Even more surprising to us, these effects were due to peripheral, not central, changes in IIS. Mice lacking IRS2 only in neurons had no change in PP2a or GSK3. We then examined the amyloid load of the brain, expecting to see no change or possibly an increase in pathology. In fact, we saw the reverse. Amyloid load was substantially decreased, in fact it was almost abolished. This decrease was not due to changes in the generation of Aβ, as APP metabolism was unaffected, or to alteration in soluble Aβ 1-40 or 1-42 but to a substantial reduction in the amount of fibrillised Aβ. The characteristic Aβ-dependent behavioural deficit of the Tg2576 animal, fear conditioning, was rescued by loss of IRS2. Finally, we were able to show that these effects on Aβ were due to an IIS effect on Aβ clearance, as the expression of transthyretin was substantially increased in Tg2576 animals lacking IRS2. The site of action of changes in transthyretin is interesting, as we saw no protein expression in cerebral cortex, as others have shown previously. Transthyetin is expressed by the choroid plexus, from where it is secreted to CSF, and in AD transthyretin is indeed altered in CSF levels (Castano et al. 2006; Davidsson et al. 2002; Gloeckner et al. 2008; Merched et al. 1998). This is not necessarily the only site of action of transthyretin, however, as we also found decreased levels in AD in plasma, suggesting a failure of a peripheral sink of amyloid either in CSF or blood.

In summary, our study, which was the first to use a precise, genetic lesion in IIS and overexpression of APP resulting in Aβ deposition, elucidated complex effects of IIS, with an enhancement of tau phosphorylation as predicted but also a substantial and unexpected decrease in Aβ-related pathology (Killick et al. 2009). These findings may explain some of the contradictory effects previously seen in animal models of IIS disruption on AD-related pathology.

## 7 Conclusions and Way Forward

There can be little doubt that insulin and IGF signalling are important in understanding AD. The evidence is overwhelming – from genes and environment, through cellular and animal models and in intervention studies in man (see Craft et al. this volume). However, although most of the evidence points to a relatively consistent and straightforward understanding whereby insulin resistance increases disease risk via a failure of intracellular signalling, some animal models suggest that caution is warranted. Moving from cellular to in vivo models, complexity is encountered, with both C. elegans and mouse studies showing that, whilst defective IIS might enhance some pathologies, it also enhances protein aggregation clearance. Given the importance of protein aggregation across neurodegeneration, this observation is troubling. Ultimately, it is intervention studies that count, and therapies based on enhancing IIS are in late stages of clinical development with more underway. Our most recent findings suggest that, whilst such therapies should continue to be developed and would be predicted to have a beneficial effect on some aspects of AD pathology, unexpected effects and even worsening of disease may occur. These cautionary notes suggest that further and deeper understanding of the precise role IIS plays in AD pathology is warranted.

## References

Adachi A, Kano F, Saido TC, Murata M (2009) Visual screening and analysis for kinase-regulated membrane trafficking pathways that are involved in extensive β-amyloid secretion. Genes Cells 14:355–369

Albanese E, Banerjee S, Dhanasiri S, Fernandez J-L, Ferri C, Knapp M, McCrone P, Prince M, Snell T, Stewart R (2007) Dementia UK: Report to the Alzheimer's Society, Kings College London and London School of Economics and Political Science

Alvarez G, Muñoz-Montaño JR, Satrústegui J, Avila J, Bogónez E, Díaz-Nido J (1999) Lithium protects cultured neurons against b-amyloid-induced neurodegeneration. FEBS Lett 453: 260–264

Alves da Costa C, Paitel E, Mattson MP, Amson R, Telerman A, Ancolio K, Checler F Wild-type and mutated presenilins 2 trigger p53-dependent apoptosis and down-regulate presenilin 1 expression in HEK293 human cells and in murine neurons. Proc Natl Acad Sci USA 99: 4043–4048

Anekonda TS, Reddy PH (2006) Neuronal protection by sirtuins in Alzheimer's disease. J Neurochem 96:305–313

Aplin AE, Gibb GM, Jacobsen JS, Gallo JM, Anderton BH (1996) In vitro phosphorylation of the cytoplasmic domain of the amyloid precursor protein by glycogen synthase kinase-3β. J Neurochem 67:699–707

Aplin AE, Jacobsen JS, Anderton BH, Gallo JM (1997) Effect of increased glycogen synthase kinase-3 activity upon the maturation of the amyloid precursor protein in transfected cells. Neuroreport 8:639–643

Avila J (2006) Tau phosphorylation and aggregation in Alzheimer's disease pathology. FEBS Lett. 580:2922-2927

Avila J, Lucas JJ, Perez M, Hernandez F (2004) Role of tau protein in both physiological and pathological conditions. Physiol Rev 84:361–384

Barroso I, Luan J, Middelberg RP, Harding AH, Franks PW, Jakes RW, Clayton D, Schafer AJ, O'Rahilly S, Wareham NJ (2003) Candidate gene association study in type 2 diabetes indicates a role for genes involved in β-cell function as well as insulin action. PLoS Biol 1: E20

Bartke A (2008) Insulin and aging. Cell Cycle 7:3338–3343

Bertram L, McQueen M, Mullin K, Blacker D, Tanzi R (2005) The AlzGene database. Alzheimer Research Forum

Biessels GJ, Staekenborg S, Brunner E, Brayne C, Scheltens P (2006) Risk of dementia in diabetes mellitus: a systematic review. Lancet Neurol 5:64–74

Bornemann KD, Staufenbiel M (2000) Transgenic mouse models of Alzheimer's disease. Ann NY Acad Sci 908:260–266

Boutajangout A, Leroy K, Touchet N, Authelet M, Blanchard V, Tremp G, Pradier L, Brion JP (2002) Increased tau phosphorylation but absence of formation of neurofibrillary tangles in mice double transgenic for human tau and Alzheimer mutant (M146L) presenilin-1. Neurosci Lett 318:29–33

Boyt AA, Taddei K, Hallmayer J, Helmerhorst E, Gandy SE, Craft S, Martins RN (2000) The effect of insulin and glucose on the plasma concentration of Alzheimer's amyloid precursor protein. Neuroscience 95:727–734

Braak H, Braak E (1991) Neuropathological stageing of Alzheimer-related changes. Acta Neuropathol Berl 82:239–259

Braak E, Braak H, Mandelkow E-M (1994) A sequence of cytoskeleton changes related to the formation of neurofibrillary tangles and neuropil threads. Acta Neuropathol (Berl ) 87: 554–567

Brownlees J, Irving NG, Brion JP, Gibb BJM, Wagner U, Woodgett J, Miller CC (1997) Tau phosphorylation in transgenic mice expressing glycogen synthase kinase-3β transgenes. Neuroreport 8:3251–3255

Brunden KR, Trojanowski JQ, Lee VM (2008) Evidence that non-fibrillar tau causes pathology linked to neurodegeneration and behavioral impairments. J Alzheimers Dis 14:393–399

Caccamo A, Oddo S, Tran LX, LaFerla FM (2007) Lithium reduces tau phosphorylation but not Aβ or working memory deficits in a transgenic model with both plaques and tangles. Am J Pathol 170:1669–1675

Cao D, Lu H, Lewis TL, Li L (2007) Intake of sucrose-sweetened water induces insulin resistance and exacerbates memory deficits and amyloidosis in a transgenic mouse model of Alzheimer disease. J Biol Chem 282:36275–36282

Carro E, Trejo JL, Gomez-Isla T, LeRoith D, Torres-Aleman I (2002) Serum insulin-like growth factor I regulates brain amyloid-β levels. Nature Med 8:1390–1397

Carro E, Trejo JL, Spuch C, Bohl D, Heard JM, Torres-Aleman I (2005) Blockade of the insulin-like growth factor I receptor in the choroid plexus originates Alzheimer's-like neuropathology in rodents: New cues into the human disease? Neurobiol Aging 27:1618-1631

Carro E, Trejo JL, Gerber A, Loetscher H, Torrado J, Metzger F, Torres-Aleman I (2006) Therapeutic actions of insulin-like growth factor I on APP/PS2 mice with severe brain amyloidosis. Neurobiol Aging 27:1250–1257

Castano EM, Roher AE, Esh CL, Kokjohn TA, Beach T (2006) Comparative proteomics of cerebrospinal fluid in neuropathologically-confirmed Alzheimer's disease and non-demented elderly subjects. Neurol Res 28:155–163

Chee FC, Mudher A, Cuttle MF, Newman TA, MacKay D, Lovestone S, Shepherd D (2005) Over-expression of tau results in defective synaptic transmission in Drosophila neuromuscular junctions. Neurobiol Dis 20:918–928

Clodfelder-Miller BJ, Zmijewska AA, Johnson GV, Jope RS (2006) Tau is hyperphosphorylated at multiple sites in mouse brain in vivo after streptozotocin-induced insulin deficiency. Diabetes 55:3320–3325

Cohen E, Dillin A (2008) The insulin paradox: aging, proteotoxicity and neurodegeneration. Nature Rev Neurosci 9:759–767

Cohen E, Bieschke J, Perciavalle RM, Kelly JW, Dillin A (2006) Opposing activities protect against age-onset proteotoxicity. Science 313:1604–1610

Cuervo AM (2008) Autophagy and aging: keeping that old broom working. Trends Genet 24: 604–612

Davidsson P, Westman-Brinkmalm A, Nilsson CL, Lindbjer M, Paulson L, Andreasen N, Sjögren M, Blennow K (2002) Proteome analysis of cerebrospinal fluid proteins in Alzheimer patients. Neuroreport 13:611–615

Dayanandan R, Van SM, Mack TG, Ko L, Yen SH, Leroy K, Brion JP, Anderton BH, Hutton M, Lovestone S (1999) Mutations in tau reduce its microtubule binding properties in intact cells and affect its phosphorylation. FEBS Lett 446:228–232

De Ferrari GV, Chacon MA, Barria MI, Garrido JL, Godoy JA, Olivares G, Reyes AE, Alvarez A, Bronfman M, Inestrosa NC (2003) Activation of Wnt signaling rescues neurodegeneration and behavioral impairments induced by β-amyloid fibrils. Mol Psychiat 8:195–208

De Ferrari GV, Papassotiropoulos A, Biechele T, Wavrant De-Vrieze F, Avila ME, Major MB, Myers A, Sáez K, Henríquez JP, Zhao A, Wollmer MA, Nitsch RM, Hock C, Morris CM, Hardy J, Moon RT (2007) Common genetic variation within the low-density lipoprotein receptor-related protein 6 and late-onset Alzheimer's disease. Proc Natl Acad Sci USA 104:9434–9439

Dore S, Kar S, Quirion R (1997) Insulin-like growth factor I protects and rescues hippocampal neurons against β-amyloid- and human amylin-induced toxicity. Proc Natl Acad Sci USA 94:4772–4777

Duff K (1998) Recent work on Alzheimer's disease transgenics. Curr Opin Biotechnol 9:561–564

Dunn N, Holmes C, Mullee M (2005) Does lithium therapy protect against the onset of dementia? Alzheimer Dis Assoc Disord 19:20–22

Engel T, Lucas JJ, Gomez-Ramos P, Moran MA, Avila J, Hernandez F (2006) Cooexpression of FTDP-17 tau and GSK-3β in transgenic mice induce tau polymerization and neurodegeneration. Neurobiol Aging 27:1258-1268

Engel T, Goni-Oliver P, Lucas JJ, Avila J, Hernandez F (2006a) Chronic lithium administration to FTDP-17 tau and GSK-3β overexpressing mice prevents tau hyperphosphorylation and neurofibrillary tangle formation, but pre-formed neurofibrillary tangles do not revert. J Neurochem 99:1445–1455

Engel T, Hernandez F, Avila J, Lucas JJ (2006b) Full reversal of Alzheimer's disease-like phenotype in a mouse model with conditional overexpression of glycogen synthase kinase-3. J Neurosci 26:5083–5090

Farias GG, Godoy JA, Hernandez F, Avila J, Fisher A, Inestrosa NC (2004) M1 muscarinic receptor activation protects neurons from β-amyloid toxicity. A role for Wnt signaling pathway. Neurobiol Dis 17:337–348

Feng Z, Hu W, Rajagopal G, Levine AJ (2008) The tumor suppressor p53: cancer and aging. Cell Cycle 7:842–847

Ferber D (2001) Neurodegenerative disease. Using the fruit fly to model tau malfunction. Science 292:1983–1984

Ferrer I, Gomez-Isla T, Puig B, Freixes M, Ribe E, Dalfó E, Avila J (2005) Current advances on different kinases involved in tau phosphorylation, and implications in Alzheimer's disease and tauopathies. Curr Alzheimer Res 2:3–18

Feulner TM, Laws SM, Friedrich P, Wagenpfeil S, Wurst SH, Riehle C, Kuhn KA, Krawczak M, Schreiber S, Nikolaus S, Förstl H, Kurz A, Riemenschneider M. (2009) Examination of the current top candidate genes for AD in a genome-wide association study. Mol Psychiat advance online publication 6 January doi:10.1038/mp.2008.141

Florez-McClure ML, Hohsfield LA, Fonte G, Bealor MT, Link CD (2007) Decreased insulin-receptor signaling promotes the autophagic degradation of β-amyloid peptide in C. elegans. Autophagy 3:569–580

Forlenza OV, Spink JM, Dayanandan R, Anderton BH, Olesen OF, Lovestone S (2000) Muscarinic agonists reduce tau phosphorylation in non-neuronal cells via GSK-3β inhibition and in neurons. J Neural Transm 107:1201–1212

Friedman DB, Johnson TE (1988) Three mutants that extend both mean and maximum life span of the nematode, Caenorhabditis elegans, define the age-1 gene. J Gerontol 43:B102–B109

Gao S, Hendrie HC, Hall KS (1998) The relationships between age, sex, and the incidence of dementia and Alzheimer disease – A meta-analysis. Arch Gen Psychiat 55:809–815

Ghribi O, Herman MM, Savory J (2003) Lithium inhibits Aβ-induced stress in endoplasmic reticulum of rabbit hippocampus but does not prevent oxidative damage and tau phosphorylation. J Neurosci Res 71:853–862

Giannakou ME, Partridge L (2007) Role of insulin-like signalling in Drosophila lifespan. Trends Biochem Sci 32:180–188
Gloeckner SF, Meyne F, Wagner F, Heinemann U, Krasnianski A, Meissner B et al (2008) Quantitative analysis of transthyretin, tau and amyloid-β in patients with dementia. J Alzheimers Dis 14:17–25
Gotz J, Barmettler R, Ferrari A, Goedert M, Probst A, Nitsch RM (2000) In vivo analysis of wild-type and FTDP-17 tau transgenic mice. Ann NY Acad Sci 920:126–133
Gotz J, Chen F, Barmettler R, Nitsch RM (2001a) Tau filament formation in transgenic mice expressing P301L tau. J Biol Chem 276:529–534
Gotz J, Chen F, Van Dorpe J, Nitsch RM (2001b) Formation of neurofibrillary tangles in P301l tau transgenic mice induced by Aβ 42 fibrils. Science 293:1491–1495
Grupe A, Abraham R, Li Y, Rowland C, Hollingworth P, Morgan A, Jehu L, Segurado R, Stone D, Schadt E, Karnoub M, Nowotny P, Tacey K, Catanese J, Sninsky J, Brayne C, Rubinsztein D, Gill M, Lawlor B, Lovestone S, Holmans P, O'Donovan M, Morris JC, Thal L, Goate A, Owen MJ, Williams J (2007) Evidence for novel susceptibility genes for late-onset Alzheimer's disease from a genome-wide association study of putative functional variants. Hum Mol Genet. 16:865–873
Haass C, Selkoe DJ (2007) Soluble protein oligomers in neurodegeneration: lessons from the Alzheimer's amyloid β-peptide. Nature Rev Mol Cell Biol 8:101–112
Hamilton G, Proitsi P, Jehu L, Morgan A, Williams J, O'Donovan MC, Owen MJ, Powell JF, Lovestone S (2007) Candidate gene association study of insulin signaling genes and Alzheimer's disease: evidence for SOS2, PCK1, and PPARγ as susceptibility loci. Am J Med Genet B Neuropsychiatr Genet 144B: 508–516
Hanger DP, Hughes K, Woodgett JR, Brion J-P, Anderton BH (1992) Glycogen synthase kinase-3 induces Alzheimer's disease-like phosphorylation of tau: Generation of paired helical filament epitopes and neuronal localisation of the kinase. Neurosci Lett 147:58–62
Hanger DP, Byers HL, Wray S, Leung KY, Saxton MJ, Seereeram A, Reynolds CH, Ward MA, Anderton BH (2007) Novel phosphorylation sites in tau from Alzheimer brain support a role for casein kinase 1 in disease pathogenesis. J Biol Chem 282:23645–23654
Hardy J (2006) Has the amyloid cascade hypothesis for Alzheimer's disease been proved? Curr Alzheimer Res 3:71–73
Hardy JA, Higgins GA (1992) Alzheimer's disease: the amyloid cascade hypothesis. Science 256:184–185
Helisalmi S, Vepsäläinen S, Hiltunen M, Koivisto AM, Salminen A, Laakso M, Soininen H (2008) Genetic study between SIRT1, PPARD, PGC-1α genes and Alzheimer's disease. J Neurol 255:668–673
Hernandez F, Borrell J, Guaza C, Avila J, Lucas JJ (2002) Spatial learning deficit in transgenic mice that conditionally over-express GSK-3β in the brain but do not form tau filaments. J Neurochem 83:1529–1533
Heutink P (2000) Untangling tau-related dementia. Hum Mol Genet 9:979–986
Ho L, Qin W, Pompl PN, Xiang Z, Wang J, Zhao Z, Peng Y, Cambareri G, Rocher A, Mobbs CV, Hof PR, Pasinetti GM (2004) Diet-induced insulin resistance promotes amyloidosis in a transgenic mouse model of Alzheimer's disease. FASEB J 18:902–904
Hong M, Lee VMY (1997) Insulin and insulin-like growth factor-1 regulate tau phosphorylation in cultured human neurons. J Biol Chem 272:19547–19553
Hooper C, Markevich V, Plattner F, Killick R, Schofield E, Engel T, Hernandez F, Anderton B, Rosenblum K, Bliss T, Cooke SF, Avila J, Lucas JJ, Giese KP, Stephenson J, Lovestone S (2007a) Glycogen synthase kinase-3 inhibition is integral to long-term potentiation. Eur J Neurosci 25:81–86
Hooper C, Meimaridou E, Tavassoli M, Melino G, Lovestone S, Killick R (2007b) p53 is upregulated in Alzheimer's disease and induces tau phosphorylation in HEK293a cells. Neurosci Lett 418:34–37
Hooper C, Killick R, Lovestone S (2008) The GSK3 hypothesis of Alzheimer's disease. J Neurochem 104:1433–1439

Howlett DR, Richardson JC (2009) The pathology of APP transgenic mice: a model of Alzheimer's disease or simply overexpression of APP? Histol Histopathol 24:83–100

Hye A, Kerr F, Archer N, Foy C, Poppe M, Brown R, Hamilton G, Powell J, Anderton B, Lovestone S (2005) Glycogen synthase kinase-3 is increased in white cells early in Alzheimer's disease. Neurosci Lett 373:1–4

Inestrosa N, De Ferrari G, Garrido J, Alvarez A, Olivares G, Barría MI, Bronfman M, Chacón MA (2002) Wnt signaling involvement in β-amyloid-dependent neurodegeneration. Neurochem Int 41:341

Ingelsson M, Ramasamy K, Russ C, Freeman SH, Orne J, Raju S, Matsui T, Growdon JH, Frosch MP, Ghetti B, Brown RH, Irizarry MC, Hyman BT (2007) Increase in the relative expression of tau with four microtubule binding repeat regions in frontotemporal lobar degeneration and progressive supranuclear palsy brains. Acta Neuropathol 114:471–479

Ishiguro K, Shiratsuchi A, Sato S, Omori A, Arioka M, Kobayashi S, Uchida T, Imahori K (1993) Glycogen synthase kinase 3β is identical to tau protein kinase I generating several epitopes of paired helical filaments. FEBS Lett 325:167–172

Jackson GR, Wiedau-Pazos M, Sang T-K, Wagle N, Brown CA, Massachi S, Geschwind DH. (2002) Human wild-type tau interacts with wingless pathway components and produces neurofibrillary tangles in Drosophila. Neuron 34:509–519

Jorm AF (2000) Is depression a risk factor for dementia or cognitive decline? A review. Gerontology 46:219–227

Kessing LV, Sondergard L, Forman JL, Andersen PK (2008) Lithium treatment and risk of dementia. Arch Gen Psychiat 65:1331–1335

Killick R, Scales G, Leroy K, Causevic M, Hooper C, Irvine EE, Choudhury AI, Drinkwater L, Kerr F, Al-Qassab H, Stephenson J, Yilmaz Z, Giese KP, Brion JP, Withers DJ, Lovestone S (2009) Deletion of IRS2 reduces amyloid deposition and rescues behavioural deficits in APP transgenic mice. Biochem Biophys Res Commun 386:257–262

Kitamura Y, Shimohama S, Kamoshima W, Matsuoka Y, Nomura Y, Taniguchi T (1997) Changes of p53 in the brains of patients with Alzheimer's disease. Biochem Biophys Res Commun 232:418–421

Kwok JB, Loy CT, Hamilton G, Lau E, Hallupp M, Williams J, Owen MJ, Broe GA, Tang N, Lam L, Powell JF, Lovestone S, Schofield PR. (2008) Glycogen synthase kinase-3β and tau genes interact in Alzheimer's disease. Ann Neurol 64:446–454

Kwon HS, Ott M (2008) The ups and downs of SIRT1. Trends Biochem Sci 33:517–525

Lanni C, Uberti D, Racchi M, Govoni S, Memo M (2007) Unfolded p53: a potential biomarker for Alzheimer's disease. J Alzheimers Dis 12:93–99

Lanni C, Racchi M, Mazzini G, Ranzenigo A, Polotti R, Sinforiani E, Olivari L, Barcikowska M, Styczynska M, Kuznicki J, Szybinska A, Govoni S, Memo M, Uberti D (2008) Conformationally altered p53: a novel Alzheimer's disease marker? Mol Psychiat 13:641–647

Lanz TA, Salatto CT, Semproni AR, Marconi M, Brown TM, Richter KE, Schmidt K, Nelson FR, Schachter JB (2008) Peripheral elevation of IGF-1 fails to alter Abeta clearance in multiple in vivo models. Biochem Pharmacol 75:1093–1103

Lesort M, Johnson GVW (2000) Insulin-like growth factor-1 and insulin mediate transient site-selective increases in tau phosphorylation in primary cortical neurons. Neuroscience 99: 305–316

Lesort M, Jope RS, Johnson GV (1999) Insulin transiently increases tau phosphorylation: involvement of glycogen synthase kinase-3β and Fyn tyrosine kinase. J Neurochem 72:576–584

Lewis J, McGowan E, Rockwood J, Melrose H, Nacharaju P, Van Slegtenhorst M, Gwinn-Hardy K, Paul Murphy M, Baker M, Yu X, Duff K, Hardy J, Corral A, Lin WL, Yen SH, Dickson DW, Davies P, Hutton M (2000) Neurofibrillary tangles, amyotrophy and progressive motor disturbance in mice expressing mutant (P301L) tau protein. Nature Genet 25:402–405

Lewis J, Dickson DW, Lin WL, Chisholm L, Corral A, Jones G, Yen SH, Sahara N, Skipper L, Yager D, Eckman C, Hardy J, Hutton M, McGowan E. (2001) Enhanced neurofibrillary degeneration in transgenic mice expressing mutant tau and APP. Science 293:1487–1491

li-Youcef N, Lagouge M, Froelich S, Koehl C, Schoonjans K, Auwerx J (2007) Sirtuins: the 'magnificent seven', function, metabolism and longevity. Ann Med 39:335–345
Liddell MB, Lovestone S, Owen MJ (2001) Genetic risk of Alzheimer's disease: advising relatives. Br J Psychiat 178:7–11
Liolitsa D, Powell J, Lovestone S (2002) Genetic variability in the insulin signalling pathway may contribute to the risk of late onset Alzheimer's disease. J Neurol Neurosurg Psychiat 73:261–266
Liu GP, Zhang Y, Yao XQ, Zhang CE, Fang J, Wang Q, Wang JZ (2008) Activation of glycogen synthase kinase-3 inhibits protein phosphatase-2A and the underlying mechanisms. Neurobiol Aging 29:1348–1358
Lovestone S, Reynolds CH (1997) The phosphorylation of tau: a critical stage in neurodevelopmental and neurodegenerative processes. Neuroscience 78:309–324
Lovestone S, Reynolds CH, Latimer D, Davis DR, Anderton BH, Gallo J-M, Hanger D, Mulot S, Marquardt B, Stabel S, Woodgett JR, Miller CC (1994) Alzheimer's disease-like phosphorylation of the microtubule-associated protein tau by glycogen synthase kinase-3 in transfected mammalian cells. Curr Biol 4:1077–1086
Lovestone S, Hartley CL, Pearce J, Anderton BH (1996) Phosphorylation of tau by glycogen synthase kinase-3β in intact mammalian cells: the effects on organisation and stability of microtubules. Neuroscience 73:1145–1157
Lowin A, Knapp M, McCrone P (2001) Alzheimer's disease in the UK: comparative evidence on cost of illness and volume of health services research funding. Int J Geriatr Psychiat 16:1143–1148
Lucas JJ, Hernandez F, Gomez-Ramos P, Moran MA, Hen R, Avila J (2001) Decreased nuclear β-catenin, tau hyperphosphorylation and neurodegeneration in GSK-3β conditional transgenic mice. EMBO J 20:27–39
Macdonald A, Briggs K, Poppe M, Higgins A, Velayudhan L, Lovestone S (2008) A feasibility and tolerability study of lithium in Alzheimer's disease. Int J Geriatr Psychiat 23:704–711
Mandelkow E-M, Drewes G, Biernat J, Gustke N, Van Lint J, Vandenheede JR, Mandelkow E (1992) Glycogen synthase kinase-3 and the Alzheimer-like state of microtubule-associated protein tau. FEBS Lett 314:315–321
Martin L, Magnaudeix A, Esclaire F, Yardin C, Terro F (2008) Inhibition of glycogen synthase kinase-3grβ downregulates total tau proteins in cultured neurons and its reversal by the blockade of protein phosphatase-2A. Brain Res 1252:66–75
Matheu A, Maraver A, Serrano M (2008) The Arf/p53 pathway in cancer and aging. Cancer Res 68:6031–6034
Mazanetz MP, Fischer PM (2007) Untangling tau hyperphosphorylation in drug design for neurodegenerative diseases. Nature Rev Drug Discov 6:464–479
McElwee JJ, Schuster E, Blanc E, Piper MD, Thomas JH, Patel DS, Selman C, Withers DJ, Thornton JM, Partridge L, Gems D (2007) Evolutionary conservation of regulated longevity assurance mechanisms. Genome Biol 8:R132
McPhee CK, Baehrecke EH (2009) Autophagy in Drosophila melanogaster. Biochim Biophys 1793:1452–1460
Merched A, Serot JM, Visvikis S, Aguillon D, Faure G, Siest G (1998)Apolipoprotein E, transthyretin and actin in the CSF of Alzheimer's patients: relation with the senile plaques and cytoskeleton biochemistry. FEBS Lett 425:225–228
MRC CFAS (2001) Pathological correlates of late-onset dementia in a multicentre, community-based population in England and Wales. Neuropathology Group of the Medical Research Council Cognitive Function and Ageing Study (MRC CFAS). Lancet 357:169–175
Mudher A, Lovestone S (2002) Alzheimer's disease – do tauists and baptists finally shake hands? Trends Neurosci 25:22–26
Mudher A, Chapman S, Richardson J, Asuni A, Gibb G, Pollard C, Killick R, Iqbal T, Raymond L, Varndell I, Sheppard P, Makoff A, Gower E, Soden PE, Lewis P, Murphy M, Golde TE, Rupniak HT, Anderton BH, Lovestone S (2001) Dishevelled regulates the metabolism of

amyloid precursor protein via protein kinase C/mitogen-activated protein kinase and c-Jun terminal kinase. J Neurosci 21:4987–4995
Mudher A, Shepherd D, Newman TA, Mildren P, Jukes JP, Squire A, Mears A, Drummond JA, Berg S, MacKay D, Asuni AA, Bhat R, Lovestone S (2004) GSK-3β inhibition reverses axonal transport defects and behavioural phenotypes in Drosophila. Mol Psychiat 9:522–530
Muyllaert D, Terwel D, Borghgraef P, Devijver H, Dewachter I, Van LF (2006) Transgenic mouse models for Alzheimer's disease: the role of GSK-3β in combined amyloid and tau-pathology. Rev Neurol (Paris) 162:903–907
Myers AJ, Pittman AM, Zhao AS, Rohrer K, Kaleem M, Marlowe L, Lees A, Leung D, McKeith IG, Perry RH, Morris CM, Trojanowski JQ, Clark C, Karlawish J, Arnold S, Forman MS, Van Deerlin V, de Silva R, Hardy J (2007) The MAPT H1c risk haplotype is associated with increased expression of tau and especially of 4 repeat containing transcripts. Neurobiol Dis 25:561–570
Nagy Z, Esiri MM, Jobst KA, Morris JH, King EM, McDonald B Litchfield S, Smith A, Barnetson L, Smith AD. (1995) Relative roles of plaques and tangles in the dementia of Alzheimer's disease: correlations using three sets of neuropathological criteria. Dementia 6:21–31
Nakashima H, Ishihara T, Suguimoto P, Yokota O, Oshima E, Kugo A, Kugo A, Terada S, Hamamura T, Trojanowski JQ, Lee VM, Kuroda S (2005) Chronic lithium treatment decreases tau lesions by promoting ubiquitination in a mouse model of tauopathies. Acta Neuropathol 110:547–556
Niikura T, Hashimoto Y, Okamoto T, Abe Y, Yasukawa T, Kawasumi M, Hiraki T, Kita Y, Terashita K, Kouyama K, Nishimoto I (2001) Insulin-like growth factor I (IGF-I) protects cells from apoptosis by Alzheimer's V642I mutant amyloid precursor protein through IGF-I receptor in an IGF-binding protein-sensitive manner. J Neurosci 21:1902–1910
Noble W, Planel E, Zehr C, Olm V, Meyerson J, Suleman F, Gaynor K, Wang L, LaFrancois J, Feinstein B, Burns M, Krishnamurthy P, Wen Y, Bhat R, Lewis J, Dickson D, Duff K (2005) Inhibition of glycogen synthase kinase-3 by lithium correlates with reduced tauopathy and degeneration in vivo. Proc Natl Acad Sci USA 102:6990–6995
Nunes PV, Forlenza OV, Gattaz WF (2007) Lithium and risk for Alzheimer's disease in elderly patients with bipolar disorder. Br J Psychiat 190:359–360
Oddo S, Caccamo A, Kitazawa M, Tseng BP, LaFerla FM (2003) Amyloid deposition precedes tangle formation in a triple transgenic model of Alzheimer's disease. Neurobiol Aging 24:1063–1070
Ogg S, Paradis S, Gottlieb S, Patterson GI, Lee L, Tissenbaum HA, Ruvkun G (1997) The Fork head transcription factor DAF-16 transduces insulin-like metabolic and longevity signals in C. elegans. Nature 389:994–999
Pedersen WA, Flynn ER (2004) Insulin resistance contributes to aberrant stress responses in the Tg2576 mouse model of Alzheimer's disease. Neurobiol Dis 17:500–506
Pei JJ, Tanaka T, Tung YC, Braak E, Iqbal K, Grundke-Iqbal I (1997) Distribution, levels, and activity of glycogen synthase kinase-3 in the Alzheimer disease brain. J Neuropathol Exp Neurol 56:70–78
Pei JJ, Braak E, Braak H, Grundke-Iqbal I, Iqbal K, Winblad B, Winblad B, Cowburn RF (1999) Distribution of active glycogen synthase kinase 3b (GSK- 3β) in brains staged for Alzheimer disease neurofibrillary changes. J Neuropathol Exp Neurol 58:1010–1019
Peineau S, Taghibiglou C, Bradley C, Wong TP, Liu L, Lu J, Lo E, Wu D, Saule E, Bouschet T, Matthews P, Isaac JT, Bortolotto ZA, Wang YT, Collingridge GL (2007) LTP inhibits LTD in the hippocampus via regulation of GSK3β. Neuron 53:703–717
Phiel CJ, Wilson CA, Lee VM, Klein PS (2003) GSK-3alpha regulates production of Alzheimer's disease amyloid-β peptides. Nature 423:435–439
Piper MD, Selman C, McElwee JJ, Partridge L (2008) Separating cause from effect: how does insulin/IGF signalling control lifespan in worms, flies and mice? J Intern Med 263:179–191
Planel E, Miyasaka T, Launey T, Chui DH, Tanemura K, Sato S, Murayama O, Ishiguro K, Tatebayashi Y, Takashima A (2004) Alterations in glucose metabolism induce hypothermia

leading to tau hyperphosphorylation through differential inhibition of kinase and phosphatase activities: implications for Alzheimer's disease. J Neurosci 24:2401–2411

Planel E, Tatebayashi Y, Miyasaka T, Liu L, Wang L, Herman M, Yu WH, Luchsinger JA, Wadzinski B, Duff KE, Takashima A (2007) Insulin dysfunction induces in vivo tau hyperphosphorylation through distinct mechanisms. J Neurosci 27:13635–13648

Plattner F, Angelo M, Giese KP (2006) The roles of cyclin-dependent kinase 5 and glycogen synthase kinase 3 in tau hyperphosphorylation. J Biol Chem 281:25457–25465

Qin W, Zhao W, Ho L, Wang J, Walsh K, Gandy S, Pasinetti GM (2008) Regulation of forkhead transcription factor FoxO3a contributes to calorie restriction-induced prevention of Alzheimer's disease-type amyloid neuropathology and spatial memory deterioration. Ann NY Acad Sci 1147:335–347

Qing H, He G, Ly PT, Fox CJ, Staufenbiel M, Cai F, Zhang Z, Wei S, Sun X, Chen CH, Zhou W, Wang K, Song W (2008) Valproic acid inhibits Aβ production, neuritic plaque formation, and behavioral deficits in Alzheimer's disease mouse models. J Exp Med 205:2781–2789

Rockenstein E, Torrance M, Adame A, Mante M, Bar-on P, Rose JB, Crews L, Masliah E (2007) Neuroprotective effects of regulators of the glycogen synthase kinase-3β signaling pathway in a transgenic model of Alzheimer's disease are associated with reduced amyloid precursor protein phosphorylation. J Neurosci 27:1981–1991

Rodier F, Campisi J, Bhaumik D (2007) Two faces of p53: aging and tumor suppression. Nucleic Acids Res 35:7475–7484

Ryder J, Su Y, Liu F, Li B, Zhou Y, Ni B (2003) Divergent roles of GSK3 and CDK5 in APP processing. Biochem Biophys Res Commun 312:922–929

Sarkar S, Ravikumar B, Floto RA, Rubinsztein DC (2009) Rapamycin and mTOR-independent autophagy inducers ameliorate toxicity of polyglutamine-expanded huntingtin and related proteinopathies. Cell Death Differ 16:46–56

Schubert M, Brazil DP, Burks DJ, Kushner JA, Ye J, Flint CL, Farhang-Fallah J, Dikkes P, Warot XM, Rio C, Corfas G, White MF (2003) Insulin receptor substrate-2 deficiency impairs brain growth and promotes tau phosphorylation. J Neurosci 23:7084–7092

Schweitzer I, Tuckwell V, O'Brien J, Ames D (2002) Is late onset depression a prodrome to dementia? Int J Geriatr Psychiat 17:997–1005

Sengupta A, Wu QL, Grundke-Iqbal I, Iqbal K, Singh TJ (1997) Potentiation of GSK-3-catalyzed Alzheimer-like phosphorylation of human tau by cdk5. Mol Cell Biochem 167: 99–105

Small SA, Duff K (2008) Linking Aβ and tau in late-onset Alzheimer's disease: a dual pathway hypothesis. Neuron 60:534–542

Smith WW, Norton DD, Gorospe M, Jiang H, Nemoto S, Holbrook NJ, Finkel T, Kusiak JW (2005) Phosphorylation of p66Shc and forkhead proteins mediates Aβ toxicity. J Cell Biol 169:331–339

Solano DC, Sironi M, Bonfini C, Solerte SB, Govoni S, Racchi M (2000) Insulin regulates soluble amyloid precursor protein release via phosphatidyl inositol 3 kinase-dependent pathway. FASEB J 14:1015–1022

Stein TD, Johnson JA (2002) Lack of neurodegeneration in transgenic mice overexpressing mutant amyloid precursor protein is associated with increased levels of transthyretin and the activation of cell survival pathways. J Neurosci 22:7380–7388

Stein TD, Anders NJ, DeCarli C, Chan SL, Mattson MP, Johnson JA (2004) Neutralization of transthyretin reverses the neuroprotective effects of secreted amyloid precursor protein (APP) in APPSW mice resulting in tau phosphorylation and loss of hippocampal neurons: support for the amyloid hypothesis. J Neurosci 24:7707–7717

Stewart R, Liolitsa D (1999) Type 2 diabetes mellitus, cognitive impairment and dementia. Diabet Med 16:93–112

Su Y, Ryder J, Li B, Wu X, Fox N, Solenberg P, Brune K, Paul S, Yan Z, Liu F, Binhui N (2004) Lithium, a common drug for bipolar disorder treatment, regulates amyloid-β precursor protein processing. Biochemistry 43:6899–6908

Tajes M, Gutierrez-Cuesta J, Folch J, Ferrer I, Caballero B, Smith MA, Casadesus G, Camins A, Pallás M (2008) Lithium treatment decreases activities of tau kinases in a murine model of senescence. J Neuropathol Exp Neurol 67:612–623

Takashima A, Honda T, Yasutake K, Michel G, Murayama O, Murayama M, Ishiguro K, Yamaguchi H (1998) Activation of tau protein kinase I glycogen synthase kinase- 3β by amyloid β peptide (25-35) enhances phosphorylation of tau in hippocampal neurons. Neurosci Res 31:317–323

Tanemura K, Akagi T, Murayama M, Kikuchi N, Murayama O, Hashikawa T, Yoshiike Y, Park JM, Matsuda K, Nakao S, Sun X, Sato S, Yamaguchi H, Takashima A (2001) Formation of filamentous tau aggregations in transgenic mice expressing V337M human tau. Neurobiol Dis 8:1036–1045

Terao T, Nakano H, Inoue Y, Okamoto T, Nakamura J, Iwata N (2006) Lithium and dementia: a preliminary study. Prog Neuropsychopharmacol Biol Psychiat 30:1125–1128

Terry RD, Masliah E, Salmon DP, Butters N, DeTeresa R, Hill R, Hansen LA, Katzman R (1991) Physical basis of cognitive alterations in Alzheimer's disease: synapse loss is the major correlate of cognitive impairment. Ann Neurol 30:572–580

Terwel D, Muyllaert D, Dewachter I, Borghgraef P, Croes S, Devijver H, Van Leuven F (2008) Amyloid activates GSK-3β to aggravate neuronal tauopathy in bigenic mice. Am J Pathol 172:786–798

Trojanowski JQ, Lee VMY (1995) Phosphorylation of paired helical filament tau in Alzheimer's disease neurofibrillary lesions: Focusing on phosphatases. FASEB J 9:1570–1576

Vijg J, Campisi J (2008) Puzzles, promises and a cure for ageing. Nature 454:1065–1071

Wang JZ, Grundke-Iqbal I, Iqbal K (2007) Kinases and phosphatases and tau sites involved in Alzheimer neurofibrillary degeneration. Eur J Neurosci 25:59–68

Williams DW, Tyrer M, Shepherd D (2000) Tau and tau reporters disrupt central projections of sensory neurons in Drosophila. J Comp Neurol 428:630–640

Withers DJ, Gutierrez JS, Towery H, Burks DJ, Ren JM, Previs S, Zhang Y, Bernal, D, Pons S, Shulman GI, Bonner-Weir S, White MF (1998) Disruption of IRS-2 causes type 2 diabetes in mice. Nature 391:900–904

Woodruff-Pak DS (2008) Animal models of Alzheimer's disease: therapeutic implications. J Alzheimers Dis 15:507–521

Zhu LQ, Wang SH, Liu D, Yin YY, Tian Q, Wang XC, Wang Q, Chen JG, Wang JZ (2007) Activation of glycogen synthase kinase-3 inhibits long-term potentiation with synapse-associated impairments. J Neurosci 27:12211–12220

# PKC and Insulin Pathways in Memory Storage: Targets for Synaptogenesis, Anti-apoptosis, and the Treatment of AD

**Miao-Kun Sun, Thomas J. Nelson, and Daniel L. Alkon**

**Abstract** Long-term memory is stored through the activation of protein kinase C (PKC) isozymes α and ε that interact with insulin pathways in the brain, as well as through the PKC-mediated formation of structurally specific synaptic transformations. These natural mechanistic pathways whereby the brain adaptively modifies itself to record experience also provide a means for the brain to repair and protect itself against neurologic disorders that destroy synapses, as occurs in Alzheimer's disease (AD), stroke, and head trauma.

The loss of synapses in the brains of AD patients is the only demonstrated pathologic change that closely correlates with the loss of cognitive function. Molecular signaling pathways controlled by PKC isozyme activation (specifically α and ε) are critically involved in the growth of new synapses, i.e., synaptogenesis, protecting against the loss of synapses, and preventing the death of neurons, i.e., apoptosis. These PKC α and ε pathways induce synaptogenesis in the fully differentiated (adult) and aging brain as well as in the developing brain and thus directly target critical pathologic synaptic changes in AD.

A principal mechanism by which PKC-mediated synaptogenesis occurs is the activation and coordination of growth factors such as brain-derived neurotrophic factor (BDNF), nerve growth factor (NGF), and insulin growth factor (IGF). PKC isozymes have reciprocal interactions with both IGF and NGF. Insulin, IGF, and PKC provide the neurotrophic signaling essential for neuronal development as well as continuous neuronal and synaptic growth and survival. Stimulation of neurotrophic activity via the PKC and insulin signaling pathways activates anti-apoptotic signaling cascades including the Ras-MAPP (mitrogen-activated protein kinase) pathway and the phosphatidyl/inositol3-kinase (PI3K). IGF-1 acts on the type I insulin receptor, resulting in the activation of PI3K/ATK and MAP kinase (MEK)/ extracellular signal-regulated kinase (ERK) pathways. These same IGF, insulin

D.L. Alkon (✉)
Blanchette Rockefeller Neurosciences Institute at West Virginia University, Morgantown, WV 26505, USA

S. Craft and Y. Christen (eds.), *Diabetes, Insulin and Alzheimer's Disease*,
Research and Perspectives in Alzheimer's Disease,
DOI 10.1007/978-3-642-04300-0_10, 

receptor (IR), and PKC interactions also have important impacts on the reduction of Aβ and hyperphosphorylated τ in AD. Similar to PKC activity deficits, insulin signaling insufficiency leads to a decreased Aβ clearance and increased hyperphosphorylated τ, as well as impaired synaptogenesis and synaptic remodeling. Increased Aβ oligomers can directly produce neuronal insulin resistance as well as directly bind to and inactivate PKC. These fundamental molecular interactions of the PKC and insulin metabolic pathways provide a basis for the close relationship between AD and diabetes. They also provide a mechanistic rationale for therapeutic interventions that activate PKC α and/or ε and, thereby, eliminate and/or prevent the loss of synapses in aging, AD, and a variety of other neurologic disorders through enhanced synaptogenesis and anti-apoptosis. A number of recent experiments with animal models of AD and stroke, in fact, have provided clear evidence of the therapeutic efficacy of activating the PKC-insulin signaling pathways and have recently lead to FDA approval of the first, clinical trial(s) of the PKC activator, Bryostatin, for the treatment of AD patients.

## 1 Introduction

In the cognitive processes of learning and memory, two signaling pathways are essential: the insulin and protein kinase C (PKC) signaling pathways. The insulin signaling pathway includes those activated by insulin and insulin-like growth factors (IGFs). PKC, a family of serine/threonine kinases, is critical for synaptic remodeling, memory storage, and neural repair, in a way similar to the insulin signaling cascade. There are currently 12 identified PKC isoforms, which are further divided into three subgroups, cPKC (α, $\beta_I$, $\beta_{II}$, and γ), nPKC (δ, ε, ε', η, θ, and μ), and aPKC (ζ and λ/ι), based on their molecular structures and thus co-factor requirements. Functional deficits of the PKC and insulin signaling pathways result in serous memory impairment, and their functional recovery dictates outcome in cognitive therapies.

The brain has the highest expression of PKC in the body but was previously considered to be an insulin-insensitive organ. While PKC functional deficits lead to impaired learning and memory, the major health concern involving insulin is diabetes, a group of metabolic disorders characterized by hyperglycemia that affects hundreds of millions of people worldwide. Recent studies, however, have shown that insulin receptors (IRs) exist in the brain (Havrankova et al. 2000), especially in the hippocampus, mediating some brain functions. Patients with either type 1 diabetes, caused by insulin deficiency, or type 2 diabetes, the most prevalent form of diabetes caused by insulin resistance, typically show impaired cognitive function (Stranahan et al. 2008a; Knight et al. 2009). Insulin signaling through insulin and IGFs has thus evolved to not only regulate glucose metabolism and energy supply but also to control brain aging and cognitive abilities (Russo et al. 2005).

Evidence indicating the convergence of molecular pathways for memory, PKC, and insulin regulation comes from studies of diabetes and Alzheimer's disease

(AD), a neurodegenerative disorder with memory loss. AD is associated with a decline in brain PKC activity. Diabetes is one of the strongest risk factors for developing AD among a number of vascular risk factors, including hypertension, heart disease, and smoking, and remains so even when vascular impact is removed (Luchsinger et al. 2005; Arvanitakis et al. 2004), revealing a two-fold increase in the risk of developing either AD or vascular dementia in patients with Type 2 diabetes mellitus (Stewart and Lilitsa 1999; Peila et al. 2002). Insulin resistance and hyperinsulinemia exert a negative influence on memory and are associated with hippocampal atrophy (Convit et al. 2003; den Heijer et al. 2003; Luchsinger et al. 2004), neuropathological markers of AD (see below; Petrovitch et al. 2000), deficits in cognition and memory (Arvanitakis et al. 2004; Peila et al. 2004), including episodic memory, and verbal and visual memory impairments (Elias et al. 1997; Helkala et al. 1995; Perlmurter et al. 1984; Strachan et al. 1997; Venhanen et al. 1999). AD patients, on the other hand, are also more vulnerable to developing Type 2 diabetes (Janson et al. 2004) and have been found to have higher rates of insulin resistance, hyperinsulinemia, and hyperglycemia compared to healthy controls (Razay and Wilcock 1994). The links between the two disorders in pathogenesis and pathophysiology are striking, and a better understanding of their pathogenic mechanisms may have important implications for pharmacological treatments of AD and its prevention.

## 2 Insulin, Insulin-like Growth Factor, and PKC Signaling

There are many similarities between pancreatic β-cells and inhibitory GABAergic neurons. Both cell types possess specialized molecular structures for vesicular secretion. Neural progenitor cells can be differentiated into insulin-secreting, glucose-responsive cells (Hori et al. 2005). Both cell types synthesize and secrete high levels of vesicular gamma-aminobutyric acid (GABA) and express a vesicular GABA transporter (Suckow et al. 2006)). The synaptic adhesion molecules neurexin, neuroligin, and SynCAM, usually thought of as synapse-specific proteins, are also present in β-cells (Suckow et al. 2008). These proteins determine whether new synapses develop into excitatory or inhibitory synapses (Chubykin et al. 2007).

There are also many similarities between diabetes and AD at the molecular level. One characteristic of type 2 diabetes is the fibrillization and deposition of islet amyloid polypeptide (IAPP; Marzban et al. 2003), a peptide that is normally co-secreted with insulin. β-cells also express high levels of amyloid precursor protein (APP), a β-secretase (BACE), and presenilin 2 (Figueroa et al. 2001). BACE and presenilins proteolytically cleave APP to produce β-amyloid, a peptide whose natural function is still unknown. In both diseases, β-amyloid polymerizes into fibrils, which are then deposited in the extracellular space. Both diseases also exhibit hyperphosphorylation of the microtubule protein τ (Miklossy et al. 2005). These many similarities may provide clues about the normal function of APP.

Insulin, a small protein of 51-amino acid residues, is known to induce plasma translocation of glucose transporters and increase glucose uptake in muscles and adipose tissues. There is very little insulin mRNA in the brain (Steen et al. 2005). IGFs are polypeptides with sequence similarity to insulin. IGF-1, for instance, is mainly produced in liver and small amounts of IGF-1 are also produced in the brain (Bondy and Lee 1993). It is, thus, not clear whether the low level of possible insulin production in the brain may play a significant functional role in neurons. Knockout of the neuronal IR gene has been shown to cause some enhanced GSK3β activity and tau hyperphosphorylation but does not cause cognitive deficits or neurodegeneration (Schubert et al. 2004). Insulin, however, can cross the blood-brain barrier (BBB) by a saturable, IR-mediated transport process (Schwartz et al. 1990; Banks et al. 1997; Banks and Kastin 1998). IGF-1 can also cross the BBB. The IR, a heterotetrameric tyrosine kinase receptor, is activated by insulin binding. Insulin binding to the extracellular IR α subunit leads to a conformational change, resulting in autophosphorylation of intracellular tyrosine residues in the β subunit, whose phosphorylation in turn leads to recognition and further tyrosine phosphorylation of phosphotyrosine-binding (PTB) domains of substrate proteins, such as IR substrates (IRS) 1-6 and the Src-homology-2-containing spinal protein (Shc). In the insulin-sensitive peripheral tissues, IR activation results in trafficking of GLUT4, the insulin-dependent glucose transporter, to the plasma membrane, thus increasing glucose uptake. Phosphorylation of IRS-1, Shc and Grb2 results in activation of ERK/MAPK and downstream transcription factors CREB and Elk-1.

IR activation also activates phosphatidylinositol 3-kinase (PI3K), resulting in formation of phosphatidylinositol triphosphate (Gasparini et al. 2002; Launer 2005; Gispen and Biessels 2000), an AKT activator. AKT in turn phosphorylates and inactivates both α and β cytosolic forms of glycogen synthase kinase 3 (GSK3; Schubert et al. 2004), BAD (pro-apoptotic) and modulates down-stream TOR (target of rapamycin) regulation of translation. Insulin-mediated AKT activation is required for insulin-induced upregulation of insulysin (insulin-degrading enzyme), whose substrates include $A\beta_{40}$ and $A\beta_{42}$. Reduction in insulin signaling and insulin resistance thus leads to decreased insulysin activity. AKT also inhibits GSK3α, which stimulates Aβ generation through γ-secretase (Phiel et al. 2003).

Insulin and PKC signaling cascades are highly correlated in functional regulation. Both of their activities are required in activity-dependent synaptic remodeling in the hippocampus (Hori et al. 2005; see below). It has been shown that PKCα regulates the IGF-1 receptor in adult and embryonic rat cardiomyocytes (Maniar et al. 2005). IGF-1 also activates PKCα, whose activity is required for IGF-1-stimulated PI3K activation (Maniar et al. 2005).

## 3 Synaptic Transmission and Synaptogenesis

Synapses are where the majority of communications between neurons occur, and their number and strength determine cognitive abilities. PKC and insulin signaling regulates synaptic transmission and function. Cholinergic and glutamatergic transmissions in the hippocampus and cortex, for instance, are essential to many types of learning and memory and are under functional regulation by insulin and PKC signaling cascades. In the prefrontal cortex, muscarinic acetylcholine receptor activation enhances $GABA_A$ receptor-mediated current in the pyramidal neurons through a PKC-dependent, Src-mediated signaling cascade that is gated by an insulin/PI3K pathway (Ma et al. 2003). NMDA receptor activation elicits the activity-dependent remodeling of postsynaptic structures in the hippocampal neurons and postsynaptic response, involving a PKC-dependent synaptic translocation of IRS p53 (Hori et al. 2005), since PKC activators induce and PKC inhibitors suppress its synaptic translocation.

Synaptic formation and remodeling regulate the function of synaptic networks and may play key roles in recovery after brain injury (Brown et al. 2007) and in learning and memory (Brown et al. 2007). Dendritic spines, the post-synaptic structures important for plasticity, are heterogeneous in shape and structure. Spines function as bridges between axons and postsynaptic contact sites for most of the excitatory synapses in the mammalian brain (Fischer et al. 1998; Yuste and Denk 1995) and are modifiable. They can be roughly divided into two groups: small spines and large spines. The former are dynamic and are generated during activity-dependent processes in acquisition of memories and change form rapidly, either disappearing or transforming into large spines. The latter, such as mushroom spines, survive for months or even years, providing a structural basis for memories. Synaptic formation requires a complex coordinated interaction between pre- and post-synaptic structures: an accumulation and organization of synaptic vesicles at the presynaptic level with signaling pathways and vesicle fusion mechanisms operable at high speed and a postysynaptic density at excitatory synapses. Synaptogenesis involves forming a contact and stabilizing into a functional and mature synapse. A dendritic process at an excitatory synapse, for example, could be either initiated by the growth of a highly motile filopodium, which stabilizes as a dendritic spine when in contact with an axon terminal (Ziv and Smith 1996), or a direct growth of a spine from the dendrite without the initial formation of a filopodium (Engert and Bonhoeffer 1999; Jourdain et al. 2003) or a presynaptic contact (De Roo et al. 2008; Knott et al. 2006). The activity-dependent remodeling of synaptic structure and strength is a fundamental process in learning and memory, involving over a hundred proteins, enzymes, and messengers as well as the interplay between insulin and PKC signaling cascades. Synaptogenesis depends on the recruitment of integrins (Webb et al. 2007) and cell adhesion molecules (CAMs), which may stabilize the cytoarchitecture in pre- and post synaptic sites. α5-Integrin signaling, for instance, regulates spine morphogenesis and synapse formation in hippocampal neurons via Src kinase, Rac, and the signaling adaptor GIT1 (Webb et al. 2007). Knockdown of α5 integrin expression with small interfering RNA decreases the number of dendritic spines and synapses (Webb et al. 2008). Migration of the integrin receptor triggers the arachidonic acid cascade, resulting in astrocytic facilitation of excitatory synaptogenesis, which can be blocked by inhibition of integrins and PKC (Hama et al. 2004). PKC activation also underlies GABAergic inhibitory synaptogenesis (Meier et al. 2003). Learning, PKC activation, and IGF-1 activation induce synaptogenesis in the dorsal hippocampal CA1 field ( Hongpaisin and Alkon 2007; O'Kusky et al. 2000). Activation of PKCα leads to an up-regulation of neuronal ELAV proteins and GAP-43 protein, GAP-43 mRNA stabilization (Pascale et al. 2005). It is also known that PI3K activation induces synaptogenesis in adult neurons and is required for synaptic maintenance (Martin-Pena et al. 2006). The synaptogenesis activity of PI3K also requires AKT. The BDNF-TrkB-PI3K-AKT pathway is involved in the NMDA receptor activation-induced delivery of PSD-95 to dendrites in synaptogenesis (Yoshii and Constantine-Paton 2007). GSK3 overexpression, on the other hand, exhibits an anti-synaptogenetic effect (Martin-Pena et al 2006).

During brain aging and cognitive diseases, synapses are gradually lost or become functionally deficient. For instance, AD is functionally characterized in early stages by memory impairment, resulting from synaptic failure (Selkoe 2002). Soluble Aβ oligomers, the neurotoxic form (Shankar et al. 2008), induce the loss of synapses (Sleary et al. 2005; Shugrue et al. 2010). Diabetes is also associated with reduced dendritic spine density (Magarinos and McEwen 2000; Martinez-Tellez et al. 2005), whereas physical exercise enhances neuronal structures (Strabahn et al. 2007; Redila and Christie 2006). The same types of synaptic changes are also consistently observed in animal studies. Axon regeneration requires MAP kinase activity (Hammarlund et al. 2009). Experimental diabetes induced by streptozotocin, a diabetogenic agent that induces a well-characterized experimental model of type I diabetes, is associated with retraction and simplification of apical dendrites of hippocampal CA3 pyramidal neurons and with an increased glucocorticoid reactivity to stress in rats (Magariños and McEwen 2000). Middle-aged rats fed a high fat/glucose diet show reduced hippocampal dendritic spine density and hippocampal BDNF expression and spatial learning impairment, associated with diabetic energy and lipid metabolism (Stranahan et al. 2008).

# 4 PKC, APP, and Synaptogenesis

There is increasing evidence that PKC is necessary for synaptogenesis. During development, synaptosomal PKCγ reaches a peak around postnatal day 14, which is a critical period of synaptogenesis (Shearman et al. 1991). Contact of neurons with astrocytes induces synaptogenesis, which is mediated in part by integrin receptors and PKC (Hama et al. 2004). During learning, maturation of dendritic spines is accelerated by PKC activators and blocked by PKC inhibitors, indicating that memory-specific synaptogenesis also requires PKC (Fig. 1; Hongpaisan and Alkon 2007). Activation of PKC by the experimental drug bryostatin can also induce synaptogenesis in inactive animals with fully developed nervous systems (Sun et al. 2008). Inhibition of MAP kinase reduces neurite outgrowth (Mundy et al. 2008), suggesting the involvement of the PKC-Ras-MAP kinase pathway.

Of course, the process of synaptogenesis involves numerous other proteins besides PKC (Fig. 2). CAMs such as SYG-1 and 2, neurexin, and neuroligin are necessary for synapse formation and stability (Dean and Dresbach 2006).

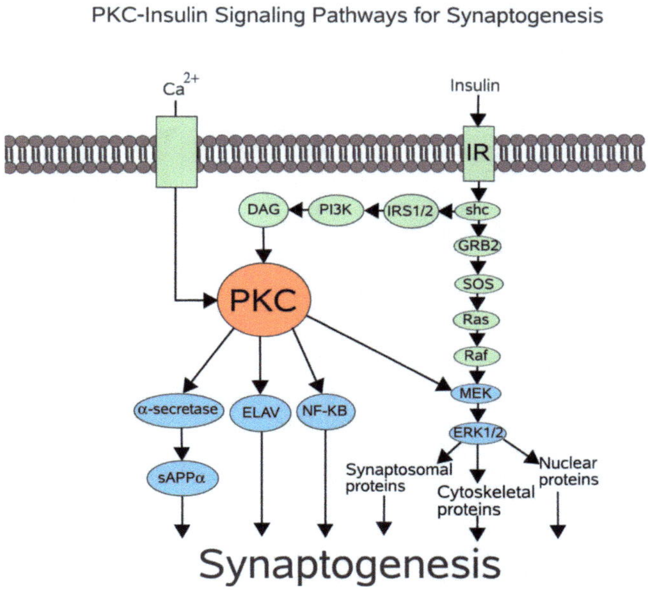


**Fig. 1** PKC-insulin signaling pathways for synaptogenesis. PKC and insulin can both affect synaptogenesis. The insulin receptor activates the Ras/Raf growth pathway by way of Shc protein. ERK 1 and 2 can modulate synaptogenesis by affecting synaptosomal proteins, inducing rearrangements of the cytoskeleton, and by changing protein expression. PKC also activates this pathway by activating MEK (mitogen-activated kinase). PKC modulates synaptogenesis by activating α-secretase, which produces extracellular sAPPα, and by activating the mRNA-stabilizing ELAV proteins and the nuclear signaling factor NF-κB. Insulin receptor substrates 1 and 2 also indirectly activate PKC by activating the synthesis of diacylglycerol (DAG)

The formation of a macromolecular complex between presynaptic neurexin and postsynaptic neuroligin is an important early step in synaptogenesis (Fabrichny et al. 2007). Neuroligin oligomerizes into multimers that function to recruit neurexin to cell-cell contact sites (Dean et al. 2003). Neurexin in turn interacts with cytoplasmic scaffolding molecules CIPP, CASK/Lin-2 and Mint/Lin-10/X11α, and possibly others, through its C-terminal PDZ binding motif (Dean et al. 2003). Stabilization of the new cell-cell contact occurs by activation of a number of pathways, including PKC, which is activated by arachidonic acid released by phospholipase A2 during cell adhesion (Hama et al. 2004). PKC and its substrate GAP-43 exert stabilizing effects on F-actin and the cytoskeleton (Larsson 2006). Interestingly, contact with astrocytes also produces a long-lasting activation of PKC, which phosphorylates a number of proteins, including GAP-43, MARCKS, and adducin, leading to actin rearrangement, spinogenesis, and synaptogenesis (Hama et al. 2004). BDNF may also indirectly activate PKC by its effect on PLCγ (Schmidt 2004). Sustained activation of PKC by pharmacological agents such as bryostatin might short-circuit this process and stabilize nascent synapses directly.

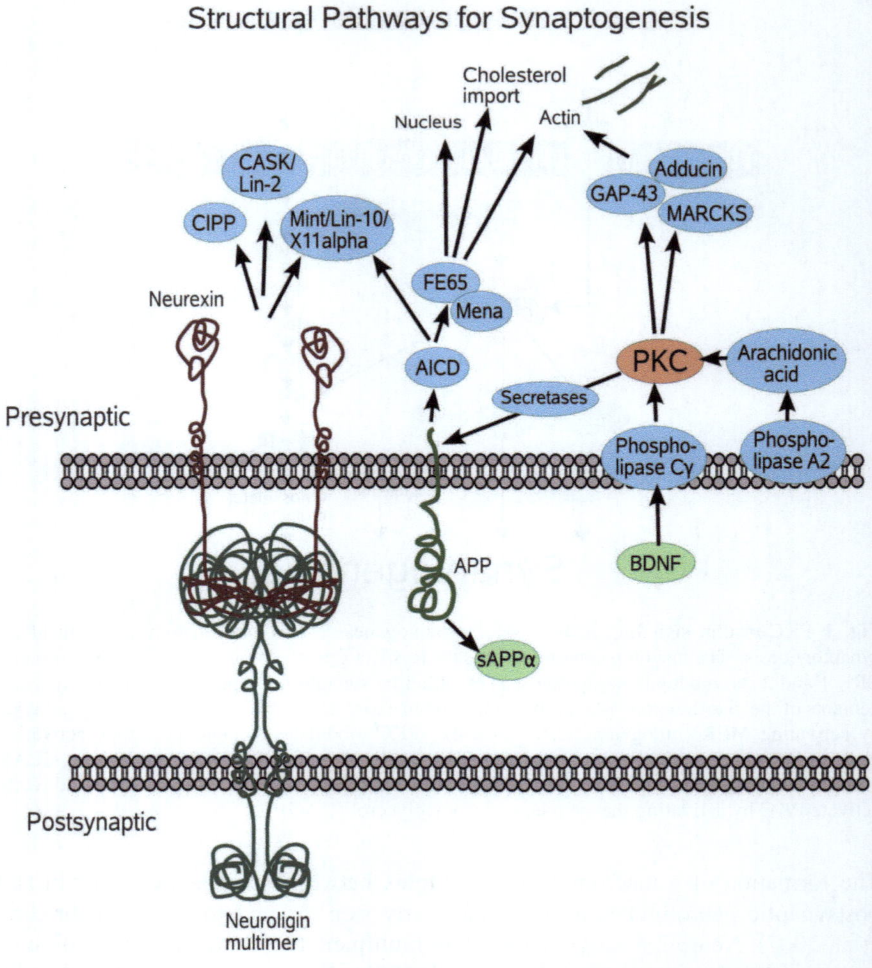


**Fig. 2** Structural pathways for synaptogenesis. During synaptogenesis, contact between the pre- and post-synaptic membranes triggers oligomerization of neuroligin, which recruits neurexin to cell-cell contact sites. This trans-synaptic protein complex interacts with cytoplasmic scaffolding molecules CIPP, CASK/Lin-2, and Mint/Lin-10/X11α. PKC's role in synaptogenesis is to stabilize the cell-cell contact by phosphorylating synaptic growth proteins GAP-43, MARCKS, and adducin, which activate actin cytoskeletal rearrangement. PKC also activates α-secretases, which release sAPPα and AICD, which interacts with FE65, Mena, and Mint to help stabilize the newly formed synapse. APP also participates in neuron-glial cell adhesion

Although APP is often considered only for its role in AD, both APP and sAPPα, the secreted extracellular domain of APP, are important in synaptogenesis (Fig. 2). APP is expressed at high levels not only on the surface of neurons but also on glial cells. Approximately 88% of the APP protein is extracellular. Increasing evidence suggests that this extracellular portion of APP acts as a contact receptor and participates in neuron-glial cell adhesion (Gralle and Ferreira 2007). The short cytoplasmic tail of APP, in contrast, binds to FE65 proteins and to Mena, forming a ternary complex that regulates actin (Sabo et al. 2001). AICD, the APP intracellular domain produced by cleavage by ε-secretase, also binds FE65. APP-induced rearrangement of the cytoskeleton is essential for dendritic arborization (Leyssen et al. 2005). In adults, APP is inserted into fast axonal transport vesicles and transported from the trans-Golgi network to the synapses (Sisodia et al. 1993), where it undergoes rapid turnover: part of it is cleaved by secretases, releasing sAPPα into the extracellular medium (Fig. 2), and the remainder is recycled by the endocytic pathway (Koo and Squazzo 1994).

Both portions of APP, the large extracellular soluble APP and the shorter cytosolic/membrane fraction that includes AICD, are involved in synaptogenesis and thereby play a central role in learning and memory. Knockout of APP causes memory deficits (Dawson et al. 1999) and intracerebral injection of sAPPα enhances memory performance (Meziane et al. 1998). The shorter, intracellular portion of APP binds FE65 in actin-rich structures in growth cones (Ikin et al. 2007). This binding occurs between the FE65 C-terminal phosphotyrosine binding (PTB) domain [sometimes called a phosphotyrosine interaction (PI) domain] and the YENPTY reinternalization motif of APP. FE65, in turn, interacts via its WW domain with Mena (Sabo et al. 2003), a protein of the Ena/VASP family that regulates cell motility, adhesion, and growth cone dynamics. A similar cytosolic protein, X11α/Mint-1, also binds to the YENPTY motif and participates in a series of protein interactions that stabilize APP and carry out the structural changes induced by APP.

The APP-binding protein FE65 also binds to the AICD fragment and translocates it to the nucleus (McLoughlin and Miller 2008). In the nucleus, FE65 may act as a transcription factor (McLoughlin and Miller 2008). FE65 also possesses a second PTB domain, which binds to the NPXY motif on the low-density lipoprotein receptor-related protein (LRP; Jaeger and Pietrzik 2008). The function of this trimeric complex of FE65, LRP, and APP is not clear, but because LRP is an ApoE receptor, it is likely to be involved in cholesterol transport or APP processing (Jaeger and Pietrzik 2008).

The other, larger fragment of APP, sAPPα, is a secreted factor with structural and functional similarities to other growth factors, such as epidermal growth factor (EGF; Caillé et al. 2004). The sAPPα protein increases neurite outgrowth (Sabo et al. 2003), acts as a proliferative factor, and inhibits apoptotic signaling by Bad, caspase 3, and caspase 9 (Caillé et al. 2004). sAPPα also stimulates the oxidation of docosahexaenoic acid to produce NPD1, a lipid that is reported to repress proinflammatory and pro-apoptotic gene expression mediated by NF-κB (nuclear factor κ B) and STAT-1 (signal transducer and activator of transcription 1; Lukiw and Bazan 2006). Thus, sAPPα may act by a variety of pathways to protect neurons and facilitate synaptogenesis.

The pathways that activate sAPPα production are unclear. sAPPα is produced by cleavage of APP by α-secretases. One such enzyme is TACE (ADAM-17), which is activated by PKC and MAP Kinase Erk 1/2 (Buxbaum et al. 1993). Thus, PKC activators could, in principle, accelerate the production of sAPPα and thereby reduce Aβ levels by competitively lowering the quantity of APP available to β-secretase. Such a strategy could simultaneously promote neurorepair and block the production of toxic Aβ, which has been shown to occur in fibroblasts (Etcheberrigaray et al. 2004) but has yet to be demonstrated in patients.

## 5 Neuronal Survival, Neurogenesis, and Anti-apoptosis

Brain growth factors are essential to neuronal survival, neurogenesis, and anti-apoptosis in adults (Stewart and Liolitsa 1999). Insulin, IGFs, and PKC provide the neurotrophic signaling that is essential for not only synaptogenesis but also neuronal development and continuous growth and survival. Apoptosis, a form of programmed cell death, is responsible for a variety of neuronal injuries and cognitive damage related to neurodegenerative disorders. Stimulation of the neurotrophic activity of the PKC and insulin signaling pathways activates anti-apoptotic signaling cascades (Sun et al. 2008, 2009), including the Ras-MARP pathway (Walker et al. 1998; Datta et al. 1997; Kurada and White 1998) and the PI3K (Klesse and Parada 1998). IGF1, for example, is a well-established neurotrophic signal for proliferation, survival, and differentiation of neuronal and glial progenitors (Aberg et al. 2000; Hsieh et al. 2004; Vicario-Abejon et al. 2003). IGF-1 acts on the type I receptor, resulting in an activation of PI3K/ATK and MEK/extracellular signal-regulated kinase (ERK) pathways. IGF-1 regulates adult neurogenesis (Aberg et al. 2000; Carro et al. 2001), enhances cell proliferation and differentiation (Aberg et al. 2000), and mediates exercise-induced neurogenesis (Trejo et al. 2001). Transgenic mice overexpressing IGF-1 show increases in cortical cell number (Popken et al. 2004). Neurogenesis involves reducing the $G_1$ phase and recruiting additional cells into the cycle, suggesting a mitogenic role for IGF-1 (Hodge et al. 2004). Activation of IGF-1 promotes $G_1$/S cell cycle progression via the PI3K/Akt pathway in developing rat cerebral cortex (Mairet-Coello et al. 2009). IRS-1 also associates with the loop domain of Bcl-2 and synergistically up-regulates the anti-apoptotic function of Bcl-2 (Ueno et al. 2000). Activation of these neurotrophic-signaling cascades results in phosphorylation of Bad and thus suppression of apoptosis (Datta et al. 1997; del Peso et al. 1997; Weinreb et al. 2004). PKCα and ε are also known to associate with Bcl-2 members to produce neuroprotection (Hevener et al. 2000).

## 6 Insulin, PKC, and Learning and Memory

Insulin signaling and PKC are important co-players in learning and memory (Fig. 1). Both are up-regulated and undergo translocation after acquiring associative memory (Zhao and Alkon 2001; Zhao et al. 1999; Bank et al. 1988; Alkon et al. 2007). Although glucose is the major energy source for the central nervous system and cognitive functions, its uptake, transport, and utilization in the brain do not depend on insulin (Zhao et al. 2004). However, hippocampal perfusion and glucose metabolism are consistently reduced in patients with mild cognitive impairment and AD (Rodriguez et al. 2000; De Santi et al. 2001). There is also a recent discovery that the hippocampus, where IRs exist (Park 2001), expresses the insulin-regulated glucose transporter, GLUT4 (Piroli et al. 2007; Fernando et al. 2008), which may represent one mechanism by which insulin affects cognition. The hippocampal GLUT4 appears to mediate glucose uptake, especially the activity-dependent uptake, into the pyramidal cells, since the uptake can be reduced by indinavir (Fernando et al. 2008), a glucose transporter inhibitor. However, insulin itself does not affect the GLUT4-mediated glucose uptake into the hippocampal neurons (Fernando et al. 2008), suggesting a different molecular control pathway.

The neuronal effects of insulin (Fig. 3) involve at least three main actions: a direct action of insulin on signal processing functions of the neurons, an action through GSK3, and its impact on amyloidosis. The first two appear beneficial in terms of learning and memory, but the third one is troublesome. Peripheral insulin administration is complicated because of its impact on glucose concentrations. In humans, intravenous insulin administration has also been reported to facilitate

**Fig. 3** 3D model of human insulin hexamer, based on the data from Chang et al. (1997). Insulin is normally found as a symmetric hexamer composed of three symmetrically arranged insulin dimers. The complex is coordinated by histidines to two central zinc ions

memory (Craft et al. 2003). Such administration may impair, rather than enhance, learning and memory, possibly due to sedative effects associated with changes in glucose levels (Akanmu et al. 2009). Intracerebrovascular administration or microinjection of insulin into the rodent hippocampal CA1 region has been shown to improve spatial memory consolidation and retrieval (Moosavi et al. 2007) and passive avoidance memory (Park et al. 2000; Babri et al. 2007). There is a learning-specific increase in the expression of the IRs and insulin signaling pathways in the hippocampus (Zhao et al. 1999, 2004; Dou et al. 2005). However, the direct actions of insulin on learning and memory are limited by insulin resistance in the brains of diabetics and in some AD patients. Furthermore, cognitive functions and neurogenesis are impaired in subjects of either type 1 or type 2 diabetes, as well as in both diabetes types of animal models (Gispen and Biessels 2000; Dou et al. 2005; Biessels et al. 1998; Kamal et al. 2006; Biessels and Gispen 2005; Lang et al. 2008), suggesting that lower insulin concentrations alone may not explain the impaired learning and memory and synaptic strength. Diabetes patients show hyperactivation of the hypothalamic-pipuitary-adrenal (HPA) axis, resulting in elevated cortisol levels (Messier 2005; Desrocher and Rovet 2004). Evidence has also been provided that the diabetes-induced impairment of cognition, neurogenesis, and synaptic strength is mediated by hyperactivation of the HPA axis and thus elevated levels of glucocorticoids (Stranahan et al. 2008a). Lowering corticosterone prevents the diabetes-induced impairment of learning and memory in insulin-deficient and insulin-resistant rodents (Stranahan et al. 2008a), consistent with the glucocorticoid effects on neural structures and functions (Gould et al. 1992; Montaron et al. 2006).

Insulin inactivates GSK-3 through a PI3K-AKT signaling pathway, resulting in phosphorylation of both GSK-3β and GSK-3α isoforms. The insulin-mediated GSK-3 inhibition may have beneficial effects in reducing tau protein hyperphosphorylation (Park 2001) and Aβ (Phiel et al. 2003; Su et al. 2004), although it remains to be determined how large a role insulin insufficiency may play in tau hyperphosphorylation and Aβ accumulation in AD brains.

Activation of PKC enhances learning and memory (Hongpaisan and Alkon 2007; Sun and Alkon 2006; Sun et al. 2008; Nelson et al. 2008). Depending on the particular isforms activated, activation of PKC results in a facilitation of signaling processing, protein synthesis and synaptogenesis, synaptic remodeling and repairing, and Aβ degradation, as well as reduction in amyloidosis and tau hyperphosphorylation (see below).

## 7 Insulin Pathways, Insulysin, PKC, and AD

The converging epidemiological and clinical evidence suggests that insulin resistance significantly increases the risk of developing AD. Patients with AD are also more likely to have insulin resistance. The main putative mechanism in the AD pathogenesis is elevation of soluble Aβ and Aβ oligomers in the brain, especially in the hippocampus and related brain structures. Recent data suggest that several actions of insulin resistance may play an important role in AD progression or even AD pathogenesis, including an inhibition of cerebral BBB Aβ clearance and insulysin-mediated degradation of Aβs, anti-synaptogenesis via GSK, in addition to a reduced expression of insulysin associated with insulin resistance. Insulysin deficits in expression are sufficient to cause both Type 2 diabetes mellitus and amyloidosis. However, the effectiveness of insulysin deficiency in inducing the two disorders does not necessarily mean that the enzyme is an actually dominant AD factor. It is, however, clear that pathophysiological processes such as cerebral Aβ imbalance between formation and clearance precede the clinical symptoms of AD. In the progression of AD, the Aβ balance in the brain shifts from Aβ clearance to deposition. Aβ accumulation damages insulin and PKC signaling pathways (Favit et al. 1998; Lee et al. 2004) and could be the direct result of an insufficiency of the insulin or PKC signaling pathways or both. The evidence points to a pathogenic role of PKC and insulin signaling insufficiencies in neurodegeneraitve disorders and suggests that restoration of the signaling pathways may have therapeutic values for AD treatment and prevention, i.e., to arrest the pathological processes before they result in clinical dementia.

A direct link between diabetes and AD is the evidence that abnormally high insulin levels or brain insulin resistance facilitates amyloidosis (Ho et al. 2004). Insulin promotes the release of intracellular Aβ in neuronal cultures (Gasparini et al. 2001) and reduces cerebral Aβ clearance. The latter involves an insulin-induced inhibition of the cerebral Aβ clearance: clearance through the BBB to circulating blood through efflux transport of intact Aβs and inhibition of degradation mediated by insulysin. One of the possible carriers/transporters for BBB Aβ clearance is the IRs expressed in the brain capillary endothelial cells. The IRs mediate transport of insulin across the BBB by transcytosis (Pardridge et al. 1995) and can bind Aβs (Xie et al. 2002). Cerebral BBB Aβ clearance is thus inhibited by insulin and is reduced in aging (by about 30% in 23-month-old rats as compared to 7-week-old rats: Shiiki et al. 2004).

Insulin inhibits Aβ degradation by insulysin (Fig. 4) in the brain (Shiiki et al. 2004; Qiu et al. 1998; Rizk et al. 2006; Perez et al. 2000; Zhu et al. 2005). Insulysin, an important regulator of extracellular Aβ levels, including the neurotoxic Aβ40 and Aβ42 (Qiu et al. 1998; Vekrellis et al. 2000; Song and Hersh 2005; Mukherjee et al. 2000), exists in the hippocampus, where its levels are reduced in AD patients who express the apolipoprotein ε4 allele (Cook et al. 2003). The enzyme is a thiol zinc metalloendopeptidase, about 110 kd, primarily located in the cytosol but also in the peroxisomes and endosomes, on the cell surface, and in the extracelluar space (secreted). Its action prevents both the neurotoxic effects of the amyloid and their ability to deposit onto amyloid plaques. Shen et al. (2006) crystallized insulin-degrading enzyme (also known as insulysin or insulinase) bound to insulin B chain and to β-amyloid and found that the enzyme consists of two domains (IDE-N and IDE-C), representing the N- and C-terminal regions. These two domains form an enclosed chamber that opens, engulfs the peptide substrate, and then closes again, chewing up the peptide at multiple sites like a molecular Pac-man. In the absence of

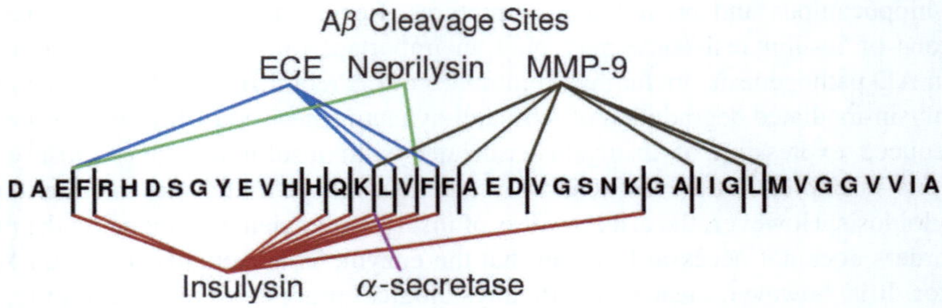


**Fig. 4** Aβ cleavage sites. Endothelin-converting enzyme (ECE), neprilysin, and insulysin (also known as insulin-degrading enzyme) have all been shown to proteolytically degrade Aβ in vivo. The cleavage sites of these three enzymes tend to cluster around Lys-16 of Aβ, which is also the cleavage site of α-secretases. However, insulyzin and ECE also cleave Aβ at several other sites. MMP-9 (matrix metallopeptidase 9) can also degrade Aβ at a number of sites. Neprilysin is regulated by nicastrin, which is a component of the γ secretase complex that produces Aβ. Neprilysin also proteolyzes a number of other signaling peptides, including atrial natriugenic factor, substance P, and enkephalins. Likewise, insulysin hydrolyzes a number of other peptides, including insulin and amylin. Activators of these enzymes have potential benefits as anti-Alzheimer drugs.
Another clearance pathway is efflux of Aβ through the blood-brain barrier (BBB). This clearance is mediated primarily by LRP1 (low-density lipoprotein receptor-related protein; Bell et al. 2007), possibly by a transcytosis mechanism. However, other possible clearance mechanisms include the receptor for advanced glycation end products (RAGE), sequestration by Aβ carrier proteins such as apolipoprotein E and α 2-macroglobulin, and endocytosis followed by degradation (Nazer et al. 2009). LRP1 at the BBB is reported to be downregulated in streptozotocin-induced diabetic mice (Hong et al. 2009), which may contribute to Aβ accumulation in diabetes

substrate, the chamber is normally closed. Mutations of IDE that promote opening of the catalytic chamber increase the catalytic rate by up to 40-fold (Shen et al. 2006). Thus, small molecules that convert IDE to a normally open configuration could behave as powerful activators.

In addition to insulin B chain and β-amyloid, insulin-degrading enzyme proteolyzes glucagon, TGF α, β-endorphin, amylin, and many other signaling peptides. IDE substrates all possess certain similarities. The peptide must have a 7- to 13-amino acid region containing a large hydrophobic group (Phe on Aβ, Tyr on insulin B) capable of forming a β-sheet. The catalytic portion of IDE (IDE-N) binds to the β-sheet and to the first three to five N-terminal amino acids of the peptide. Since the interior of the cavity is positively charged, the peptide must also be neutral or acidic. The peptide must also be smaller than about 6 kDa to fit inside the digestive cavity of IDE. These X-ray crystallographic results will be of great benefit in designing modulators of IDE and other β-amyloid-degrading enzymes.

Interestingly, insulin-degrading enzyme is also an important receptor for varicella-zoster virus (VZV) glycoprotein E (Li et al. 2006). Binding of this protein to IDE is important for cell-to-cell spread and infectivity (Ali et al. 2009). VZV is a neurotrophic herpes virus similar to herpes simplex virus type 1 (HSV1), which has been suggested as a possible etiological factor in AD (Itzhaki and Wozniak 2008).

Evidence suggests that homozygous deletion of insulysin in mice leads to a ~65% elevation of endogenous cerebral $A\beta_{40}$ levels (Rizk et al. 2006; Song and

Hersh 2005; Miller et al. 2003). Partial loss-of-function mutations (about 15 to 30% deficits) in insulysin sufficient to cause Type 2 diabetes also impair Aβ degradation (Farris et al. 2004), suggesting that other Aβ-degrading enzymes are insufficient to compensate for the reduced catalytic efficiency of insulysin in neurons. Insulysin is sensitive to oxidative damage (Shinall et al. 2005) and impaired insulin degradation is related to late-life verbal memory decline (Okereke et al. 2008). Excess insulin inhibits the degradation of $A\beta_{40}$ and $A\beta_{42}$, largely because of its higher binding affinity for the insulysin ($k_m \sim 0.1$ μM; Aberg et al. 2000), whereas both Aβ derivatives weakly inhibit insulin degradation in a dose-dependent manner ($k_m >$ 0.2 μM). In humans, peripheral infusion of insulin leads to increases Aβ levels in the cerebrospinal fluid (Watson et al. 2003), and insulin treatment in diabetes has been shown to accelerate AD (Pardridge et al. 1995). In rats, cerebral insulin at 300 ng/ml significantly increases Aβ levels in the brain (Shiiki et al. 2004). Thus, it remains to be investigated extensively whether intranasal administration of insulin (Luchsinger et al. 2004), which bypasses the BBB and thus avoids the systemic side effects of insulin, may be beneficial to AD patients, although some studies have already shown cognitive benefits (Okereke et al. 2008; Watson et al. 2003; Hallschmid et al. 2007; Hanson and Frey 2008; Reger et al. 2008).

## 8 Treatment of AD via Targeting the PKC and Insulin Signaling Pathways

AD, a devastating neurodegenerative disorder with impaired memory and cognition, is pathologically characterized by extracellular senile plaques of mainly aggregated fibrillar insoluble Aβ and intracellular neurofibrillary tangles of hyperphosphorylated tau protein, pathophysiologically by synaptic dysfunction, and clinically by a progressive dementia. Abnormal processing and accumulation of the neurotoxic amyloid are believed by many to be the primary causative AD factors, responsible for synaptic dysfunction and impaired synaptogenesis in the early disease stages and neuronal loss/death at a late stage. It should be noted, however, that the Aβs, including the major neurotoxic forms $A\beta_{40}$ and $A\beta_{42}$, are normally and continually produced in the brain (Zou et al. 2005) and may have physiological functions (Pusso et al. 2008), suggesting that eliminating Aβ may not be entirely therapeutically beneficial. Although overproduction of the neurotoxic Aβs has been implicated in some early-onset familial AD cases, which account for only $<5\%$ of the entire AD population, the Aβ accumulation in the sporadic AD brains occurring in the majority of AD cases most likely results from an insufficient Aβ clearance, since the rate of neurotoxic Aβ production in most sporadic AD brain is unchanged (Selkoe 2001; Wang et al. 2003). One strategy, therefore, for AD treatment is to restore the Aβ clearance/removal. The proteases capable of degrading Aβ include insulysin, neprilysin, endothelin-converting enzyme, the uPA/tPA/plasminogen system, and angiotensin-converting enzyme. Their deficits in functions, especially insulysin and neprilysin, shift the Aβ balance towards accumulation and represent important therapeutic targets for AD treatment. Evidence suggests that reductions in the insulin signaling cascade, such as those occurring in insulin resistance, provides a signal for a reduced expression of insulysin (Zhao et al. 2004). Transgenic overexpression of insulysin in neurons has been shown to significantly reduce brain Aβ levels, retard or completely prevent amyloid plaque formation, and rescue the premature lethality in transgenic AD mice (Leissring et al. 2003). Importantly, PKCε activation markedly enhances Aβ degradation (see below).

Deficits of PKC and insulin signaling pathways appear, on the other hand, to play essential roles in AD pathogenesis (Alkon et al. 2007; Watson and Craft 2004). Dysfunction of PKC and insulin signaling is associated with the development of AD and other neurodegenerative disorders (Alkon et al. 2007). Brain insulin and IRs are reduced in AD (Steen et al. 2005; Rivera et al. 2005; Frolich et al. 1998; Craft et al. 1998), with higher levels of plasma insulin. IGF-1 receptor blockers have been shown to produce amyloidosis and hyperphosphorylated tau (Carro et al. 2006). Deficits in one signaling pathway can facilitate signaling insufficiency of the other, initiating a deteriorating pathological cycle. PKC signal deficits downregulate IR gene transcription (Yamamoto et al. 2000). In a recent study, induction of a PKC-dependent (blocked with calphostin C) up-regulation of IR expression was shown to reduce insulin resistance (Kong et al. 2009). Similar to PKC activity deficits, insulin signaling insufficiency leads to impaired synaptogenesis and synaptic remodeling due to the lack of neurotrophic activity, a decreased Aβ clearance, and an increased hyperphosphorylated tau. The increased Aβ oligomers can directly produce neuronal insulin resistance. Dendritic IRs are virtually eliminated after 30 min of exposure of cultured hippocampal neurons to Aβ oligomers (Zhao et al. 2008), suggesting that insulin resistance in AD brains may be the result of Aβ oligomers-induced IR deficits (redistribution from the dendrites). Patients with diabetes have insulin resistance in the brains, judging by a functional decrease in IR-mediated signal transduction in the brain (Ho et al. 2004), and have increased amyloid plaques and neurofibrillary tangles in the hippocampus at autopsy (Peila et al. 2002). Intracerebroventricular administration of streptozotocin (STZ) to deplete brain insulin produces brain insulin resistance and progressive neurodegeneration and hyperphosphorylated tau (Grunblatt et al. 2007; de la Monte et al. 2006; Zhang et al. 2008). Peripheral STZ administration has also been found to impair adult hippocampal neurogenesis (Stranahan et al. 2008a; Ali et al. 2009). Insulin resistance caused by a high fat diet is associated with reduced insulysin levels, decreased PI3K and AKT activation and increased amyloidosis (Ho et al. 2004), probably the results of reduced insulysin and increased GSK3α. Animals with selective deletion of the insulysin gene show the hallmark phenotypic characteristics of AD and type 2 diabetes mellitus, i.e., chronic elevation of cerebral Aβ, hyperinsulinemia, and glucose intolerance (Rizk et al. 2006), indicating strong links in pathogenesis between the two disorders (Watson and Craft 2004). Consistent with this linkage is the observation that aging is associated with reduced insulysin activity, especially in the hippocampus (Caccamo et al. 2005).

Insulin/IGF-1 and PKC signaling pathways may thus serve as potential therapeutic targets in AD. PKC activators reduce the rate of Aβ formation (Etcheberrigarary et al. 2004), in part via activating α-secretase and thus producing soluble APP fragment α (sAPPα) instead of the neurotoxic Aβs through β and γ-secretases. Activation of PKCε, in addition, increases the activity of endothelin-converting-energy (ECE), which catalyzes the degradation of Aβ (Nelson and Alkon 2009). Over-expression of PKCε was also previously shown to markedly reduce Aβ plaques in AD transgenic mice. (Choi et al. 2006) Synaptogenesis and synaptic remodeling are impaired in AD and other dementia (Sun et al. 2008). Enhancing insulin signaling in the central nervous system is a potential strategy for AD intervention to correct the insulysin deficits in AD (Zhao et al. 2004). CNS insulin can be a protective agent against memory impairment induced by ischemia (Rizk et al. 2006), lesions (De Castro et al. 1976), chronic stress (Moosavi et al. 2007), and some pharmacological agents. Expression of IGF-1 increases after injury, protecting and rescuing hippocampal neurons from neurotoxic amyloid (Dore et al. 1997). In a recent study, Kuhad et al. (2007) reported that inhibition of NF-κβ with tocotrienol prevented spatial memory deficits in STZ-induced diabetic rats. Improving insulin sensitivity pharmacologically is expected to benefit both AD and diabetes patients. Removing insulin resistance restores the impaired neurotrophic activity and GSK inhibition and increases the expression and function of insulysin for Aβ clearance and BBB Aβ clearance. Insulin itself, however, is not a good therapeutic candidate, since it enhances Aβ release from neurons, competes with neurotoxic Aβs for cerebral BBB Aβ clearance and available insulysin, and produces hypoglycemia if administered systemically.

Oligomerized Aβ (also known as Aβ-derived diffusible ligands, or ADDLs) produces a downregulation of insulin receptors in the plasma membrane. When the insulin receptors are occupied by insulin, this downregulation is prevented (De Felice et al. 2009). The downregulation also can be blocked by inhibitors of calcium calmodulin-dependent kinase II (CaMKII) and casein kinase II (CK2), suggesting that these kinases may play important roles in mediating the synaptic loss caused by Aβ.

Other neurohormones related to insulin are less beneficial than insulin. IGF-1 promotes phosphorylation of APP at Thr668 (Araki et al. 2009). This site appears to be critical in determining the amyloidogenic fate of APP. Phosphorylation of this site after treatment with IGF-1 increases the rate of cleavage by γ-secretase, increasing the levels of extracellular Aβ (Feyt et al. 2007). Phosphorylation of Thr668 is increased in AD patients (Lee et al. 2003), suggesting that this mechanism may contribute to AD. Phosphorylation of APP can be mediated in vivo by a number of protein kinases, including GSK3β, SAPK1b/JNK3, Cdc2, and Cdk5 (Lee et al. 2003). However, the precise roles of IGF-1 and Thr668 are still not completely understood.

Improving insulin sensitivity through diet, exercise (Hevener et al. 2000; De Angelis et al. 1999; Mayer-Davis et al. 1998), and drugs is important for diabetes treatment and may also benefit those patients at risk for AD. IR modulators that are not substrates of insulysin may possess such therapeutic potential, especially those that lead to a selective insulysin activation towards Aβ relative to insulin (Song et al. 2003). Small peptide substrates have been shown to increase the activity of insulysin toward $A\beta_{40}$ hydrolysis without affecting its catalytic activity to insulin (Song et al. 2005). Consistent with this line of therapeutic hypothesis is the evidence that IGF-1 blocks amyloid toxicity by increasing survival signaling through ERK and PI3K (Wei et al. 2002).

Particularly interesting are the so-called insulin sensitizers, a class of agents named the thiazolidinediones, including rosiglitazone and pioglitazone. These are agonists for the peroxisome proliferator-activated receptor-gamma (PPAR-γ), whose activation maintains an appropriate level of expression of key gluco- and liporegulatory molecules and protein involved in the transduction of insulin signaling, thus improving insulin sensitivity (Schmidt 2004; Malinowski and Bolesta 2000). Rats fed a high-fat diet exhibit characteristic features of insulin resistance (hyperglycemia, hypertriglyceridemia, hypercholestrolemia, and hyperinsulinemia) and severe spatial memory deficit. Peripheral rosiglitazone administration (5 mg/kg, p.o. for seven days) lowers the plasma glucose, triglycerides, cholesterol, and insulin levels and improves spatial memory in these rats (Pathan et al. 2008). Chronic rosiglitazone (5 mg/kg, p.o., for 10 weeks) has also been shown to prevent cognitive impairment in an object recognition test in transgenic AD mice (Escribano et al. 2009). These observations suggest that peripheral insulin sensitization is sufficient to improve learning and memory, since the compound does not cross the intact BBB (Pederden et al. 2006). One possibility is that the drug acts on the down-regulation of insulin transport associated with insulin resistance and chronic peripheral hyperinsulinemia (Baura et al. 1996). Pioglitazone treatment has also been reported to restore PKC signaling from deficiency (Yamagishi et al. 2008). PKC signaling appears to be involved in insulin resistance but reports differ in which direction. There is a report that PKCθ-dependent phosphorylation of phosphoinositide-dependent protein kinase-1 (PDK1) contributes to palmitate-induced insulin resistance (Wang et al. 2009). On the other hand, atypical protein kinase Cλdeficiency has been shown to be sufficient to induce systemic insulin resistance, reduced glucose tolerance, abdominal obesity and dyslipidemia in mice (Farese et al. 2007). The impact of glucocorticoids is also worth careful consideration. Glucocorticoid exposure reduces insulin action, probably through inhibition of IR binding and alteration of several components of the insulin signaling pathway, resulting in insulin resistance. Glucocorticoids inhibit insulysin activity, its expression in the hippocampus (Kulstad et al. 2005), and binding to insulin.

At the core of cognitive impairment and therapy is the ability of the neural network to support the synaptogenesis and synaptic strength changes required by memory demands. PKC and insulin neurotrophic signaling insufficiency, increased Aβ oligomers and GSK activity all exert potent activity of anti-synaptogesis and damage the neural ability to maintain synaptic strength. Restoration of the cerebral Aβ balance is the therapeutic goal in AD treatment and prevention and may be achieved effectively through enhancing neurotrophic and insulysin activity. People at risk for AD especially benefit from treatment for insulin resistance. Development of the agents that activate or dis-inhibit insulysin would be a viable therapeutic strategy for the treatment of AD and is worth more intensive research efforts and resources.

# 9 Concluding Remarks

An emerging view arising from the results reviewed above suggests that:

1. AD is a disease of synapses that involves the molecular dysregulation of synaptogenesis in the aging brain;
2. The amyloidogenic-pathophysiologic pathway in AD serves in a physiologic capacity to regulate synaptogenesis;
3. PKCα and ε pathways play central roles in synaptogenesis in development, learning and memory in the mature nervous system, in protecting against the degenerating nervous system, and neurorepair; and
4. insulin signaling pathways also regulate synaptogenesis in physiologic and pathophysiologic conditions that contribute to AD.

According to this view, PKC signaling regulates normal and abnormal synaptogenesis, particularly the abnormalities that occur in AD. Similarly, insulin signaling regulates normal and abnormal glucose utilization, particularly as the abnormality occurs in diabetes. Because these two phenomena, synaptogenesis and glucose utilization, involve important interaction and interdependence in brain metabolism, it is not surprising that AD and diabetes appear to show mutual vulnerabilities. This emerging view is also consistent with new therapeutic strategies that facilitate synaptogenesis by means of PKC activation as well as optimized insulin signaling that cannot involve only exogenous insulin itself.

# References

Aberg MA, Aberg ND, Hedbacker H, Oscarsson J, Eriksson PS (2000) Peripheral infusion of IGF-1 selectively induces neurogenesis in the adult rat hippocampus. J Neurosci 20:2896–2903

Akanmu MA, Nwabudike N, Ilesanmi OR (2009) Analgesic, learning and memory and anxiolytic effects of insulin in mice. Beh Brain Res 196:237–241

Ali MA, Li Q, Fischer ER, Cohen JI (2009) The insulin degrading enzyme binding domain of varicella-zoster virus (VZV) glycoprotein E is important for cell-to-cell spread and VZV infectivity, while a glycoprotein I binding domain is essential for infection Virology 386:270–279

Alkon DL, Sun M-K, Nelson TJ (2007) PKC signaling deficits: a mechanistic hypothesis for the origins of Alzheimer's disease. Trends Pharmacol Sci 28:51–60

Araki W, Kume H, Oda A, Tamaoka A, Kametani F (2009) IGF-1 promotes β-amyloid production by a secretase-independent mechanism. Biochem Biophys Res Commun 380:111–114

Arvanitakis Z, Wilson RS, Bienias JL, Evans DA, Bennett DA (2004) Diabetes mellitus and risk of Alzheimer disease and decline in cognitive function. Arch Neurol 61:661–666

Babri S, Badie HG, Khamenei S, Seyedlar MO (2007) Intrahippocampal insulin improves memory in a passive-avoidance task in male wistar rats. Brain Cogn 64:86–91

Bank B, DeWeer A, Kuzirian AM, Rasmussen H, Alkon DL (1988) Classical conditioning induces long-term translocation of protein kinase C in rabbit hippocampal CA1 cells. Proc Natl Acad Sci USA 85:1988–1992

Banks WA, Jaspan JB, Kastin AJ (1997) Selective physiological transport of insulin across blood-brain barrier: novel demonstration by species specific radioimmunoassay. Peptides 28:1257–1262

Banks WA, Kastin AJ (1998) Differential permeability of the blood-brain barrier to two pancreatic peptides: insulin and amylin Peptides 19:883–889

Baura GD, Foster DM, Kaiyala K, Porte D Jr, Kahn SE, Schwartz MW (1996) Insulin transport from plasma into the central nervous system is inhibited by dexamethasone in dogs. Diabetes 45:86–90

Bell RD, Sagare AP, Friedman AE, Bedi GS, Holtzman DM, Deane R, Zlokovic BV (2007a) Transport pathways for clearance of human Alzheimer's amyloid β-peptide and apolipoproteins E and J in the mouse central nervous system. J Cereb Blood Flow Metab 27:909–918

Biessels GJ, Kamal A, Urban IJA, Spruijt BM, Erkelens DW, Gispen WH (1998b) Water maze learning and hippocampal synaptic plasticity in streptozotocin-diabetic rats: effects of insulin treatment. Brain Res 800:125–135

Biessels GJ, Gispen WH (2005) The impact of diabetes on cognition: what can be learned from rodent models? Neurobiol Aging 26 (Suppl 1):36–41

Bondy CA, Lee WH (1993) Patterns of insulin-like growth factor and IGF receptor gene expression in the brain Functional implications. Ann NY Acad Sci 692:33–43

Brown CE, Li P, Boyd JD, Delaney KR, Murphy TH (2007) Extensive turnover of dendritic spines and vascular remodeling in cortical tissues recovering from stroke. J Neurosci 27: 4101–4109

Buxbaum JD, Koo EH, Greengard P (1993) Protein phosphorylation inhibits production of Alzheimer amyloid β/A4 peptide. Proc Natl Acad Sci USA 90:9195–9198

Caccamo A, Oddo S, Sugarman MC, Akbari Y, LaFerla FM (2005) Age- and region-dependent alterations in Aβ-degrading enzyme: implications for Aβ-induced disorders. Neurobiol Aging 26:645–654

Caillé I, Allinquant B, Dupont E, Bouillot C, Langer A, Müller U, Prochiantz A (2004) Soluble form of amyloid precursor protein regulates proliferation of progenitors in the adult subventricular zone. Development 131:2173–2181

Carro E, Trejo JL, Busiguina S, Torres-Aleman S (2001) Circulating insulin-like growth factor 1 mediates the protective effects of physical exercise against brain insults of different etiology and anatomy. J Neurosci 21:5678–5684

Carro E, Trejo JL, Spuch C, Bohl D, Heard JM, Torres-Aleman I. (2006) Blockade of the insulin-like growth factor-1 receptor in the choroids plexus originates Alzheimer's-like neuropathology in rodents: new cues into the human disease? Neurobiol Aging 27:16279–16284

Chang X, Jorgensen AMM, Bardrum P, Led JJ (1997) Solution structures of the r6 human insulin hexamer. Biochemistry 36:9409–9422

Choi DS, Wang D, Yu GQ, Zhu G, Kharazia VN, Paredes JP, Chang WS, Deitchman JK, Mucke L, Messing RO (2006) PKCε increases endothelin converting enzyme activity and reduces amyloid plaque pathology in transgenic mice. Proc Natl Acad Sci USA 103:8215–8220

Chubykin AA, Atasoy D, Etherton MR, Brose N, Kavalali ET, Gibson JR, Sudhof TC (2007) Activity-dependent validation of excitatory versus inhibitory synapses by neuroligin-1 versus neuroligin-2. Neuron 54:919

Cleary JP, Walsh DM, Hofmeister JJ, Shankar GM, Kuskowski MA, Selkoe DJ, Ashe KH (2005) Natural oligomers of the amyloid- protein specifically disrupt cognitive function. Nature Neurosci 8:79–84

Convit A, Wolf OT, Tarshish C, de Leon MJ (2003) Reduced glucose tolerance is associated with poor memory performance and hippocampal atrophy among normal elderly. Proc Natl Acad Sci USA 100:2019–2022

Cook DG, Leverenz JB, McMillan PJ, Kulstad JJ, Ericksen S, Roth RA, Schellenberg GD, Jin LW, Kovacina KS, Craft S (2003) Reduced hippocampal insulin-degrading enzyme in late-onset Alzheimer's disease is associated with the apolipoprotein E- ε4 allele. Am J Pathol 162:313–319

Craft S, Peskind E, Schwartz MW, Schellenberg GD, Raskind M, Porte D Jr (1998) Cerebrospinal fluid and plasma insulin levels in Alzheimer's disease: relationship to severity of dementia and apolipoprotein E genotype Neurology 50:164–168

Craft S, Asthana S, Cook DG , Baker LD, Cherrier M, Purganan K, Wait C, Petrova A, Latendresse S, Watson GS, Newcomer JW, Schellenberg GD, Krohn AJ (2003) Insulin dose response effects on memory and amyloid precursor protein in Alzheimer's disease: interactions with apolipoprotein E genotype. Psychoneuroendocrinology 28:809–822

Datta SR, Dudek H, Tao X, Masters S, Fu H, Gotoh Y, Greenberg ME (1997) Akt phosphorylation of BAD couples survival signals to the cell-intrinsic death machinery. Cell 91:231–241

Dawson GR, Seabrook G, Zheng H, Smith DW, Graham S, ODowd G, Bowery BJ, Boyce S, Trumbauer ME, Chen HY, Van der Ploeg LH, Sirinathsinghji DG (1999) Age-related cognitive deficits, impaired long-term potentiation and reduction in synaptic marker density in mice lacking the β-amyloid precursor protein. Neuroscience 90:1–13

De Angelis Lobo d'Avila K, Gadonski G, Fang J, Dall'Ago P, Albuquerque VL, Peixoto LR, Fernandes TG, Irigoyen MC (1999) Exercise reverses peripheral insulin resistance in trained L-NAME-hypertensive rats. Hypertension 34:768–772

De Castro JM, Balagura S (1976) Insulin pretreatemnt facilitates recovery after dorsal hippocampal lesions. Physiol Behav 16:517–520

De Felice FG, Vieira MN, Bomfim TR, Decker H, Velasco PT, Lambert MP, Viola KL, Zhao WQ, Ferreira ST, Klein WL (2009) Protection of synapses against Alzheimer's-linked toxins: insulin signaling prevents the pathogenic binding of Aβ oligomers. Proc Natl Acad Sci USA 106:1971–1976

de la Monte S, Tong M, Lester-Coll N, Plater M Jr, Wands JR (2006) Therapeutic rescue of neurodegeneration in experimental type 3 diabetes: Relevance to Alzheimer's disease. J Alzheimer's Dis 10:89–109

De Roo M, Klauser P, Mendez P, Poglia L, Muller D (2008) Activity-dependent PSD formation and stabilization of newly formed spines in hippocampal slice cultures. Cereb Cortex 18:151–161

De Santi S, de Leon, MJ, Rusinek H, Convit A, Tarshish CY, Roche A, Tsui WH, Kandil E, Boppana M, Daisley K, Wang GJ, Schlyer D, Fowler J (2001) Hippocampal formation glucose metabolism and volume losses in MCI and Alzheimer's disease. Neurobiol Aging 22:529–539

Dean C, Dresbach T (2006) Neuroligins and neurexins: linking cell adhesion, synapse formation and cognitive function. Trends Neurosci 9:21–29

Dean C, Scholl FG, Choih J, DeMaria S, Berger J, Isacoff E, Scheiffele P (2003) Neurexin mediates the assembly of presynaptic terminals. Nature Neurosci 6:708–716

del Peso L, González-García M, Page C, Herrera R, Nuñez G (1997) Interleukin-3-induced phosphorylation of BAD through the protein kinase Akt. Science 278:687–689

den Heijer T, Vermeer SE, van Dijk EJ, Prins ND, Koudstaal PJ, Hofman A, Breteler MM (2003) Type 2 diabetes and atrophy of medial temporal lobe structures on brain MRI. Diabetologia 46:1604–1610

Desrocher M, Rovet J (2004) Neurocognitive correlates of type 1 diabetes mellitus in childhood. Child Neuropsychol 10:36–52

Dore S, Kar S, Quiron R (1997) Insulin-like growth factor 1 protects and rescues hippocampal neurons against β-amyloid- and human amylin-induced toxicity. Proc Natl Acad Sci USA 94:4772–4777

Dou JT, Chen M, Dufour F, Alkon DL, Zhao WQ (2006) Insulin receptor signaling in long-term memory consolidation following spatial learning. Learn Mem 12:646–655
Elias PK, Elias MF, D'Agostino RB, Cupples LA, Wilson PW, Silbershatz H, Wolf PA (1997) NIDDM and blood pressure as risk factor for poor cognitive performance: the Framingham Study. Diabetes Care 20:1388–1395
Engert F, Bonhoeffer T (1999) Dendritic spine changes associated with hippocampal long-term synaptic plasticity. Nature 399:66–70
Escribano L, Simón AM, Pérez-Mediavilla A, Salazar-Colocho P, Del Río J, Frechilla D (2009) Rosiglitazone reverses memory decline and hippocampal glucocorticoid receptor down-regulation in an Alzheimer's disease mouse model. Biochem Biophys Res Comm 379:406–410
Etcheberrigaray R, Tan M, Dewachter I, Kuipri C, Van der Auwera I, Wera S, Qiao L, Bank B, Nelson TJ, Kozikowski AP, Van Leuven F, Alkon DL (2004) Therapeutic effects of PKC activators in Alzheimer's disease transgenic mice. Proc Natl Acad Sci USA 101:11141–11146
Fabrichny IP, Leone P, Sulzenbacher G, Comoletti D, Miller MT, Taylor P, Bourne Y, Marchot P (2007) Structural analysis of the synaptic protein neuroligin and its β-neurexin complex: determinants for folding and cell adhesion. Neuron 56:979–991
Farese RV, Sajan MP, Yang H, Li P, Mastorides S, Gower WR Jr, Nimal S, Choi CS, Kim S, Shulman GI, Kahn CR, Braun U (2007) Muscle-specific knockout of PKC-lambda impairs glucose transport and induces metabolic and diabetic syndromes. J Clin Invest 117:2289–2301
Farris W, Mansourian S, Leissring MA, Eckman EA, Bertram L, Eckman CB, Tanzi RE, Selkoe DJ (2004) Partial loss-of-function mutations in insulin-degrading enzyme that induce diabetes also impair degradation of amyloid β-protein. Am J Pathol 164:1425–1434
Favit A, Grimaldi M, Nelson TJ, Alkon DL (1998) Alzheimer's-specific effects of soluble β-amyloid on protein kinase C-α and -γ degradation in human fibroblasts. Proc Natl Acad Sci USA 95:5562–5567
Fernando RN, Albiston AL, Chai SY (2008) The insulin-regulated aminopeptidase IRAP is colocalised with GLUT4 in the mouse hippocampus – potential role in modulation of glucose uptake in neurons? Eur J Neurosci 28:588–598
Feyt C, Pierrot N, Tasiaux B, Van Hees J, Kienlen-Campard P, Courtoy PJ, Octave JN (2007) Phosphorylation of APP695 at Thr668 decreases γ-cleavage and extracellular Aβ. Biochem Biophys Res Commun 357:1004–1010
Figueroa DJ, Shi XP, Gardell SJ, Austin CP (2001) AβPP secretases are co-expressed with AβPP in the pancreatic islets. J Alzheimers Dis 4:393–396
Fischer M, Kaech S, Knutti D, Matus A (1998) Rapid actin-based plasticity in dendritic spines. Neuron 20:847–854
Frölich L, Blum-Degen D, Bernstein HG, Engelsberger S, Humrich J, Laufer S, Muschner D, Thalheimer A, Türk A, Hoyer S, Zöchling R, Boissl KW, Jellinger K, Riederer P (1998) Brain insulin and insulin receptors in aging and sporadic Alzheimer's disease. J Neural Transm 105:423–438
Gasparini L, Gouras GK, Wang R Gross RS, Beal MF, Greengard P, Xu H (2001) Stimulation of β-amyloid precursor protein trafficking by insulin reduces intraneuronal β-amyloid and required mitogen-activated protein kinase signaling. J Neurosci 21:2561–2570
Gasparini L, Netzer WJ, Greengard P, Xu H (2002) Does insulin dysfunction play a role in Alzheimer's disease? Trends Pharmacol Sci 23:288–293
Gispen WH, Biessels GJ (2000) Cognition and synaptic plasticity in diabetes mellitus. Trends Neurosci 23:542–549
Gould E, Carmeron HA, Daniels DC, Woolley CS, McEwen BS (1992) Adrenal hormones suppress cell division in the adult rat dentate gyrus. J Neurosci 12:3642–3650
Gralle M, Ferreira ST (2007) Structure and functions of the human amyloid precursor protein: the whole is more than the sum of its parts. Prog Neurobiol 82:11–32
Grunblatt E, Salkovic-Petrisic M, Osmanovic J, Riederer P, Hoyer S (2007) Brain insulin system dysfunction in streptozocin intracerebroventricularly treated rats generates hyperphosphorylated tau protein. J Neurochem 101:757–770

Hallschmid M, Benedict C, Born J, Kern W (2007) Targeting metabolic and cognitive pathways of the CNS by intranasal insulin administration. Expert Opin Drug Deliv 4:319–322,
Hama H, Hara C, Yamaguchi K, Miyawaki A (2004) PKC signaling mediates global enhancement of excitatory synaptogenesis in neurons triggered by local contact with astrocytes Neuron 41:405–415
Hammarlund M, Nix P, Hauth L, Jorgensen EM, Bastiani M (2009) Axon regeneration requires a conserved MAP kinase pathway. Science 323:802–806
Hanson LR, Frey II WH (2008) Intranasal delivery bypass the blood-brain barrier to target therapeutic agents to the central nervous system and treat neurodegenerative disease. EMC Neurosci 9 (Suppl 3):S5
Havrankova J, Roth J, Brownstein M (1978) Insulin receptors are widely distributedin the central nervous system of the rat. Nature. 272:827–829
Helkala EL, Niskanen L, Viinamaki H, Partanen J, Uusitupa M (1995) Short-term and long-term memory in elderly patients with NIDDM. Diabetes Care 18:681–685
Hevener AL, Reichart D, Olefsky J (2000) Exercise and thiazolidinedione therapy normalize insulin action in the obese Zucker fatty rat. Diabetes 49:2154–2159
Ho L, Qin W, Pompl PN, Xiang Z, Wang J, Zhao Z, Peng Y, Cambareri G, Rocher A, Mobbs CV, Hof PR, Pasinetti GM (2004) Diet-induced insulin resistance promotes amyloidosis in a transgenic mouse model of Alzheimer's disease. FASEB J 18:902–904
Hodge RD, D'Ercole AJ, O'Kusky JR (2004) Insulin-like growth factor-1 accelerates the cell cycle by decreasing G1 phase length and increases cell cycle reentry in the embryonic cerebral cortex. J Neurosci 24:10201–10210
Holtmaat A, Wilbrecht L, Knott GW, Welker E, Svoboda K (2006) Experience-dependent and cell-type-specific spine growth in the neocortex. Nature 441:979–983
Hong H, Liu LP, Liao JM, Wang TS, Ye FY, Wu J, Wang YY, Wang Y, Li YQ, Long Y, Xia YZ (2007) Downregulation of LPR1 at the blood-brain barrier in streptozotocin-induced diabetic mice. Neuropharmacology 56:1054–1059
Hongpaisan J, Alkon DL (2007) A structural basis for enhancement of long-term associative memory in single dendritic spines regulated by PKC. Proc Natl Acad Sci USA 104: 19571–19576
Hori K, Yasuda H, Konno D, Maruoka H, Tsumoto T, Sobue K (2005) NMDA receptor-dependent synaptic translocation of insulin receptor substrate p53 via protein kinase C signaling. J Neurosci 25:2670–2681
Hori Y, Gu X, Xie X, Kim SK (2005) Differentiation of insulin-producing cells from human neural progenitor cells. PLoS Med 2:e103
Hsieh J, Aimone JB, Kasper BK, Kuwabara T, Nakashima K, Gage FH 2004) IGF-1 instructs multipotent adult progenitor cells to become oligodendrocytes. J Cell Biol 164:111–122
Ikin AF, Sabo SL, Lanier LM, Buxbaum JD (2007) A macromolecular complex involving the amyloid precursor protein (APP) and the cytosolic adapter FE65 is a negative regulator of axon branching. Mol Cell Neurosci 35:57–63
Itzhaki RF, Wozniak MA (2008) Herpes simplex virus type 1 in Alzheimer's disease: the enemy within. J Alzheimers Dis 13:393–405
Jaeger S, Pietrzik CU (2008) Functional role of lipoprotein receptors in Alzheimer's disease. Curr Alzheimer Res 5:15–25
Janson J, Laedtke T, Patisi JE, O'Brien P, Petersen RC, Butler PC (2004) Increased risk of type 2 diabetes in Alzheimer disease. Diabetes 53:474–481
Jourdain P, Fukunaga K, Muller D (2003) Calcium/calmodulin-dependent protein kinase II contributes to activity-dependent filopodia growth and spine formation. J Neurosci 23:10645–10649
Kamal A, Biessels GJ, Gispen WH, Ramakers GM (2006) Synaptic transmission changes in the pyramidal cells of the hippocampus in streptozotocin-induced diabetes mellitus in rats. Brain Res 1073-1074:276–280
Klesse LJ, Parada LF (1998) p21 ras and phosphatidylinositol-3 kinase are required for survival of wild-type and NF1 mutant sensory neurons. J Neurosci 18:10420–10428

Knight S Northam E, Donath S, Gardner A, Harkin N, Taplin C, Joy P, Cameron FJ, Ambler GR (2009) Improvements in cognition, mood and behaviour following commencement of continuous subcutaneous insulin infusion therapy in children with type 1 diabetes mellitus: a pilot study. Diabetologia 52:193–198

Knott GW, Holtmaat A, Wilbrecht L, Welker E, Svoboda K (2006) Spine growth precedes synapse formation in the adult neocortex in vivo. Nature Neurosci 9:1117–1124

Kong W-J, Zhang H, Song DQ, Xue R, Zhao W, Wei J, Wang YM, Shan N, Zhou ZX,Yang P, You XF, Li ZR, Si SY, Zhao LX, Pan HN, Jiang JD (2009) Berberine reduces insulin resistance through protein kinase C-dependent up-regulation of insulin receptor expression. Metabolism 58:109–119

Koo EH, Squazzo SL (1994) Evidence that production and release of amyloid β-protein involves the endocytic pathway. J Biol Chem 269:17386–17389

Kuhad A, Bishnoi M, Tiwari V, Chopra K (2009) Suppresion of NF-κβ signaling pathway by tocotrienol can prevent diabetes associated cognitive deficits. Pharmacol Biochem Behav 92:251–259

Kulstad JJ, McMillan PJ, Leverenz JB, Cook DG, Green PS, Peskind ER, Wilkinson CW, Farris W, Mehta PD, Craft S (2005) Effects of chronic glucocorticoid administration on insulin-degrading enzyme and amyloid-β peptide in the aged Macaque. J Neuropathol Exp Neurol 64:139–146

Kurada P, White K (1998) Ras promotes cell survival in Drosophila by downregulating hid expression. Cell 95:319–329

Lang BT, Yan Y, Dempsey RJ, Vemuganti R (2008) Impaired neurogenesis in adult type-2 diabetic rats. Brain Res 1258:25–33

Larsson C (2006) Protein kinase C and the regulation of the actin cytoskeleton. Cell Signal 18:276–284

Launer LJ (2005) Diabetes and brain aging: Epidemiologic evidence. Curr Diabetes Rep 5:59–63

Lee MS, Kao SC, Lemere CA, Xia W, Tseng HC, Zhou Y, Neve R, Ahlijanian MK, Tsai LH (2003) APP processing is regulated by cytoplasmic phosphorylation. J Cell Biol 163:83–95

Lee W, Boo JH, Jung MW, Park SD, Kim YH, Kim SU, Mook-Jung I (2004) Amyloid beta peptide directly inhibits PKC activation. Mol Cell Neurosci 26:222–231

Leissring M, Farris W, Chang AY, Walsh DM, Wu X, Sun X, Frosch MP, Selkoe DJ (2003) Enhanced proteolysis of β-amyloid in APP transgenic mice prevents plaque formation, secondary pathology, and premature death. Neuron 40:1087–1093

Leyssen M, Ayaz D, Hebert SS, Reeve S, De Strooper B, Hassan BA (2005) Amyloid precursor protein promotes post-developmental neurite arborization in the Drosophila brain. EMBO J 24:2944–2955

Li Q, Ali MA, Cohen JI (2006) Insulin degrading enzyme is a cellular receptor mediating varicella-zoster virus infection and cell-to-cell spread. Cell 127:305–316

Luchsinger JA, Tang M-X, Shea A, Mayeux R (2004) Hyperinsulinemia and risk of Alzheimer's disease. Neurology 63:1187–1192

Luchsinger JA, Reitz C, Honig LS, Tang M-X, Shea S, Mayeux R (2005) Aggregation of vascular risk factors and risk of incident Alzheimer's disease. Neurology 65:545–551

Lukiw WJ, Bazan NG (2006) Survival signalling in Alzheimer's disease. Biochem Soc Trans 34:1277–1282

Ma X-H, Zhong P, Gu Z, Feng J, Yan Z (2003) Muscarinic potentiation of GABAA receptor currents is gated by insulin signaling in the prefrontal cortex. J Neurosci 23:1159–1168

Magariños AM, McEwen BS (2000) Experimental diabetes in rats causes hippocampal dendritic and synaptic reorganization and increased glucocortocoid reactive stress. Proc Natl Acad Sci USA 97:11056–11061

Mairet-Coello G, Tury A, DiCicco-Bloom E (2009) Insulin-like growth factor-1 promotes G1/S cell cycle progression through bi-directional regulation of cyclins and cyclin-dependent kinase inhibitors via the phosphatidylinositol 3-kinase/Akt pathway in developing rat cerebral cortex. J Neurosci 29:775–788

Malinowski J, Bolesta S (2000) Rosiglitazone in the treatment of type 2 diabetes mellitus: a critical review. Clin Ther 22:1151–1168

Maniar R, Pecherskaya A, Ila R, Solem M (2005) PKC α-dependent regulation of IGF1 receptor in adult and embryonic rat cardiomyocytes. Mol Cell Biochem 275:15–24

Martin-Pena A, Acebes A, Rodríguez JR, Sorribes A, de Polavieja GG, Fernández-Fúnez P, Ferrús A (2006) Age-independent synaptogenesis by phosphoinositide 3 kinase. J Neurosci 26:10199–10208

Martinez-Tellez R, Goemz-Villalobos Mde J, Flores G (2005) Alteration in dendritic morphology of cortical neurons in rats with diabetes mellitus induced by streptozotocin. Brain Res 1048:108–115

Marzban L, Park K, Verchere CB (2003) Islet amyloid polypeptide and type 2 diabetes. Exp Gerontol 38:347–51

Mayer-Davis E, D'Agostino R, Karter A, Haffner SM, Rewers MJ, Saad M, Bergman RN (1998) Intensity and amount of physical activity in relation to insulin sensitivity: the insulin resistance atherosclerosis study. JAMA 279:669–74

McLoughlin DM, Miller CC (2008) The FE65 proteins and Alzheimer's disease. J Neurosci Res 86:744–754

Meier J, Akyeli J, Kirischuk S, Grantyn R (2003) GABAA receptor activity and PKC control inhibitory synaptogenesis in CNS tissue slices. Mol Cell Neurosci 23:600–613

Messier C (2005) Impact of impaired glucose tolerance and type 2 diabetes on cognitive aging. Neurobiol Aging 26 (Suppl 1):S26–S30

Meziane H, Dodart JC, Mathis C Little S, Clemens J, Paul SM, Ungerer A (1998) Memory-enhancing effects of secreted forms of the β-amyloid precursor protein in normal and amnestic mice. Proc Natl Acad Sci USA 95:12683–12688

Miklossy J, Qing H, Radenovic A, Kis A, Vileno B, Lszl F, Miller L, Martins RN, Waeber G, Mooser V, Bosman F, Khalili K, Darbinian-Sarlissian N, McGeer PL (2008) β amyloid and hyperphosphorylated tau deposits in the pancreas in type 2 diabetes. Neurobiol Aging doi:10.1016/j.neurobiolaging.2008.08.019

Miller BC, Eckman EA, Sambamurti K, Dobbs N, Chow KM, Eckman CB, Hersh LB,Thiele DL (2003) Amyloid-beta peptide levels in brain are inversely correlated with insulysin activity levels in vivo. Proc Natl Acad Sci USA 100:6221–6226

Montaron MF, Drapeau E, Dupret D, Kitchener P, Aurousseau C, Le Moal M, Piazza PV, Abrous DN (2006) Lifelong corticosterone level determines age-related decline in neurogenesis and memory. Neurobiol Aging 27:645–654

Moosavi M, Naghdi N, Choopani S (2007a) Intra CA1 insulin microinjection improves memory consolidation and retrieval. Peptides 28:1029–1034

Moosavi M, Naghdi N, Maghsoudi N, Asl SZ (2007b) Insulin protects against stress-induced impairments in water maze performance. Behav Brain Res 176:230–236

Mukherjee A, Song E, Kihiko-Ehmann M, Goodman JP Jr, Pyrek JS, Estus S, Hersh LB. (2000) Insulysin hydrolyzes amyloid β peptides to products that are neither neurotoxic nor deposit on amyloid plaques. J Neurosci 20:8745–8749

Mundy WR, Robinette B, Radio NM, Freudenrich TM (2008) Protein biomarkers associated with growth and synaptogenesis in a cell culture model of neuronal development. Toxicology 249:220–229

Nazer B, Hong S, Selkoe DJ (2008) LRP promotes endocytosis and degradation, but not transtcytosis, of the amyloid-β peptide in a blood-brain barrier in vitro model. Neurobiol Dis 30:94–102

Nelson TJ, Alkon DL (2009) Neuroprotective versus tumorigenic protein kinase C activators. Trends Biochem Sci 34: 136-45

Nelson TJ, Sun M-K, Hongpaisan J, Alkon DL (2008) Insulin, PKC signaling pathways and synaptic remodeling during memory storage and neuronal repair. Eur J Pharmacol 585:76–87

Okereke OI, Selkoe DJ, Pollak MN, Stampfer MJ, Hu FB, Hankinson SE, Grodstein F (2008) A profile of impaired insulin degradation in relation to late-life cognitive decline: a preliminary investigation Int J Geriat Psychiat 24:178–182

O'Kusky JR, Ye P, D'Ercole AJ (2000) Insulin-like growth factor-1 promotes neurogenesis and synaptogenesis in the hippocampal dendrite gyrus during postnatal development. J Neurosci 20:8435–8442

Pardridge WM, Kang YS, Buciak JL, Yang J (1995) Human insulin receptormonoclonal antibody undergoes high affinity binding to human brain capillaries in vitro and rapid transcytosis through the blood-brain barrier in vivo in the primate. Pharm Res 12:807–816

Park CR (2001) Cognitive effects of insulin in the central nervous system. Neurosci Biobeh Res 25:311–323

Park CR, Seeley RJ, Craft S, Seeley RJ, Craft S, Woods SC (2000) Intracerebroventricular insulin enhances memory in a passive-avoidance task. Physiol Behav 68:509–514

Pascale A, Amadio M, Scapagnini G, Lanni C, Racchi M, Provenzani A, Govoni S, Alkon DL, Quattrone A (2005) Neuronal ELAV proteins enhance mRNA stability by a PKCα-dependent pathway. Proc Natl Acad Sci USA 102:12065–12070

Pathan AR, Gaikwad AB, Viswanad B, Ramarao P (2008) Rosiglitazone attenuates the cognitive deficits induced by high fat diet feeding in rats. Eur J Pharmacol 589:176–179

Pedersen WA, McMillan PJ, Kulstad JJ, Leverenz JB, Craft S, Haynatzki GR (2006) Rosiglitazone attenyates learning and memory deficits in Tg2576 Alzheimer mice. Exp Neurol 199:265–273

Peila R, Rodriguez BL, Launer LJ (2002) Type 2 diabetes, APOE gene, and the risk for dementia and related pathologies: the Honolulu-Asia aging study. Diabetes 51:1256–1262

Peila R, Rodriguez BL, White LR, Lauer LJ (2004) Fasting insulin and incident dementia in an elderly population of Japanese-American men. Neurology 63:228–233

Perez A, Morelli L, Cresto JC, Castano EM (2000) Degradation of soluble amyloid β-peptide 1-40, 1-42, and the Dutch variant 1-40Q by insulin degrading enzyme from Alzheimer's disease and control brains. Neurochem Res 25:247–255

Perlmurter LC, Kakami MK, Hodgson-Harrington C, Ginsberg J, Katz J, Singer DE, Nathan MD (1984) Decreased cognitive function in aging non-insulin-dependent diabetic patients. Am J Med 77:1043–1048

Petrovitch H, White LR, Izmirlian G, Ross GW, Havlik RJ, Markesbery W, Nelson J, Davis DG, Hardman J, Foley DJ, Launer LJ (2000) Midlife blood pressure and neuritic plaques, neurofibrillary tangles, and brain weight at death: the HAAS. Neurobiol Aging 21:57–62

Phiel CJ, Wilson CA, Lee VM, Klein PS (2003) GSK-3α regulates production of Alzheimer's disease amyloid-β peptides. Nature 423:435–439

Piroli GG, Grillo CA, Reznikov LR, Adams S, McEwen BS, Charron MJ, Reagan LP (2007) Corticosterone impairs insulin-stimulated translocation of GLUT4 in the rat hippocampus. Neuroendocrinology 85:71–80

Popken GJ, Hodge RD, Ye P, Zhang J, Ng W, O'Kusky JR, D'Ercole AJ (2004) In vivo effects of insulin-like growth factor-1 (IGF-1) on prenatal and early postnatal development of the central nervous system. Eur J Neurosci 19:2056–2068

Puzzo D, Privitera L, Leznik E, Fà M, Staniszewski A, Palmeri A, Arancio O (2008) Picomolar amyloid-b positively modulates synaptic plasticity and memory in hippocampus. J Neurosci 28:14537–14545

Qiu WQ, Walsh DM, Ye Z, Vekrellis K, Zhang J, Podlisny MB, Rosner MR, Safavi A, Hersh LB, Selkoe DJ (1998) Insulin-degrading enzyme regulates extracellular levels of amyloid β-protein by degradation J Biol Chem 273:32730–32738

Razay G, Wilcock GK (1994) Hyperinsulinemia and Alzheimer's disease. Age Ageing 23:396–399

Redila VA, Christie BR (2006) Exercise-induced changes in dendritic structure and complexity in the adult hippocampal dendritic gyrus. Neuroscience 137:1299–1307

Reger MA, Watson GS, Green PS, Baker LD, Cholerton B, Fishel MA, Plymate SR,Cherrier MM, Schellenberg GD, Frey WH 2nd, Craft S (2008) Intranasal insulin administration dose-dependently modulates verbal memory and plasma amyloid-β in memory-impaired older adults. J Alzheimer's Dis 13:323–331

Rivera EJ, Goldin A, Fulmer N, Tavares R, Wands JR, de la Monte SM (2005) Insulin and insulin-like growth factor expression and function deteriorate with progression of Alzheimer's disease: link to brain reductions in acetylcholine. J Alzheim Dis 8:247–268

Rizk NN, Rafols JA, Dunbar JC (2006) Cerebral ischemia-induced apoptosis and necrosis in normal and diabetic rats: Effects of insulin and C-peptide. Brain Res 1096:204–212

Rodriguez, G, Vitali P, Calvini P, Bordoni C, Girtler N, Taddei G, Mariani G, Nobili F, (2000) Hippocampal perfusion in mild Alzheimer's disease. Psychiat Res 100:65–74

Russo VC, Gluckman PD, Feldman EL, Werther GA (2005) The insulin-like growth factor system and its pleiotropic functions in brain. Endocrine Rev 26:916–943

Sabo SL, AF Ikin AF, Buxbaum JD, Greengard P (2001) The Alzheimer amyloid precursor protein (APP) and FE65, an APP-binding protein, regulate cell movement. J Cell Biol 153:1403–1414

Sabo SL, Ikin AF, Buxbaum JD, Greengard P (2003) The amyloid precursor protein and its regulatory protein, FE65, in growth cones and synapses in vitro and in vivo. J Neurosci 23:5407–5415

Schmidt JT (2004) Activity-driven sharpening of the retinotectal projection: the search for retrograde synaptic signaling pathways. J Neurobiol 59:114–133

Schubert M, Gautam D, Surjo D, Ueki K, Baudler S, Schubert D, Kondo T, Alber J, Galldiks N, Küstermann E, Arndt S, Jacobs AH, Krone W, Kahn CR, Brüning JC (2004) Role for neuronal insulin resistance in neurodegenerative diseases. Proc Natl Acad Sci USA 101:3100–3105

Schwartz MW, Sipols A, Kahn SE, Lattemann DF, Taborsky GJ Jr, Bergman RN, Woods SC, Porte D Jr (1990) Kinetics and specificity of insulin uptake from plasma into cerebrospinal fluid. Am J Physiol 259:E378–E383

Selkoe D (2001) Clearing the brain's amyloid cobwebs. Neuron 32:177–180

Selkoe DJ (2002) Alzheimer's disease is a synaptic failure Science 298:789–791

Shankar GM, Li S, Mehta TH, Garcia-Munoz A, Shepardson NE, Smith I, Brett FM, Farrell MA, Rowan MJ, Lemere CA, Regan CM, Walsh DM, Sabatini BL, Selkoe DJ (2008) Amyloid-β protein dimers isolated directly from Alzheimer's brains impair synaptic plasticity and memory. Nature Med 14:837–842

Shearman MS, Shinomura T, Oda T, Nishizuka Y (1991) Synaptosomal protein kinase C subspecies: A. Dynamic changes in the hippocampus and cerebellar cortex concomitant with synaptogenesis. J Neurochem 56:1255–1262

Shen Y, Joachimiak A, Rosner MR, Tang WJ (2006) Structures of human insulin-degrading enzyme reveal a new substrate recognition mechanism. Nature 443:870

Shiiki T, Ohtsuki S, Kurihara A, Naganuma H, Nishimura K, Tachikawa M, Hosoya K, Terasaki T (2004) Brain insulin impairs amyloid-β(1-40) clearance from the brain. J Neurosci 24:9632–9637

Shinall H, Song ES, Hersh LB (2005) Susceptibility of amyloid β peptide degrading enzyme to oxidative damage: a potential Alzheimer's disease spiral Biochemistry 44:15345–15350

Shughrue PJ, Acton PJ, Breese RS, Zhao WQ, Chen-Dodson E, Hepler RW, Wolfe AL, Matthews M, Heidecker GJ, Joyce JG, Villarreal SA, Kinney GG (2010) Anti-ADDL antibodies differentially block oligomer binding to hippocampal neurons. Neurobiol Aging 31:189–202

Sisodia SS, Koo EH, Hoffman PN, Perry G, Price DL (1993) Identification and transport of full-length amyloid precursor proteins in rat peripheral nervous system. J Neurosci 13: 3136–3142

Song ES, Hersh LB (2005) Insulysin: an allosteric enzyme as a target for Alzheimer's disease. J Mol Neurosci 25:210–216

Song E-S, Juliao MA, Juliao L, Hersh LB (2003) Substrate activation of insulin-degrading enzyme (insulysin). J Biol Chem 278:49789–49794

Steen E, Terry BM, Rivera EJ, Cannon JL, Neely TR, Tavares R, Xu XJ, Wands JR, de la Monte SM (2005) Impaired insulin and insulin-like growth factor expression and signaling mechanisms in Alzheimer's disease – is this type 3 diabetes? J Alzheimer's Dis 7:63–80

Stewart R, Liolitsa D (1999) Type 2 diabetes mellitus, cognitive impairment and dementia. Diabet Med 16:93–112

Strabahn AM, Khalil D, Gould E (2007) Running induces widespread structural alterations in the hippocampus and entorhinal cortex. Hippocampus 17:1017–1022

Strachan MW, Deary IJ, Ewing FM, Frier BM (1997) Is type II diabetes associated with an increased risk of cognitive dysfunction? A critical review of published studies. Diabetes Care. 20:438–445

Stranahan AM, Arumugam TV, Cutler RG, Lee K, Egan JM, Mattson MP (2008a) Diabetes impairs hippocampal function through glucocorticoid-mediated effects on new and mature neurons. Nature Neurosci 11:309–317

Stranahan AM, Norman ED, Lee K, Cutler RG, Telljohann RS, Egan JM, Mattson MP . (2008b) Diet-induced insulin resistance impairs hippocampal synaptic plasticity and cognition in middle-aged rats. Hippocampus 18:1085–1088

Su Y, Ryder J, Li B, Wu X, Fox N, Solenberg P, Brune K, Paul S, Zhou Y, Liu F, Ni B (2004) Lithium, a common drug for bipolar disorder treatment, regulates amyloid-β precursor protein processing. Biochemistry 43:6899–6908

Suckow AT, Sweet IR, Van Yserloo B, Rutledge EA, Hall TR, Waldrop M, Chessler SD (2006) Identification and characterization of a novel isoform of the vesicular γ-aminobutyric acid transporter with glucose-regulated expression in rat islets. J Mol Endocrinol 36:187–199

Suckow AT, Comoletti D, Waldrop MA, Mosedale M, Egodage S, Taylor P, Chessler SD (2008) Expression of neurexin, neuroligin, and their cytoplasmic binding partners in the pancreatic beta-cells and the involvement of neuroligin in insulin secretion. Endocrinology 149:6006–6017

Sun M-K, Alkon DL (2006) Bryostatin-1: pharmacology and therapeutic potential as a CNS drug. CNS Drug Reviews 12:1–8

Sun M-K, Hongpaisan J, Nelson TJ, Alkon DL (2008) Poststroke neuronal rescue and synaptogenesis mediated in vivo by protein kinase C in adult brains. Proc Natl Acad Sci USA 105:13620–13625

Sun M-K, Hongpaisan J, Alkon DL (2009) Postischemic PKC activation rescues retrograde and anterograde long-term memory. Proc Natl Acad Sci USA 106:14676–14680

Trejo JL, Carro E, Torres-Aleman I (2001) Circulating insulin-like growth factor 1 mediates exercise-induced increases in the number of new neurons in the adult hippocampus. J Neurosci 21:1628–1634

Ueno H, Kondo E, Yamamoto-Honda R, Tobe K, Nakamoto T, Sasaki K, Mitani K, Furusaka A, Tanaka T, Tsujimoto Y, Kadowaki T, Hirai H (2000) Association of insulin receptor substrate proteins with Bcl-2 and their effects on its phosphorylation and antiapoptotiv function. Mol Biol Cell 11:735–746

Vekrellis K, Ye Z, Qiu WQ, Ye Z, Qiu WQ, Walsh D, Hartley D, Chesneau V, Rosner MR, Selkoe DJ (2000) Neurons regulate extracellular levels of amyloid β-protein via proteolysis by insulin-degradating enzyme J Neurosci 20:1657–1665

Vanhanen M, Kuusisto J, Koivisto K, Mykkänen L, Helkala EL, Hänninen T, Riekkinen P Sr, Soininen H, Laakso M (1999) Type-2 diabetes and cognitive function in a non-demented population. Acta Neurol Scand 100:97–101

Vicario-Abejon C, Yusta-Boyo MJ, Fernandez-Moreno C, de Pablo F (2003) Locally born olfactory bulb stem cells proliferate in response to insulin-related factors and require endogenous insulin-like factor-1 for differentiation into neurons and glia. J Neurosci 23:895–906

Walker F, Kato A, Gonez LJ, Hibbs ML, Pouliot N, Levitzki A, Burgess AW (1998) Activation of the Ras/mitogen-activated protein kinase pathway by kinase-defective epidermal growth factor receptors results in cell survival but not proliferation. Mol Cell Biol 18:7192–7204

Wang C, Liu M, Riojas RA, Xin X, Gao Z, Zeng R, Wu J, Dong LQ, Liu F. (2009) Protein kinase C θ(PKCθ)-dependent phosphorylation of PDK1 at Ser504 and Ser532 contributes to palmitate-induced insulin resistance. J Biol Chem 284:2038–2044

Wang DS, Iwata N, Hama E, Saido TC, Dickson DW (2003) Oxidized neprilysin in aging and Alzheimer's disease brains. Biochem Biophys Res Commun 310:236–241

Watson GS, Craft S (2004) Modulation of memory by insulin and glucose:neuropsychological observations in Alzheimer's disease Eur J Pharmacol 490:97–113

Watson GS, Peskind ER, Asthana S, Purganan K, Wait C, Chapman D, Schwartz MW, Plymate S, Craft S (2003) Insulin increases CFS Aβ42 levels in normal older adults. Neurology 60:1899–1903

Webb D, Zhang H, Majumdar D, Horwitz AF (2007) α5 Integrin signaling regulates the formation of spines and synapses in hippocampal neurons. J Biol Chem 282:6929–6935

Wei W, Wang X, Kusiak JW (2002) Signaling events in amyloid β-peptide-induced death and insulin-like growth factor 1 protection. J Biol Chem 277:17649–17656

Weinreb O, Bar-Am O, Amit T, Chillag-Talmor O, Youdim, MBH (2004) Neuroprotection via pro-survival protein kinase C isoforms associated with Bcl-2 family members. FASEB J 18:1472–1473

Xie L, Helmerhorst E, Taddei K, Pelwright B, Van Bronswijk W, Martins R (2002) Alzheimer's amyloid peptides compete for insulin binding to the insulin receptor. J Neurosci 22: RC221 (1–5)

Yamagishi SI, Ogasawara S, Mizukami H, Yajima N, Wada R, Sugawara A, Yagihashi S (2008) Correction of protein kinase C activity and macrophage migration in peripheral nerve by pioglitazone, peroxisome proliferator activated-c-ligand, in insulin-deficient diabetic rats. J Neurochem 104:491–499

Yamamoto R, Kobayashi H, Yanagita T, Yokoo H, Kurose T, Shiraishi S, Minami S, Matsukura S, Wada A (2000) Up-regulation of cell surface insulin receptor by protein kinase C-α in adrenal chromaffin cells: involvement of transcriptional and translational events. J Neurochem 75:672–682

Yoshii A, Constantine-Paton M (2007) BDNF induces transport of PSD-95 to dendrites through PI3K-AKT signaling after NMDA receptor activation. Nature Neurosci 10:702–711

Yuste R, Denk W (1995) Dendritic spines as basic functional units of neuronal integration. Nature 375:682–684

Zhang W-J, Tan Y-F, Yue JTY, Wojtowicz JM (2008) Impairment of hippocampal neurogenesis in streptozotocin-treated diabetic rats. Acta Neurol Scand 117:205–210

Zhao L, Teter B, Morihara T, Lim GP, Ambegaokar SS, Ubeda OJ, Frautschy SA, Cole GM (2004) Insulin-degrading enzyme as a downstream target of insulin receptor signaling cascade: implications for Alzheimer's disease intervention. J Neurosci 24:11120–11126

Zhao W, Chen H, Quon MJ, Alkon DL (2004) Insulin and insulin receptors in experimental models of learning and memory. Eur J Pharmacol 490:71–81

Zhao WQ, Chen H, Xu H, Moore E, Meiri N, Quon MJ, Alkon DL (1999) Brain insulin receptors and spatial memory Correlated changes in gene expression, tyrosine phosphorylation, and signaling molecules in the hippocampus of water maze trained rats. J Biol Chem 274:34893–34902

Zhao WQ, Alkon DL (2001) Role of insulin and insulin receptor in learning and memory. Mol Cell Endocrinol 177:125–134

Zhao WQ, Chen H, Quon MJ, Alkon DL (2004) Insulin and the insulin receptor in experimental models of learning and memory. Eur J Pharmacol 490:71–81

Zhao W-Q, De Felice FG, Fernandez S, Chen H, Lambert MP, Quon MJ, Krafft GA, Klein WL. (2008) Amyloid β oligomers induce impairment of neuronal insulin receptors. FASEB J 22:246–260

Zhu X-W, Perry G, Smith MA (2005) Insulin signaling, diabetes mellitus and risk of Alzheimer's disease. J Alzheim Dis 7:81–84

Ziv NE, Smith SJ (1996) Evidence for a role of dendritic filopodia in synaptogenesis and spine formation. Neuron 17:91–102

Zou L-B, Mouri A, Iwata N, Saido TC, Wang D, Wang MW, Mizoguchi H, Noda Y, Nabeshima T (2005) Inhibition of neprilysin by infusion of thiorphan into the hippocampus causes an accumulation of amyloid β and impairment of learning and memory. J Pharmacol Exp Ther 317:334–340

# Diet, Abeta Oligomers and Defective Insulin and Neurotrophic Factor Signaling in Alzheimer's Disease

**Greg M. Cole, Qiu-Lan Ma, Fusheng Yang, Atul Deshpande, Oliver Ubeda, and Sally A. Frautschy**

**Abstract** Epidemiology suggests that the risk for developing Alzheimer's disease (AD) is reduced by the dietary omega-3 fatty acid docosahexaenoic acid (DHA) and increased by type II diabetes, corresponding to diet-induced obesity (metabolic syndrome). The brains of AD patients show evidence of insulin resistance, including insulin and neurotrophic factor signaling defects. We investigated whether the insulin signaling defects play a causal role in the loss of dendritic spines and arbor, which is associated with memory loss in AD brain. We focused on evaluating increases in cytosolic hyperphosphorylated insulin receptor substrate 1 (IRS-1) and losses of total cytosolic IRS-1, which are signatures of insulin resistance. IRS is critical as an adaptor protein, coupling insulin/trophic receptors to survival signaling via Akt (a serine threonine protein kinase). We demonstrated in cultured neurons that hyperphosphorylation of IRS-1 accompanied the Aβ oligomer (AβO)-induced loss of dendritic spines and arbor and the increases in the levels of phosphorylated tau (pTau) and activated c-Jun N-terminal kinase (JNK), a known tau kinase; these effects were antagonized by JNK-specific inhibitors as well as by omega-3 fatty acid DHA, a non-specific inhibitor of JNK. We observed IRS-1 signaling defects in three trangenic models of AD. In a Tg model of severe neuron loss, the 5x FAD APP/PS1 mice (Robert Vassar, Northwestern University, Chicago, IL) showed a loss of nuclear and stronger granular perikaryal and dendritic inclusion pIRS-1 in neurons from the cortex and CA1 region of the hippocampus by six months, mirroring changes seen in the AD brain. In the cortex of 22-month-old APPsw Tg2576 mice, total IRS-1 levels were reduced, consistent with its degradation by Aβ and the loss of insulin-like signaling. In the 3xTg-AD model, mice were placed on a high fat (21%) omega-3-depleting "diabetogenic" diet for four months, and they showed elevated phospho-IRS-1 (Ser 616) and

G.M. Cole (✉)
Greater Los Angeles Veteran's Administration Healthcare, Geriatric Research Education and Clinical Core

S. Craft and Y. Christen (eds.), *Diabetes, Insulin and Alzheimer's Disease*, Research and Perspectives in Alzheimer's Disease,
DOI 10.1007/978-3-642-04300-0_11, © Springer-Verlag Berlin Heidelberg 2010

JNK-sensitive phospho-tau (Ser 422) in the hippocampus. Four months of dietary treatment with fish oil or curcumin corrected hyperphosphorlation of JNK, IRS-1 and tau, but the combination of fish oil and curcumin was most effective at restoring cognitive defects (in Y maze). Thus, treatment with fish oil/DHA, curcumin or a combination, which both limit amyloid accumulation and/or JNK activation, has the potential to improve insulin/trophic signaling and downstream synaptic and cognitive deficits that cause AD. Consistent with these observations, several small clinical trials suggest that omega-3 supplements can arrest the progression of AD at early stages.

## 1 Introduction

Diet and exercise are modifiable risk factors that influence obesity and metabolic syndrome, contributing to insulin-resistant diabetes, a risk factor for Alzheimer's disease (AD). AD is the most common cause of cognitive decline leading to dementia in the aged. Diabetes is predicted to at least double the risk of developing dementia (Peila et al. 2002) and AD (Ott et al. 1999), independent of whether patients carry the major genetic risk factor for AD (ApoE4; Profenno and Faraone 2008); the risk increases to 4.5-fold for those carrying ApoE4 (Irie et al. 2008). In this review, we will elucidate how defects in insulin signaling pathways contribute to AD pathogenesis and whether dietary measures influencing these pathways may reduce AD risk. Here we illustrate the complex interactions between insulin signaling and AD pathogenesis (Fig. 1); it is important to summarize the relevant details of AD pathogenesis.

## 2 AD Pathogenesis

In AD brains, cognitive loss correlates with selective neuron and synapse loss. The pathological hallmark of AD is the accumulation of a 40-42 amino acid amyloid beta protein (Aβ) in fibrillar extracellular plaques and vascular deposits and the intraneuronal accumulations of the microtubule-associated protein tau in neurofibrillary tangles (NFTs; Cummings and Cole 2002). Aβ has been implicated in initiating AD by the discovery of multiple pathogenic mutations in autosomal dominant, early onset forms of the disease. These mutations are in the source of the Aβ (β-amyloid precursor or protein, APP) or in presenilin, a component of the g-secretase enzyme γ-secretase enzyme complex that cuts Aβ out of APP. Mutations that cause AD increase the production and/or aggregation of Aβ, suggesting that these Aβ aggregates initiate AD. Much of the bulk Aβ accumulation in extracellular deposits in plaques and vessels occurs years before the onset of symptoms of cognitive decline (Mormino et al. 2008; Price and Morris 1999), which is one reason why the focus has shifted from studying plaques and their impact to elucidating the role of the smaller amyloid peptide aggregates or Aβ

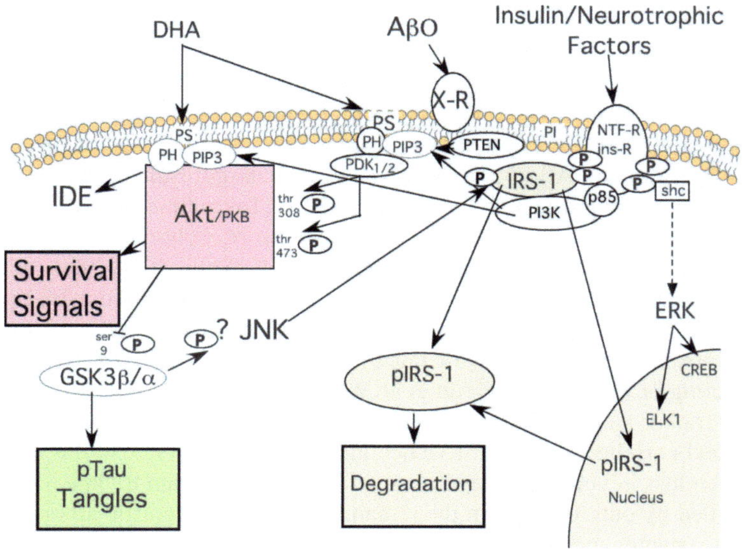


**Fig. 1** Intersection of Aβ oligomers (AβO) and insulin/neurotrophic factor (NTF) signaling. Insulin or NTF binds membrane NTF (NTF-R) or insulin (ins-R) receptors that autophosphorylate to generate tyrosine phosphate sites (P) that bind adaptor proteins. These include insulin receptor substrate-1 (IRS-1), which couples to the p85a subunit of phosphatidylinositol 3-kinase (PI3-K) to generate PIP3 at the membrane, which helps dock and activate PDK1/2 and Akt through pleckstrin homology (PH) domains. These PH domains on PDK and Akt bind both PIP3 and phosphatidylserine (PS) on the inner leaflet of the plasma membrane. One effect of DHA is to increase PS to facilitate this activation of Akt by PDK phosphorylation at residues 308 and 473. Active Akt is upstream of an Aβ-protease, insulin-degrading enzyme (IDE) as well as survival signals (inhibition of BAD) and inhibition of the tau and IRS kinase, GSK3β, which appears to be upstream of c-jun N-terminal kinase (JNK). JNK and GSK3β act sequentially to phosphorylate tau to promote tangles and IRS-1 to promote its uncoupling from receptors and rapid proteasomal degradation. Ins-R and NTF-R also bind SHC coupling to ERK and then CREB and ELK1 nuclear transcription factors. AβO interact with an unidentified surface receptor (X-R), possibly NR2B or other components of the glutamate NMDA receptor to activate fyn kinase and downstream signaling through p21-activated kinases (PAKs; Ma et al. 2008). How AβOs reduce Akt and activate both GSK3β and JNK is not entirely clear, but one possibility is through downregulation of PAK1, which is a candidate for PDK2. AβO uncoupling and reduction of IRS-1 are hypothesized to limit trophic factor and insulin signaling to further reduce Akt, resulting in further increased GSK3β and JNK in a positive feedback loop that drives tau pathology and loss of neuronal arbor and synapses

oligomers (AβO) that occur in different cellular compartments, including intraneuronal accumulations in multivesicular bodies (Takahashi et al. 2004), lipid rafts (Yanagisawa et al. 1995). In soluble fractions, Aβ oligomers have been called micelles (Lomakin et al. 1996; Soreghan et al. 1994) or amyloid-derived diffusible ligands (ADDLs), which represent oligomer species with selective binding to synapses, notably at excitatory synaptic sites (Lacor et al. 2004). While pathogenic mutations in tau protein have not been found in AD, they do occur in other forms of dementia, and when these tau mutations are introduced into animal models, they can produce AD-like synapse and neuron loss (Ballatore et al. 2007; Gotz and Ittner 2008). Lastly, some species of Aβ appear to induce NFTs (Götz et al. 2001).

Connections between Aβ and tau, and between insulin signaling and AD pathogenesis, merge in the extensive literature demonstrating Aβ activation of tau kinases. For example, Aβ inhibits PI3-K/ Akt signaling, resulting in activation of tau kinases, including direct activation of GSK3β (Takashima 2006); Aβ-induced disruptions in insulin signaling can result in aberrant phosphorylation of tau. Aβ is also known to activate MAP kinases, including JNK (Shoji et al. 2000; Wei et al. 2002), which triggers caspase cleavage of tau, and tau phosphorylation by both JNK and GSK3β, which increases tau aggregation (Sahara et al. 2008). Aβ may also inhibit proteasomal degradation of tau by reducing levels of heat shock protein 70 interacting protein (CHIP; Oddo et al. 2008).

In AD, rather than tangles or plaques or neuron loss, synapse loss is the closest pathological correlate with cognitive decline (Terry et al. 1991). In the absence of AD, neuron loss in vulnerable regions, for example, after brain trauma or stroke, is accompanied by outgrowth from the remaining neuronal dendritic arbor and compensatory synaptogenesis, but in AD and APP transgenic AD model mice, surviving (CA1) neurons show reduced dendrite length and spine density (Coleman 1987; Moolman et al. 2004; Spires et al. 2005), which is consistent with a selective loss of the dendritic spine marker, drebrin (Calon et al. 2004).

# 3 Insulin Signaling and AD Pathogenesis

Figure 1 illustrates the relationship between insulin/trophic factor signaling with AD pathogenesis. Following activation by insulin or trophic factors, receptors to these trophic factors autophosphorylate at tyrosine residues, resulting in recognition by the phosphotyrosine-binding domains of adaptor proteins, including SHC and Insulin Receptor Substrates 1 and 2 (IRS-1 and IRS-2; White 2002). IRS then tyrosine phosphorylates itself, permitting their binding and recruitment of Src homology 2 (SH2)-domain–containing proteins, notably the p85 subunit of phosphatidylinositol (PI) 3-kinase (Sun et al. 1995). PI3-K activation further leads to signaling pathways responsible for many of insulin's pleiotropic actions, including activation of Akt (protein kinase B; Virkamaki et al. 1999). The primary cause of insulin resistance in insulin-resistant diabetes and its experimental models is the uncoupling of the adaptor protein, IRS-1, following its phosphorylation at Ser312 (307 in mouse) and other sites, including ser616 (D'Alessandris et al. 2007; Morino et al. 2006). This uncoupling can occur via insulin-stimulated kinases as a negative feedback step, for example, via PI3-K/Akt/mTOR (Pederson et al. 2001; Rui et al. 2001b) or can result from stress-activated serine/threonine kinases. For example, IRS-1 can be phosphorylated at different serine/threonine residues by multiple kinases, including GSK3β (Eldar-Finkelman and Krebs 1997), JNK1 (Aguirre et al. 2000; Rui et al. 2001a), Rho kinase (Begum et al. 2002) and PKC ζ (Liu et al. 2001). The uncoupled serine/threonine-phosphorylated IRS then undergoes rapid proteasomal degradation (Rui et al. 2001b; Sun et al. 1999), resulting in a deficient signal transduction response (Pederson et al. 2001). Thus, phosphorylation of IRS and subsequent downstream deficits contribute to insulin resistance in animal models and diabetics (Rondinone et al. 1997; Saad et al. 1992).

## 4 IRS-1 Defects in AD

Mounting evidence implicates insulin pathway defects in AD (reviewed by Cole and Frautschy 2007). Some of these effects may be specific to loss of insulin mRNA (Steen et al. 2005) or of insulin receptors on dendrites (Zhao et al. 2008). However, defects in insulin signaling may also may result via more generalized changes in the adaptor proteins, which are used by multiple neurotrophic signaling pathways. Evidence for adaptor protein defects in AD include reports that total IRS-1 and IRS-2 are reduced in AD brain, accompanied by elevations in cytosolic phospho-IRS-1 (Ser312 and Ser616), which co-localizes with NFTs (Moloney et al. 2010; Talbot 2006). These data also demonstrate a direct link between "insulin resistance" and tau pathology.

Although most evidence suggests reduced insulin or trophic factor signaling in AD brain (Steen et al. 2005), there is a seemingly contradictory report of elevation of downstream active phospho-Akt (pAkt) and mTOR in pellet fractions of AD brain, particularly in relation to co-localization with pSer214 on tau, an Akt phosphorylation epitope (Griffin et al. 2005), which could also explain the observed elevations in pIRS-1. Alternatively, regional elevations in pAkt, which is normally ubiquitinated for degradation (Dickey et al. 2008), could be explained by defects in pAkt proteosomal degradation or defects in dephosphorylation (reduced phosphatases). Since active Akt binds and phosphorylates tau, the accumulation of pAkt in pellet fractions in AD brain, where tau aggregates accumulate, could also be secondary to sequestration by tau aggregates, as is the case with active pGSK3 (Ishizawa et al. 2003). Arguing for defective insulin signaling in AD, PI3-K/Akt signaling is strongly neuroprotective (as discussed in Calon et al. 2004). The brains of the Tg2576 mouse model of AD, which have been depleted of DHA by diet and transgene, showed significant reductions in pAkt in the TBS fractions (Tg+ 5.01 + 1.72 optical density units) compared to Tg− (12.25 + 2.31) or to Tg+ supplemented with DHA (10.5 + 1.90) but no changes in pAKT in pellet fractions. Consistent with the above observations, Aβ toxicity reduces PI3-K/Akt (Takashima et al. 1996), and conversely, direct PI3-K activation protects against Aβ toxicity in vitro (Lee et al. 2008). Akt accumulation could also be secondary to PTEN deficits because PTEN is a lipid phosphatase that degrades the PIP3 required to activate Akt, thus negatively regulating PI3/Akt signaling (Zhang et al. 2006; see Fig. 1).

Another possible explanation for the co-localization of aberrant phospho-tau and phospho-IRS-1 is that both IRS-1 and tau are substrates for common AβO-activated kinases. Of the repertoire of common IRS and tau kinases, Aβ aggregates are well-known to induce activated GSK3b (Ma et al. 2006; Takashima et al. 1996), GSK3Í is implicated in Aβ toxicity, elevated in AD and AD models and one of the well- established tau kinases (Sahara et al. 2008). Aβ aggregates also activate JNK (Shoji et al. 2000; Troy et al. 2001), which plays a role as a priming tau kinase (Sahara et al. 2008). Activation of JNK is also induced by reactive oxygen species (Shen and Liu 2006) and observed in neurons and dystrophic neurites of AD model mice and AD brain, where it progressively overlaps tau-positive neurofibrillary pathology (Shoji et al. 2000; Zhu et al. 2001). In the APPsw Tg2576 model of AD, active JNK localizes to plaque neurites around plaques containing phosphorylated tau (Puig et al. 2004), and in the APPsw x $PS1^{P264L}$ β-amyloidosis model mice (Savage et al. 2002), JNK activation parallels age-dependent amyloid deposition, tau phosphorylation, and synaptophysin loss. Thus, both JNK and GSK3β are activated in AD brain and AD models and are potential mediators of aberrant IRS-1 phosphorylation, uncoupling and degradation. JNK priming is required for subsequent GSK3β phosphorylation of tau (Sahara et al. 2008) and for GSK3β inactivation of IRS-2 (Sharfi and Eldar-Finkelman 2008).

# 5 IRS-1 Defects in Preclinical Models Relevant to AD

As discussed above, aberrant pIRS-1 frequently co-localizes with tau pathology in AD brain and could be caused by tau aggregates that sequester one or more serine/threonine tau kinases or by pro-inflammatory cytokines like TNFα (Aguirre et al. 2000) or by Aβ activation of tau/IRS-1 kinases. To examine the latter hypothesis, we tested the impact of AβO treatment on IRS-1 in primary rat neuronal cultures and found a rapid (within five hrs) elevation of cytosolic phospho-IRS-1 accompanied by parallel increases in p422Ser tau (Ma et al. submitted for publication). Both AβO effects were entirely blocked by a specific inhibitor of JNK (SP-600125) that also blocked other Aβ-induced ptau epitopes and synaptic marker loss. These data in primary neurons suggested that direct oligomer effects are sufficient to phosphorylate and inactivate IRS-1 and its downstream effects. Our in vitro results on IRS-1 phosphorylation are novel but consistent with other evidence that soluble Aβ aggregates can diminish the trophic factor signal transduction response to both NGF (Chromy et al. 2003) and brain-derived neurotrophic factor (BDNF; Tong et al. 2004).

We then examined phospho-IRS-1 in brains of an animal model that lacks tangle pathology but develops significant neurodegeneration, the aggressive 5xFAD mouse line, which contains three familial AD mutations in APP and two mutations in PS1 (Oakley et al. 2006). By six months, 5xFAD mice show a loss of the predominantly nuclear labeling seen in transgene negative controls and elevated neuronal cytosolic phospho-IRS-1 312, similar to that found in AD brain (Fig. 2). Neurons in the cortex and the CA1 neurons of the hippocampus showed a loss of nuclear and stronger granular perikaryal and dendritic inclusion pIRS-1, mirroring changes seen in the AD brain. Similar observations were found in APPsw Tg2576

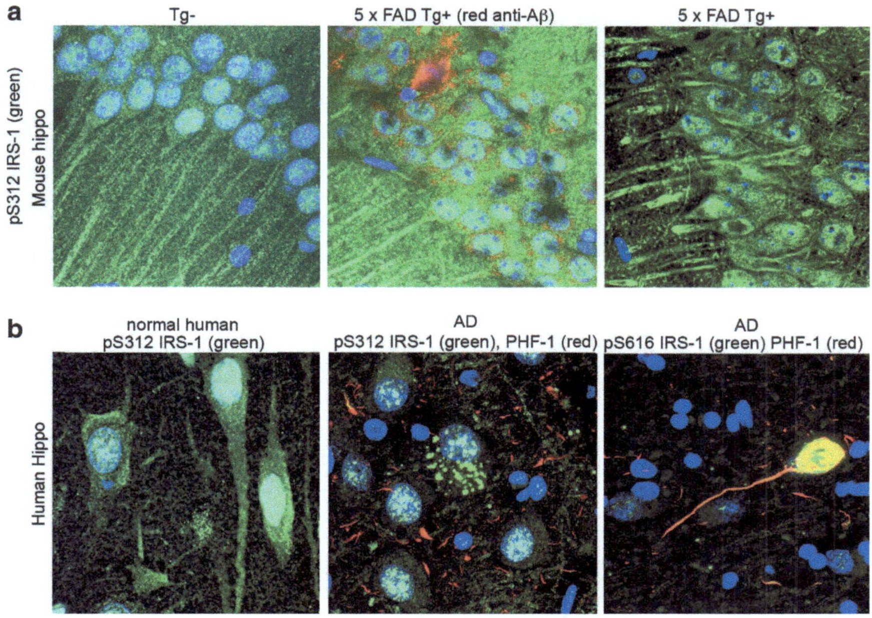


**Fig. 2** Increased cytosolic phospho-IRS-1 (ser 312) in CA1 hippocampus of transgene-positive 5xFAD AD mice and AD. **a.** Sections from Tg negative controls (left) show prominent nuclei (blue DAPI) that are colabeled with pS312 IRS-1 (green), but sections from six-month-old 5xFAD mice (middle) show reduced pS312 IRS-1 staining in nuclei, which appears to be redistributed to granular perikaryal cytosolic and focal neurites (green) or with intraneuronal Aβ (6E10, red). As shown in the right panel, the intraneuronal pIRS-1 is reduced in many nuclei and increased in dendrites. **b.** Sections from human hippocampus stained for IRS-1 show difference in the pattern of pS312 IRS-1 staining in normal (left) and AD (middle), with normal AD IRS-1 staining being largely nuclear whereas in AD, the staining appears to be granular and cytosolic. Staining for pS616 IRS-1 in AD demonstrates colocalization with PHF-1 (right)

mice, where transgene-positive mice showed increased cytosolic pIRS-1 312 and a reduction in nuclear fractions relative to transgene negative controls (Fig.3A). This loss of normal nuclear pIRS-1was confirmed by Western analysis of nuclear fractions (Fig. 3B). Total IRS-1 was reduced by 54% in Western blots from the hippocampus of aged (22-month-old) APPsw Tg2576 mice, consistent with an amyloid peptide-induced degradation and loss of insulin-like signaling (Fig. 3C). Together with the direct effects of AβO in vitro, our in vivo data suggested that Aβ aggregate accumulation is sufficient to cause AD-like elevations of cytosolic p-IRS-1 in the absence of tau aggregate or tangle accumulation.

IRS-1 defects were also examined in a triple transgenic (3xTg-AD) model (APPsw, P301L tau, PS1) that develops intraneuronal Aβ and tau aggregates and eventually plaques and tangles (Oddo et al. 2003). By nine months of age, we found that these mice showed similar deficits in total IRS-1 analyzed by Western blot analysis, corresponding to an increased cytosolic phospho IRS-1 Ser 616

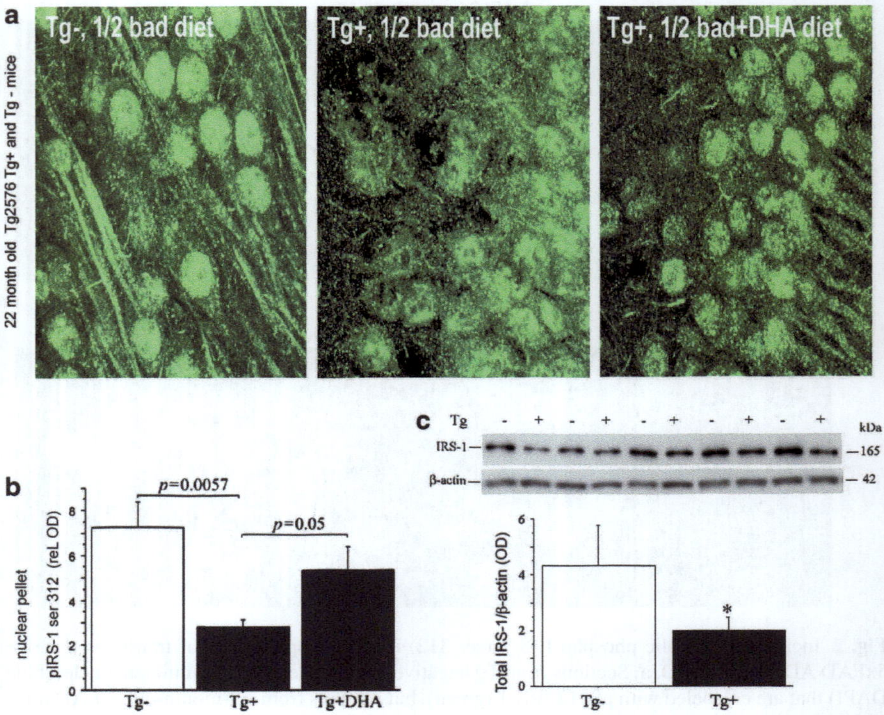


**Fig. 3** Dietary DHA protects loss of nuclear and increased cytosolic phospho-IRS-1 (ser 312) in CA1 hippocampus of transgene-positive Tg2576 AD model mice. **a**. Predominantly nuclear phospho-IRS-1 312 in CA1 neurons in aged control Tg2576 APPsw transgene negative (Tg−, left panel) was markedly shifted toward cytosol in APPsw transgene-positive (Tg+, middle panel) mice, but the normal pattern was largely restored in APPsw Tg+ mice with diets supplemented with 0.6% DHA. To exacerbate the transgene-dependent phenotype, mice were aged to 17 months on breeder chow (PMI5015) and then switched to a less severe modification of an omega-3-depleting n-6 (safflower oil)-based diet (Calon et al. 2004) with half the DHA-depleting safflower oil replaced with oleic acid; and they were aged to 22 months before euthanizing for immunocytochemistry and biochemistry. **b**. The loss of normal nuclear pIRS-1 ser 312 and restoration with DHA were confirmed in Western blots of hippocampal nuclear pellet fractions with densitometry quantifying the major (~75 kD) pIRS-1 ser 312-immunoreactive band. **c**. Loss of full-length total IRS-1 (~165kD band) in aged Tg+ mice and its prevention with dietary DHA was confirmed using Western blot of hippocampal nuclear pellet fractions. Densitometric quantification was normalized by β-actin

(Ma et al. 2009). Collectively, our results support the hypothesis that Aβ aggregates activate kinases that drive aberrant IRS-1 phosphorylation and IRS-1 deficits in vitro and in multiple AD transgenic mouse models. The resulting loss of IRS-1 coupling to PI3-K/Akt is predicted to activate another major tau kinase, GSK3β, which is also activated by Aβ (Ma et al. 2006; Takashima 2006) and appears to be upstream from JNK in AβO toxicity in vivo (Hu et al. 2008). This process further deregulates insulin signaling, since GSK3 itself directly phosphorylates IRS-1 at Ser332 to help inactivate IRS-1 (Liberman and Eldar-Finkelman 2005) and may also activate JNK (Hu et al. 2008), suggesting a positive feedback loop where Aβ activation of JNK and GSK3β initiates an uncoupling of the IRS-1/PI3-K/Akt/GSK3β pathway.

Defects in IRS-1 induced by AβO would be predicted to impair not only signaling of insulin but also the activity of neurotrophic factors like NGF, BDNF, IGF-1 and bFGF. Loss of trophic factor signaling would cause loss of neuronal arbor and dendritic spines and shrinkage of vulnerable remaining neurons, in addition to causing increased tau phosphorylation by JNK and GSK3, which would contribute to tau pathology and further neurodegeneration. Normal Akt signaling also induces the C-terminus of heat shock protein 70 interacting protein (CHIP), a tau ubiquitin ligase that facilitates degradation of misfolded tau (Dickey et al. 2008). Thus, reduction in Akt levels reduces levels of CHIP, which could exacerbate tau pathogenesis (Dickey et al. 2008). These important consequences suggest that this more generalized trophic factor resistance beyond the specific insulin/insulin receptor interaction may play a critical role in AD pathogenesis.

## 6 Western Diet Versus Omega 3 Fatty Acids

Western diets with high saturated fat and cholesterol fed to rats induce insulin resistance and cognitive deficits within a few months (Greenwood and Winocur 2005). The cognitive deficits from high saturated fat are accompanied by reduced hippocampal BDNF (Molteni et al. 2002), fewer dendrites (Granholm et al. 2008) and a reduction in the number of dendritic spines (Stranahan et al. 2008). These cognitive deficits can be ameliorated by the insulin-sensitizing drug rosaglitazone (Pathan et al. 2008), suggesting that some defect in "insulin" signaling may be downstream and apply to other trophic factor functions, for example, the phospho-IRS-1 defect discussed above. Dietary omega 3 fatty acids can partially suppress high fat diet-induced insulin resistance in peripheral tissues, for example, muscle (Taouis et al. 2002), suggesting that they may also be relevant to insulin signaling in the brain. In fact, in the aging APPsw Tg2576 transgenic mice, feeding them dietary omega 3 docosahexaenoic acid (DHA) increased activity in the PI3-K/Akt pathway and protected the anti-caspase Akt target phospho-BAD, caspase activation and postsynaptic markers (Calon et al. 2004, 2005). DHA may potentiate the activation of neurotrophic factor (BDNF, NGF, bFGF) or insulin/IGF-1 signaling via the phosphatidylinositol-3 kinase (PI3-K) > Akt pathway by increasing neuronal phosphatidylserine (PS). Higher PS content increases the rate of Akt membrane docking via a pleckstrin homology domain pocket that binds both the PI3-K product PIP3 and PS (Akbar et al. 2005). We find that, in primary neurons, DHA behaves similarly to the JNK inhibitor, protecting from AβO-induced JNK activation and downstream phosphorylation of tau and IRS-1 (Ma et al. 2009). Similar to these in vitro data on neuroprotection by DHA, feeding 3xTg-AD mice from five to nine months of age diets enriched in 2% fish oil (on a DHA-depleting base diet) corrected transgene-dependent defects in pJNK, cytosolic pIRS-1, total IRS-1 and cognition (Ma et al. 2009).

## 7 Curcumin

In the same study with 3xTg-AD on a DHA-depleting diet, we observed similar protective results with 500 ppm curcumin, a phenolic antioxidant with anti-inflammatory and anti-amyloid properties that is capable of reducing JNK activation in preclinical models of AD (Begum et al. 2008). Interestingly, the combination of fish oil plus curcumin produced more robust and significant protection of IRS-1 and cognitive function (Ma et al., 2009). This synergism is consistent with both protection against peroxidation of polyunsaturated fatty acids like DHA and curcumin's additional direct anti-amyloid (Yang et al. 2005) and other neuroprotective activities (Cole et al. 2007).

Clearly not all DHA activity can be attributed to direct effects on IRS-1. Beyond enhancing Akt signaling, DHA has additional pleiotropic protective mechanisms (reviewed in Calon and Cole 2007); for example, DHA may increase the production of BDNF (Rao et al. 2007) and is a precursor for a neuroprotective lipoxygenase metabolite of DHA, neuroprotectin D1 (NPD1), that has multiple beneficial activities (Bazan 2005). DHA can also reduce Aβ accumulation, possibly by altering APP processing (Lim et al. 2005; Oksman et al. 2006) by NPD1 modulation (Bazan 2005), lowering presenilin-1 and γ-secretase (Green et al. 2007) or raising the anti-Aβ chaperone, SorLA/ LR11 (Ma et al. 2007). Because DHA protects IRS-1 from direct inactivation by exogenous AβO in vitro, it is likely that both reducing the Aβ toxin and more direct neuroprotective effects contribute to the in vivo improvements in insulin/ neurotrophic factor signaling defects.

## 8 Omega-3 Protection in Humans

A robust protective impact of omega-3 fatty acids against some of the negative effects of high fat diet in people is suggested by observations that higher plasma omega-3 fatty acid levels are associated with less decline in verbal fluency, particularly in hypertensive and dyslipidemic patients (Beydoun et al. 2008). Omega-3 from marine sources has a strong positive epidemiology, associating higher DHA/ EPA intake with lower risk for dementia and AD (Maclean et al. 2005). Subsequent evidence from the prospective Framingham study found that those with higher quartile DHA blood levels taken 10 years prior to ascertainment of cognitive status showed protection from dementia or AD (Schaefer et al. 2006). Three studies have found that dietary DHA or fish oil protection against dementia and AD has only been observed in the non-ApoE4 carrier groups (Barberger-Gateau et al. 2007; Huang et al. 2005; Whalley et al. 2008). Consistent with an "insulin-sensitizing" effect of omega-3 fatty acids, this ApoE-dependence strongly resembles clinical trial results with insulin-sensitizing rosiglitazone (Risner et al. 2006) and intranasal insulin (Reger et al. 2008), where only the ApoE4 non-carriers show positive treatment responses.

# 9 Clinical trials

Initial results from small clinical trials suggest some utility stabilizing cognitive decline with fish oil or DHA in very early stage AD (minimally cognitively impaired) but not with established AD. This finding was true for a Japanese trial (Kotani et al. 2006), a Swedish trial (Freund-Levi et al. 2006) and a Taiwanese trial (Chiu et al. 2008). DHA or fish oil appeared to stabilize MMSE in the MCI patients but not in those with established AD. We have argued that the highly unsaturated and peroxidizable DHA should be combined with an antioxidant (Calon et al. 2004). In a 12-month pilot trial where AD subjects were given the antioxidant alpha lipoate and/or fish oil (NCT000904029), MMSE declined in subjects given alpha lipoate or fish oil alone but not in subjects given the combination (Shinto and Oken 2008). While each reported trial has been too small to be compelling, these initial results in preclinical models and clinical trials with early stage patients strongly encourage the design of larger and more definitive trials with omega 3 for primary prevention and with combinations of omega 3 with antioxidants like curcumin and α-lipoate for early stage AD treatment.

# 10 Conclusions

AD is characterized by selective neuron loss, for example in the CA1 of the hippocampus, that under normal conditions would be accompanied by compensatory dendritic outgrowth and synaptic plasticity; but in AD, dendritic arbor and spine density of surviving neurons are decreased. This decrease could be caused by a loss of trophic stimulation or a diminished response to trophic factors, as discussed above. Accumulating evidence suggests that an improvement in insulin/trophic factor signaling, either by insulin or insulin-sensitizing drugs, could be of therapeutic value for AD. The data reviewed here support the argument that supplementation with dietary fatty acids and antioxidants, an approach suitable for primary prevention, can mediate a significant improvement in AD-related defects in insulin receptor substrate, an adaptor protein common to insulin and neurotrophic factor signaling to Akt and a critical component of the neuroprotective pathway.

## References

Aguirre V, Uchida T, Yenush L, Davis R, White MF (2000) The c-Jun NH(2)-terminal kinase promotes insulin resistance during association with insulin receptor substrate-1 and phosphorylation of Ser(307). J Biol Chem 275:9047–9054

Akbar M, Calderon F, Wen Z, Kim HY (2005) Docosahexaenoic acid: a positive modulator of Akt signaling in neuronal survival. Proc Natl Acad Sci USA 102:10858–10863

Ballatore C, Lee VM, Trojanowski JQ (2007) Tau-mediated neurodegeneration in Alzheimer's disease and related disorders. Nature Rev Neurosci 8:663–672

Barberger-Gateau P, Raffaitin C, Letenneur L, Berr C, Tzourio C, Dartigues JF, Alperovitch A (2007) Dietary patterns and risk of dementia: the three-City cohort study. Neurology 69:1921–1930

Bazan NG (2005) Neuroprotectin D1 (NPD1): a DHA-derived mediator that protects brain and retina against cell injury-induced oxidative stress. Brain Pathol 15:159–166

Begum AN, Jones MR, Lim GP, Morihara T, Kim P, Heath DD, Rock CL, Pruitt MA, Yang F, Hudspeth B, Hu S, Faull KF, Teter B, Cole GM, Frautschy SA (2008) Curcumin structure-function, bioavailability, and efficacy in models of neuroinflammation and Alzheimer's disease. JPET 326:196–208

Begum N, Sandu OA, Ito M, Lohmann SM, Smolenski A (2002) Active Rho kinase (ROK-alpha) associates with insulin receptor substrate-1 and inhibits insulin signaling in vascular smooth muscle cells. J Biol Chem 277:6214–6222

Beydoun MA, Kaufman JS, Sloane PD, Heiss G, Ibrahim J (2008) n-3 Fatty acids, hypertension and risk of cognitive decline among older adults in the Atherosclerosis Risk in Communities (ARIC) study. Public Health Nutr 11:17–29

Calon F, Cole G (2007) Neuroprotective action of omega-3 polyunsaturated fatty acids against neurodegenerative diseases: evidence from animal studies. Prostaglandins Leukot Essent Fatty Acids 77:287–293

Calon F, Lim GP, Yang F, Morihara T, Teter B, Ubeda O, Rostaing P, Triller A, Salem N, Jr., Ashe KH, Frautschy SA, Cole GM (2004) Docosahexaenoic acid protects from dendritic pathology in an Alzheimer's disease mouse model. Neuron 43:633–645

Calon F, Lim GP, Morihara T, Yang F, Ubeda O, Salem NJ, Frautschy SA, Cole GM (2005) Dietary n-3 polyunsaturated fatty acid depletion activates caspases and decreases NMDA receptors in the brain of a transgenic mouse model of Alzheimer's disease. Eur J Neurosci 22:617–626

Chiu CC, Su KP, Cheng TC, Liu HC, Chang CJ, Dewey ME, Stewart R, Huang SY (2008) The effects of omega-3 fatty acids monotherapy in Alzheimer's disease and mild cognitive impairment: a preliminary randomized double-blind placebo-controlled study. Prog Neuropsychopharmacol Biol Psychiat 32:1538–1544

Chromy BA, Nowak RJ, Lambert MP, Viola KL, Chang L, Velasco PT, Jones BW, Fernandez SJ, Lacor PN, Horowitz P, Finch CE, Krafft GA, Klein WL (2003) Self-assembly of Aβ(1-42) into globular neurotoxins. Biochemistry 42:12749–12760

Cole GM, Frautschy SA (2007) The role of insulin and neurotrophic factor signaling in brain aging and Alzheimer's disease. Exp Gerontol 42:10–21

Cole GM, Teter B, Frautschy SA (2007) Neuroprotective effects of curcumin. Adv Exp Med Biol 595:197–212

Coleman PD (1987) Neuron numbers and dendritic extent in normal aging and Alzheimer's disease. Neurobiol Aging 8:521–545

Cummings JL, Cole G (2002) Alzheimer disease. JAMA 287:2335–2338

D'Alessandris C, Lauro R, Presta I, Sesti G (2007) C-reactive protein induces phosphorylation of insulin receptor substrate-1 on Ser307 and Ser 612 in L6 myocytes, thereby impairing the insulin signalling pathway that promotes glucose transport. Diabetologia 50:840–849

Dickey CA, Koren J, Zhang YJ, Xu YF, Jinwal UK, Birnbaum MJ, Monks B, Sun M, Cheng JQ, Patterson C, Bailey RM, Dunmore J, Soresh S, Leon C, Morgan D, Petrucelli L (2008) Akt and

CHIP coregulate tau degradation through coordinated interactions. Proc Natl Acad Sci USA 105:3622–3627

Eldar-Finkelman H, Krebs EG (1997) Phosphorylation of insulin receptor substrate 1 by glycogen synthase kinase 3 impairs insulin action. Proc Natl Acad Sci USA 94:9660–9664

Freund-Levi Y, Eriksdotter-Jonhagen M, Cederholm T, Basun H, Faxen-Irving G, Garlind A, Vedin I, Vessby B, Wahlund LO, Palmblad J (2006) Omega-3 fatty acid treatment in 174 patients with mild to moderate Alzheimer disease: OmegAD study: a randomized double-blind trial. Arch Neurol 63:1402–1408

Gotz J, Ittner LM (2008) Animal models of Alzheimer's disease and frontotemporal dementia. Nature Rev Neurosci 9:532–544

Götz J, Chen F, van Dorpe J, Nitsch RM (2001) Formation of neurofibrillary tangles in P301L tau transgenic mice induced by Aβ42 fibrils. Science 293:1491–1495

Granholm AC, Bimonte-Nelson HA, Moore AB, Nelson ME, Freeman LR, Sambamurti K (2008) Effects of a saturated fat and high cholesterol diet on memory and hippocampal morphology in the middle-aged rat. J Alzheimers Dis 14:133–145

Green KN, Martinez-Coria H, Khashwji H, Hall EB, Yurko-Mauro KA, Ellis L, LaFerla FM (2007) Dietary docosahexaenoic acid and docosapentaenoic acid ameliorate amyloid-β and tau pathology via a mechanism involving presenilin 1 levels. J Neurosci 27: 4385–4395

Greenwood CE, Winocur G (2005) High-fat diets, insulin resistance and declining cognitive function. Neurobiol Aging 26 Suppl 1:42–45

Griffin RJ, Moloney A, Kelliher M, Johnston JA, Ravid R, Dockery P, O'Connor R, O'Neill C (2005) Activation of Akt/PKB, increased phosphorylation of Akt substrates and loss and altered distribution of Akt and PTEN are features of Alzheimer's disease pathology. J Neurochem 93:105–117

Hu S, Begum AN, Jones MR, Oh MS, Beech WK, Beech BH, Yang F, Chen P, Ubeda OJ, Kim PC, Davies P, Ma Q, Cole GM, Frautschy SA (2008) GSK3 inhibitors show benefits in an Alzheimer's disease (AD) model of neurodegeneration but adverse effects in control animals. Neurobiol Dis 33:193–206

Huang TL, Zandi PP, Tucker KL, Fitzpatrick AL, Kuller LH, Fried LP, Burke GL, Carlson MC (2005) Benefits of fatty fish on dementia risk are stronger for those without APOE epsilon4. Neurology 65:1409–1414

Irie F, Fitzpatrick AL, Lopez OL, Kuller LH, Peila R, Newman AB, Launer LJ (2008) Enhanced risk for Alzheimer disease in persons with type 2 diabetes and APOE epsilon4: the Cardiovascular Health Study Cognition Study. Arch Neurol 65:89–93

Ishizawa T, Sahara N, Ishiguro K, Kersh J, McGowan E, Lewis J, Hutton M, Dickson DW, Yen SH (2003) Co-localization of glycogen synthase kinase-3 with neurofibrillary tangles and granulovacuolar degeneration in transgenic mice. Am J Pathol 163:1057–1067

Kotani S, Sakaguchi E, Warashina S, Matsukawa N, Ishikura Y, Kiso Y, Sakakibara M, Yoshimoto T, Guo J, Yamashima T (2006) Dietary supplementation of arachidonic and docosahexaenoic acids improves cognitive dysfunction. Neurosci Res 56:159–164

Lacor PN, Buniel MC, Chang L, Fernandez SJ, Gong Y, Viola KL, Lambert MP, Velasco PT, Bigio EH, Finch CE, Krafft GA, Klein WL (2004) Synaptic targeting by Alzheimer's-related amyloid β oligomers. J Neurosci 24:10191–10200

Lee KY, Koh SH, Noh MY, Kim SH, Lee YJ (2008) Phosphatidylinositol-3-kinase activation blocks amyloid β-induced neurotoxicity. Toxicology 243:43–50

Liberman Z, Eldar-Finkelman H (2005) Serine 332 phosphorylation of insulin receptor substrate-1 by glycogen synthase kinase-3 attenuates insulin signaling. J Biol Chem 280:4422–4428

Lim GP, Calon F, Morihara T, Yang F, Teter B, Ubeda O, Salem N, Jr., Frautschy SA, Cole GM (2005) A diet enriched with the omega-3 fatty acid docosahexaenoic acid reduces amyloid burden in an aged Alzheimer mouse model. J Neurosci 25:3032–3040

Liu YF, Paz K, Herschkovitz A, Alt A, Tennenbaum T, Sampson SR, Ohba M, Kuroki T, LeRoith D, Zick Y (2001) Insulin stimulates PKCzeta -mediated phosphorylation of insulin receptor substrate-1 (IRS-1). A self-attenuated mechanism to negatively regulate the function of IRS proteins. J Biol Chem 276:14459–14465

Lomakin A, Chung DS, Benedek GB, Kirschner DA, Teplow DB (1996) On the nucleation and growth of amyloid β-protein fibrils: detection of nuclei and quantitation of rate constants. Proc Natl Acad Sci USA 93:1125–1129

Ma QL, Lim GP, Harris-White ME, Yang F, Ambegaokar SS, Ubeda OJ, Glabe CG, Teter B, Frautschy SA, Cole GM (2006) Antibodies against β-amyloid reduce Aβ oligomers, glycogen synthase kinase-3β activation and tau phosphorylation in vivo and in vitro. J Neurosci Res 83:374–384

Ma QL, Teter B, Ubeda OJ, Morihara T, Dhoot D, Nyby MD, Tuck ML, Frautschy SA, Cole GM (2007) Omega-3 fatty acid docosahexaenoic acid increases SorLA/LR11, a sorting protein with reduced expression in sporadic Alzheimer's disease (AD): relevance to AD prevention. J Neurosci 27:14299–14307

Ma QL, Yang F, Calon F, Ubeda OJ, Hansen JE, Weisbart RH, Beech W, Frautschy SA, Cole GM (2008) p21-activated kinase-aberrant activation and translocation in Alzheimer disease pathogenesis. J Biol Chem 283:14132–14143

Ma QL, Yang F, Rosario ER, Ubeda OJ, Beech W, Gant DJ, Chen PP, Hudspeth B, Chen C, Zhao Y, Vinters HV, Frautschy SA, Cole GM (2009) Beta-amyloid oligomers induce phosphorylation of tau and inactivation of insulin receptor substrate via c-Jun N-terminal kinase signaling: suppression by omega-3 fatty acids and curcumin. J Neurosci Jul 15; 29(28): 9078–89

Maclean C, Issa A, Newberry S, Mojica W, Morton S, Garland R, Hilton L, Traina S, Shekelle P (2005) Effects of omega-3 fatty acids on cognitive function with aging, dementia, and neurological diseases. Evid Rep Technol Assess (Summ):1–88

Moloney AM, Griffin RJ, Timmons S, O'Connor R, Ravid R, O'Neill C (2010) Defects in IGF-1 receptor, insulin receptor and IRS-1/2 in Alzheimer's disease indicate possible resistance to IGF-1 and insulin signalling. Neurobiol Aging May 12; 31:224–243

Molteni R, Barnard RJ, Ying Z, Roberts CK, Gomez-Pinilla F (2002) A high-fat, refined sugar diet reduces hippocampal brain-derived neurotrophic factor, neuronal plasticity, and learning. Neuroscience 112:803–814

Moolman DL, Vitolo OV, Vonsattel JP, Shelanski ML (2004) Dendrite and dendritic spine alterations in Alzheimer models. J Neurocytol 33:377–387

Morino K, Petersen KF, Shulman GI (2006) Molecular mechanisms of insulin resistance in humans and their potential links with mitochondrial dysfunction. Diabetes 55 Suppl 2: S9–S15

Mormino EC, Kluth JT, Madison CM, Rabinovici GD, Baker SL, Miller BL, Koeppe RA, Mathis CA, Weiner MW, Jagust WJ (2008) Episodic memory loss is related to hippocampal-mediated β-amyloid deposition in elderly subjects. Brain 28 132:1310–1323

Oakley H, Cole SL, Logan S, Maus E, Shao P, Craft J, Guillozet-Bongaarts A, Ohno M, Disterhoft J, Van Eldik L, Van Eldik L, Berry R, Vassar R (2006) Intraneuronal β-amyloid aggregates, neurodegeneration, and neuron loss in transgenic mice with five familial Alzheimer's disease mutations: potential factors in amyloid plaque formation. J Neurosci 26:10129–10140

Oddo S, Cacccamo A, Shepherd JD, Murphy MP, Golde TE, Metherate R, Mattson MP, Akbari Y, LaFerla FM (2003) Triple transgenic model of Alzheimer's disease that develops age-dependent Aβ and tau pathology and synaptic dysfunction. Neuron 39:409–421

Oddo S, Caccamo A, Tseng B, Cheng D, Vasilevko V, Cribbs DH, LaFerla FM (2008) Blocking Aβ42 accumulation delays the onset and progression of tau pathology via the C terminus of heat shock protein70-interacting protein: a mechanistic link between Aβ and tau pathology. J Neurosci 28:12163–12175

Oksman M, Iivonen H, Hogyes E, Amtul Z, Penke B, Leenders I, Broersen L, Lutjohann D, Hartmann T, Tanila H (2006) Impact of different saturated fatty acid, polyunsaturated fatty acid and cholesterol containing diets on β-amyloid accumulation in APP/PS1 transgenic mice. Neurobiol Dis 23:563–572

Ott A, Stolk RP, van Harskamp F, Pols HA, Hofman A, Breteler MM (1999) Diabetes mellitus and the risk of dementia: The Rotterdam Study. Neurology 53:1937–1942

Pathan AR, Gaikwad AB, Viswanad B, Ramarao P (2008) Rosiglitazone attenuates the cognitive deficits induced by high fat diet feeding in rats. Eur J Pharmacol 589:176–179

Pederson TM, Kramer DL, Rondinone CM (2001) Serine/threonine phosphorylation of IRS-1 triggers its degradation: possible regulation by tyrosine phosphorylation. Diabetes 50:24–31

Peila R, Rodriguez BL, Launer LJ (2002) Type 2 diabetes, APOE gene, and the risk for dementia and related pathologies: The Honolulu-Asia Aging Study. Diabetes 51:1256–1262

Price JL, Morris JC (1999) Tangles and plaques in nondemented aging and "preclinical" Alzheimer's disease. Ann Neurol 45:358–368

Profenno LA, Faraone SV (2008) Diabetes and overweight associate with non-APOE4 genotype in an Alzheimer's disease population. Am J Med Genet B Neuropsychiatr Genet 147B: 822–829

Puig B, Gomez-Isla T, Ribe E, Cuadrado M, Torrejon-Escribano B, Dalfo E, Ferrer I (2004) Expression of stress-activated kinases c-Jun N-terminal kinase (SAPK/JNK-P) and p38 kinase (p38-P), and tau hyperphosphorylation in neurites surrounding Aβ plaques in APP Tg2576 mice. Neuropathol Appl Neurobiol 30:491–502

Rao JS, Ertley RN, Lee HJ, DeMar JC, Jr., Arnold JT, Rapoport SI, Bazinet RP (2007) n-3 polyunsaturated fatty acid deprivation in rats decreases frontal cortex BDNF via a p38 MAPK-dependent mechanism. Mol Psychiat 12:36–46

Reger MA, Watson GS, Green PS, Baker LD, Cholerton B, Fishel MA, Plymate SR, Cherrier MM, Schellenberg GD, Frey WH 2nd, Craft S (2008) Intranasal insulin administration dose-dependently modulates verbal memory and plasma amyloid-β in memory-impaired older adults. J Alzheimer's Dis 13:323–331

Risner ME, Saunders AM, Altman JF, Ormandy GC, Craft S, Foley IM, Zvartau-Hind ME, Hosford DA, Roses AD (2006) Efficacy of rosiglitazone in a genetically defined population with mild-to-moderate Alzheimer's disease. Pharmacogenomics J 6:246–254

Rondinone CM, Wang LM, Lonnroth P, Wesslau C, Pierce JH, Smith U (1997) Insulin receptor substrate (IRS) 1 is reduced and IRS-2 is the main docking protein for phosphatidylinositol 3-kinase in adipocytes from subjects with non-insulin-dependent diabetes mellitus. Proc Natl Acad Sci USA 94:4171–4175

Rui L, Aguirre V, Kim JK, Shulman GI, Lee A, Corbould A, Dunaif A, White MF (2001a) Insulin/IGF-1 and TNF-alpha stimulate phosphorylation of IRS-1 at inhibitory Ser307 via distinct pathways. J Clin Invest 107:181–189

Rui L, Fisher TL, Thomas J, White MF (2001b) Regulation of insulin/insulin-like growth factor-1 signaling by proteasome-mediated degradation of insulin receptor substrate-2. J Biol Chem 276:40362–40367

Saad MJ, Araki E, Miralpeix M, Rothenberg PL, White MF, Kahn CR (1992) Regulation of insulin receptor substrate-1 in liver and muscle of animal models of insulin resistance. J Clin Invest 90:1839–1849

Sahara N, Murayama M, Lee B, Park JM, Lagalwar S, Binder LI, Takashima A (2008) Active c-jun N-terminal kinase induces caspase cleavage of tau and additional phosphorylation by GSK-3β is required for tau aggregation. Eur J Neurosci 27:2897–2906

Savage MJ, Lin YG, Ciallella JR, Flood DGSRW (2002) Activation of c-Jun N-terminal kinase and p38 in an Alzheimer's disease model is associated with amyloid deposition. J Neurosci 22:3376–3385

Schaefer EJ, Bongard V, Beiser AS, Lamon-Fava S, Robins SJ, Au R, Tucker KL, Kyle DJ, Wilson PW, Wolf PA (2006) Plasma phosphatidylcholine docosahexaenoic acid content and risk of dementia and Alzheimer disease: the Framingham Heart Study. Arch Neurol 63:1545–1550

Sharfi H, Eldar-Finkelman H (2008) Sequential phosphorylation of insulin receptor substrate-2 by glycogen synthase kinase-3 and c-Jun NH2-terminal kinase plays a role in hepatic insulin signaling. Am J Physiol Endocrinol Metab 294:E307–315

Shen HM, Liu ZG (2006) JNK signaling pathway is a key modulator in cell death mediated by reactive oxygen and nitrogen species. Free Radic Biol Med 40:928–939

Shinto L, Quinn J, Montine T, Baldauf-Wagner S, Oken DB, Kaye J (2008) Omega-3 fatty acids and lipoic acid in Alzheimer's disease. Neurology 70:A393

Shoji M, Iwakami N, Takeuchi S, Waragai M, Suzuki M, Kanazawa I, Lippa CF, Ono S, Okazawa H (2000) JNK activation is associated with intracellular β-amyloid accumulation. Brain Res Mol Brain Res 85:221–233

Soreghan B, Kosmoski J, Glabe C (1994) Surfactant properties of Alzheimer's Ab peptides and the mechanism of amyloid aggregation. J Biol Chem 269:28551–28554

Spires TL, Meyer-Luehmann M, Stern EA, McLean PJ, Skoch J, Nguyen PT, Bacskai BJ, Hyman BT (2005) Dendritic spine abnormalities in amyloid precursor protein transgenic mice demonstrated by gene transfer and intravital multiphoton microscopy. J Neurosci 25:7278–7287

Steen E, Terry BM, Rivera EJ, Cannon JL, Neely TR, Tavares R, Xu XJ, Wands JR, de la Monte SM (2005) Impaired insulin and insulin-like growth factor expression and signaling mechanisms in Alzheimer's disease-is this type 3 diabetes? J Alzheimer's Dis 7:63–80

Stranahan AM, Norman ED, Lee K, Cutler RG, Telljohann RS, Egan JM, Mattson MP (2008) Diet-induced insulin resistance impairs hippocampal synaptic plasticity and cognition in middle-aged rats. Hippocampus 18:1085–1088

Sun XJ, Wang LM, Zhang Y, Yenush L, Myers MG, Jr., Glasheen E, Lane WS, Pierce JH, White MF (1995) Role of IRS-2 in insulin and cytokine signalling. Nature 377:173–177

Sun XJ, Goldberg JL, Qiao LY, Mitchell JJ (1999) Insulin-induced insulin receptor substrate-1 degradation is mediated by the proteasome degradation pathway. Diabetes 48:1359–1364

Takahashi RH, Almeida CG, Kearney PF, Yu F, Lin MT, Milner TA, Gouras GK (2004) Oligomerization of Alzheimer's β-amyloid within processes and synapses of cultured neurons and brain. J Neurosci 24:3592–3599

Takashima A (2006) GSK-3 is essential in the pathogenesis of Alzheimer's disease. J Alzheimer's Dis 9:309–317

Takashima A, Noguchi K, Michel G, Mercken M, Hoshi M, Ishiguro K, Imahori K (1996) Exposure of rat hippocampal neurons to amyloid β peptide (25-35) induces the inactivation of phosphatidyl inositiol-3 kinase and the activation of tau protein kinase I/glycogen synthase kinase-3 β. Neurosci Lett 203:33–36

Talbot K, Han LY, Schneider JA, Wilson RS, Bennett DA, Arnold SA (2006) Expression of PIRS-1 (S312 and S616) is elevated in MCI and AD and correlates with cognitive impairment and neurofibrillary pathology. Neurobiol Aging 27:S54

Taouis M, Dagou C, Ster C, Durand G, Pinault M, Delarue J (2002) N-3 polyunsaturated fatty acids prevent the defect of insulin receptor signaling in muscle. Am J Physiol Endocrinol Metab 282:E664–671

Terry RD, Masliah E, Salmon DP, Butters N, DeTeresa R, Hill R, Hansen LA, Katzman R (1991) Physical basis of cognitive alterations in Alzheimer's disease: synapse loss is the major correlate of cognitive impairment. Annals Neurol 30:572–580

Tong L, Balazs R, Thornton PL, Cotman CW (2004) β-amyloid peptide at sublethal concentrations downregulates brain-derived neurotrophic factor functions in cultured cortical neurons. J Neurosci 24:6799–6809

Troy CM, Rabacchi SA, Xu Z, Maroney AC, Connors TJ, Shelanski ML, Greene LA (2001) β-Amyloid-induced neuronal apoptosis requires c-Jun N-terminal kinase activation. J Neurochem 77:157–164

Virkamaki A, Ueki K, Kahn CR (1999) Protein-protein interaction in insulin signaling and the molecular mechanisms of insulin resistance. J Clin Invest 103:931–943

Wei W, Norton DD, Wang X, Kusiak JW (2002) Aβ 17-42 in Alzheimer's disease activates JNK and caspase-8 leading to neuronal apoptosis. Brain 125:2036–2043

Whalley LJ, Deary IJ, Starr JM, Wahle KW, Rance KA, Bourne VJ, Fox HC (2008) n-3 Fatty acid erythrocyte membrane content, APOE varepsilon4, and cognitive variation: an observational follow-up study in late adulthood. Am J Clin Nutr 87:449–454

Yanagisawa K, Odaka A, Suzuki N, Ihara Y (1995) GM1 ganglioside-bound amyloid β-protein (Aβ): a possible form of preamyloid in Alzheimer's disease. Nature Med 1:1062–1066

Yang F, Lim GP, Begum AN, Ubeda OJ, Simmons MR, Ambegaokar SS, Chen PP, Kayed R, Glabe CG, Frautschy SA, Cole GM (2005) Curcumin inhibits formation of amyloid β oligomers and fibrils, binds plaques, and reduces amyloid in vivo. J Biol Chem 280:5892–5901

Zhang X, Li F, Bulloj A, Zhang YW, Tong G, Zhang Z, Liao FF, Xu H (2006) Tumor-suppressor PTEN affects tau phosphorylation, aggregation, and binding to microtubules. FASEB J 20:1272–1274

Zhao WQ, De Felice FG, Fernandez S, Chen H, Lambert MP, Quon MJ, Krafft GA, Klein WL (2008) Amyloid β oligomers induce impairment of neuronal insulin receptors. FASEB J 22:246–260

Zhu X, Castellani RJ, Takeda A, Nunomura A, Atwood CS, Perry G, Smith MA (2001) Differential activation of neuronal ERK, JNK/SAPK and p38 in Alzheimer disease: the 'two hit' hypothesis. Mech Ageing Dev 123:39–46

# Serum IGF-I, Life Style, and Risk of Alzheimer's disease

**Joaquin Piriz, Takeshi Nishijima, Jose Luis Trejo, and Ignacio Torres Aleman**

**Abstract** The ancient insulin-like family of peptides gave rise in higher organisms to insulin-growth factors (IGFs) and insulin. Formerly considered functionally divergent, new evidence suggests that IGFs and insulin probably share a close functional relationship. These links are still poorly defined but may eventually turn out to be of great relevance in the development of Alzheimer´s pathology. IGF-I in the circulation acts as a neuroprotective hormone, entering into the brain through a transport system at the blood-brain barriers. The neuroactive role of serum IGF-I is modulated by environmental factors and behavior. Importantly, both environmental factors and life style are increasingly recognized to impact the development of Alzheimer´s disease (AD). Risk factors classically associated with cardiovascular disease, such as unhealthy diets, lack of physical exercise, or stress, are now also related to AD. The molecular underpinnings of these links are starting to be unveiled. There is evidence pointing to serum IGF-I in this regard. Circumstantial observations, such as that serum IGF-I declines with aging, the single most important risk factor for AD, or that serum IGF-I correlates with cognitive status in humans, have hinted at this connection. Further, diet, and physical or mental activity influence serum IGF-I input to the brain. In addition, stress and general health status may influence brain input of serum IGF-I. All these factors have been linked to a risk of AD. Analysis of the molecular and cellular pathways involved in serum IGF-I traffic at the blood-brain interfaces indicates that pathogenic disturbances at these sites may be of great relevance in the development of AD. Indeed, reduced brain IGF-I input elicits all the neuropathological changes associated with AD. As all the above-mentioned life style factors impinge on the transport of IGF-I at the barriers, a molecular understanding of their role as risk factors is now within reach.

---

I.T. Aleman (✉)
Cajal Institute, CSIC, and Ciberned, Avenida, Doctor Arce 37, Madrid 28002, Spain

S. Craft and Y. Christen (eds.), *Diabetes, Insulin and Alzheimer's Disease*,
Research and Perspectives in Alzheimer's Disease,
DOI 10.1007/978-3-642-04300-0_12, 

# 1 Introduction

Mammalian insulin peptides (insulin, IGF-I, and IGF-II) originated from a common insulin-like ancestor present in ancient organisms around 600 million years ago (Chan et al. 1990; Luo and Murphy 1989). While they are found in the circulation and display typical hormonal traits, the IGFs are also considered ubiquitous growth factors, acting as local messengers in body tissues (Daughaday and Rotwein 1989). The classical role assigned to IGF-I as a hormone is to translate growth hormone (GH) signals to the periphery. The hormonal significance of IGF-II is still poorly understood, with a reported role during fetal growth (DeChiara et al. 1990). At any rate, the actions of IGFs as growth modulators and of insulin as a metabolic modulator have been clearly differentiated. However, in primitive organisms wherein dozens of insulin-like peptides have been described (Pierce et al. 2001), the functional separation between them is still uncertain. New evidence may also blur this functional distinction between insulin peptides in higher organisms, at least for the closely related IGFs and insulin.

Insulin-like metabolic actions of IGFs at high concentrations were described when they were originally characterized in mammals (Daughaday and Rotwein 1989). In turn, the growth factor-like features of insulin during development and in specific tissues have been recognized (de Pablo and de la Rosa 1995). However, new data might lead to considering these three peptides to be functionally interconnected in a more direct way. An important first clue was the promiscuous signalling of the three peptides on their receptors. An additional clue was that IGF-I and insulin receptors dimerize and form functional receptor dimers (Moxham et al. 1989) of pathological consequence when dysregulated (Fernandez et al. 2001). More recently, it has been shown that serum IGF-I modulates insulin sensitivity (Clemmons 2004) and that IGF-binding protein 1 (which does not bind insulin) may be a reliable marker of insulin sensitivity (Lewitt et al. 2008). Simultaneous changes in circulating levels of IGF-I and insulin are present in many human diseases, not only in diabetes and other endocrine illnesses, but also in different neurodegenerative conditions (Busiguina et al. 2000). Collectively, these data point to a functional connection between the IGFs and insulin at the brain level as well (Davila et al. 2007).

# 2 Neuroprotection by Brain- and Serum-derived IGF-I

IGF-I is probably the prototype neuroprotective factor. Its wide array of beneficial actions on brain cells is now beyond dispute. In recent years, we have hypothesized that attenuation of IGF-I neuroprotection is involved in most, if not all, pathological processes in brain tissue that eventually lead to neuronal death/dysfunction (Davila et al. 2007). Loss of function of IGF-I may underlie the local increases in IGF-I, its receptor and its binding proteins found in all types of brain lesions (Torres-Aleman 2005).

Remarkably, different elements of the canonical IGF-I signalling pathway appear to be dysregulated in neurological impairments of widely different etiology. Thus, in Huntington´s disease, the kinase Akt appears to be involved (Humbert et al. 2002); the same is true for schizophrenia (Emamian et al. 2004), spinocerebellar ataxia 1 (Chen et al. 2003) or Parkinson´s disease (Malagelada et al. 2008). Various other proteins of the IGF-I pathway have been shown to be involved in diseases such as dentatorubral-pallydoluysian atrophy (Okamura-Oho et al. 1999), ataxia-telangectasia (Peretz et al. 2001), amyotrophic lateral sclerosis (Wilczak et al. 2003), or Rett syndrome, an autistic-like brain disorder (Itoh et al. 2007). In the case of AD, we and others have specifically pointed to IGF-I/insulin resistance status, which we consider to be instrumental in the pathogenesis of AD (Carro and Torres-Aleman 2004). An insulin-resistant state was long ago claimed to be a key metabolic disturbance in AD (Blum-Degen et al. 1995), with a possible prognostic value, as recently proposed (Ronnemaa et al. 2008). The processes leading to this connection are starting to be unveiled (De Felice et al. 2009; Gasparini et al. 2002). On the other hand, we recently postulated that the initial disturbance leading to AD is a resistant state to IGF-I initiated at the blood-brain-barriers (BBBs) that eventually leads to insulin resistance in brain cells and all the neuropathological and clinical features of AD (see Carro and Torres-Aleman 2004 for further detail).

Because brain cells synthesize IGF-I (Bondy and Lee 1993) and IGF-I expression increases in response to many types of brain insults (Walter et al. 1997), it was assumed that locally produced IGF-I underlies its neuroprotective effects. Indeed, both mice and human IGF-I null mutants display profound brain disturbances (Baker et al. 1993; Woods et al. 1997). However, this phenotype could be ascribed to a general lack of IGF-I throughout development. At least in rodents, expression of IGF-I in the brain peaks perinatally, decreasing to almost negligible levels in the adult (Bondy and Lee 1993). Conversely, levels of IGF-I protein in the brain remain more stable from birth to adulthood (Torres-Aleman et al. 1994). This mismatch between IGF-I mRNA and protein content points to an extracerebral source for brain IGF-I (Busiguina et al. 2000). A logical source is the circulation, as the blood contains very high levels of IGF-I. Against this possibility is the existence of the BBBs, which impede the free-passage of blood proteins into the brain (Dziegielewska and Saunders 2002). However, it has been shown that circulating IGF-I can cross the BBB (Pardridge 1993). This transcytosis process involves a specific transport system at the choroid plexus (Carro et al. 2005) and brain vessels (unpublished observations). Yet, passage of serum IGF-I to the brain is still not widely recognized and warrants further study. At any rate, blood-borne IGF-I is a key neuroprotective signal, actively involved in adult neurogenesis, cognition, angiogenesis, or resilience against injury, and underlies many of the beneficial effects of physical exercise on the brain (Carro et al. 2000). Changes in circulating IGF-I levels have been shown to run in parallel with changes in brain IGF-I (Busiguina et al. 2000). Absence of IGF-I receptors in the brain not only produces profound brain deficits but also results in high serum IGF-I levels (Kappeler et al. 2008). The latter finding may be explained as a typical homeostatic response to IGF-I resistance due to lack of receptors. Altogether these data indicate that serum and brain IGF-I levels share a common regulatory link, regardless of the BBBs.

## 3 Factors Influencing Serum IGF-I Levels

Site-specific genetic ablation in mice determined that the liver is the principal source of circulating IGF-I (Yakar et al. 1999). The major known regulator of serum IGF-I is growth hormone (GH). Serum levels of IGF-I are mostly controlled by GH; in turn, IGF-I decreases GH pituitary output to regulate its own levels through a classical neuroendocrine feedback loop (Daughaday and Rotwein 1989). More recently, it has been shown that IGF-I controls its serum levels acting on fat tissue, opening a new level of regulation for serum IGF-I involving energy stores (Kloting et al. 2008). Other hormones, such as estrogens, thyroid hormone or glucocorticoids, also affect basal or GH-stimulated liver IGF-I output. Glucocorticoids exert a complex modulatory role (Luo and Murphy 1989; Miell et al. 1993), but there is a clear depressing action of various types of stress on serum IGF-I (Gomez-Merino et al. 2005). Other physiological stimuli known to modulate serum IGF-I are sleep (Everson and Crowley 2004), nutrition (Thissen et al. 1994), or physical exercise (Schwarz et al. 1996). Regulation of IGF-I levels appears to be highly dependent on type of exercise, physical fitness, sex, and species. For instance, exercise has been reported to either exert no effect (Eliakim et al. 1997), reduce serum IGF-I (Eliakim et al. 1998), or increase it (Schwarz et al. 1996). A genetic influence on circulating IGF-I levels is also well established (Harrela et al. 1996), with various known polymorphisms involved (Jernstrom et al. 2001). More recently, a correlation between social status and serum IGF-I levels has also been noted (Kumari et al. 2008), Intriguingly, a relationship between social status, brain volume decline and preclinical AD has also been noted (Fotenos et al. 2008).

Infections, sepsis, severe burns and almost any type of critical illness is associated with decreased serum IGF-I levels (Dimopoulou et al. 2005; Lorenzo-Zuniga et al. 2007; Schwab et al. 1997; Sierra-Johnson et al. 2008). Diseases such as diabetes and the majority of neurological dysfunctions are also associated with decreased serum IGF-I (Busiguina et al. 2000). In several specific instances, such as in rare inherited diseases like ataxia-telangectasia (Busiguina et al. 2000), or at distinct disease stages (i.e., AD; see Vardy et al. 2007), high serum IGF-I levels are reported, which may be interpreted as a response to tissue resistance (Jain et al. 1998). One may conclude that general health status is an important determinant of serum IGF-I levels.

Because we postulate that the extent of peripheral IGF-I input to the brain is a key factor in the development of AD (Carro and Torres-Aleman 2004), all of the physiological and pathological processes affecting serum IGF-I levels would have a potential influence on this disease (Table 1). In other words, environmental and behavioral influences on AD risk will be those affecting serum IGF-I levels.

**Table 1** Factors influencing serum IGF-I levels

| Factor | Effect | Reference |
|---|---|---|
| Growth hormone | ↑ | Daughaday et al. (1989) |
| Sleep deprivation | ↓ | Everson et al. (2004) |
| Meal protein | ↑ | Thissen et al. (1994) |
| Stress | ↓ | Gomez-Merino et al. (2005) |
| Critical illness | ↓ | Dimopoulou et al. (2005); Lorenzo-Zuniga et al. (2007); Schwab et al. (1997); Sierra-Johnson et al. (2008) |
| Brain diseases | ↓↑ | Busiguina et al. (2000) |
| Social status | ↑ | Kumari et al. (2008) |
| Endocrine milieu | ↓↑ | Daughaday et al. (1989); Luo et al. (1989); Miell et al. (1993) |
| Exercise | = ↓↑ | Eliakim et al. (1997, 1998); Schwarz et al. (1996) |
| Inheritance | ↓↑ | Harrela et al. (1996); Jernstrom et al. 2001) |

Although this is a simplistic approach to a complex disease, we propose to use it as a surrogate marker in search of the factors modulating risk of AD. Specifically, longitudinal analysis of serum IGF-I levels in individuals at risk for AD (i.e., >65 years old, diabetic or metabolic syndrome patients, etc.) should be conducted on a regular basis as an additional indicator of risk.

## 4 Influence of Environmental Factors on the Passage of Serum IGF-I into the Brain

Following this reasoning, a second potential way to identify factors involved in risk forAD may be determining those impacting on the input of serum IGF-I onto the brain. Deficient serum IGF-I uptake by the brain may be due not only to low serum levels but also to impaired passage. Analysis of the mechanisms involved in this traffic and of their regulation is necessary for a full understanding of the causes underlying low IGF-I input to the brain.

Increasing the levels of IGF-I in the circulation after injection of recombinant IGF-I resulted in increased brain IGF-I content (Fernandez et al. 1998). In search of a physiological context for this experimental observation, we analyzed the influence of physical exercise on serum-derived IGF-I passage to the brain, as exercise reportedly increased blood GH/IGF-I levels in humans (Schwarz et al. 1996). While acute treadmill running did induce passage of peripheral IGF-I into the rat brain, it was not accompanied by increased serum IGF-I levels (Carro et al. 2000). Building on this observation we went further to explore the potential beneficial effects of exercise and the possible role played by serum IGF-I. We found that serum IGF-I is crucial in exercise neuroprotection (Carro et al. 2001; Lopez-Lopez et al. 2004; Trejo et al. 2001; Trejo et al. 2007). From these studies we concluded that physical exercise enhances brain entrance of serum IGF-I as part of its beneficial actions on brain function (Trejo et al. 2002). Based on this conclusion, we could readily provide a molecular explanation for the protective effects of exercise on progression of AD-like symptoms in mouse models of AD (Adlard et al. 2005). Thus, as a result of enhanced brain IGF-I input after exercise, amyloidosis, tauopathy, gliosis, and cognitive loss would all be attenuated, as IGF-I counteracts all these pathological changes. Accessibility to exercise is usually included in "environmental enrichment" laboratory protocols aimed at determining the role of enhanced brain activity as a neuroprotective behavior. Experimental environmental enrichment, which may be translated into engaged social, mental and physical activity within a human context, has been shown to be beneficial in AD models (Cracchiolo et al. 2007). As neuronal activity regulates the entrance of serum IGF-I into the brain (Davila et al. 2007 and unpublished observations),we can again readily accommodate this observation by invoking a mediatory role for serum IGF-I. In this regard, it has been documented that the protective effect of environmental enrichment on spinal cord injury involves serum IGF-I (Koopmans et al. 2006). In our view, the fact that brain activity modulates the entrance of serum IGF-I into the brain is a very significant observation, as it provides a molecular frame for cognitive reserve theory (Richards and Deary 2005). Keeping an active mind, as a result of social engagement, higher education, and other environmental factors, is postulated to protect against AD (Grant et al. 2002; Katzman 1993). A proposed explanation is the elaboration of a cognitive reserve in accordance with overall brain activity. This reserve would provide the brain with additional resources to cope with insults (van Praag et al. 2000), and activity-dependent entrance of serum IGF-I would be a means to generate these defences.

Obesity, imbalanced diets and other cardiovascular risk factors are also associated with a risk forAD (Grant et al. 2002). We explored the influence of diet on the blood-to-brain traffic of IGF-I. A high-fat diet accelerating AD pathology in animal models (Refolo et al. 2000) also led to reduced brain IGF-I input (Sharma et al. 2008; Dietrich et al. 2007). Specifically, the resultant high tryglyceride levels in blood interfered with transport of serum IGF-I at the choroid plexus through a process involving the co-receptor IGF-I cargo protein megalin (Dietrich et al. 2007). As mentioned above, compromised entrance of serum IGF-I greatly contributes to AD pathology (Carro et al. 2006). An established risk factor for AD is general health status and what is now collectively known as the metabolic syndrome. This condition includes not only obesity but also a pro-inflammatory status, glucose intolerance and dyslipidemia (Milionis et al. 2008). In this regard, we found that systemic pro-inflammatory cytokines such as TNF$\alpha$ reduce IGF-I input (Carro et al. 2002). As already mentioned, insulin dysfunction is always associated with impaired IGF-I activity (Davila et al. 2007). Because there is a bi-directional relationship between insulin and IGF-I function, we can consider that insulin resistance in metabolic syndrome leads to impaired IGF-I function and the resultant lower IGF-I input to the brain. Conversely, impaired serum-to-brain IGF-I traffic in metabolic syndrome could lead to insulin resistance as central IGF-I modulates insulin sensitivity (Muzumdar et al. 2006; Foster et al. 1991). Our data indicate that impaired serum IGF-I entrance to the brain as a result of unhealthy diets precedes development of glucose intolerance (Dietrich et al. 2007), suggesting that the initial disruption may be insufficient brain IGF-I input followed by reduced insulin sensitivity.

## 5 Insulin, IGF-I and Late-onset AD

We postulate that a functional inter-dependence between IGF-I and insulin is critical in the development of AD. Because the role of insulin as a glucose regulator seems to be less important in brain than in other tissues (Baskin et al. 1987), and because many other effects of insulin have been reported in brain cells (Davila et al. 2007), it is particularly important to better define the specific role played by insulin in brain physiology. For instance, insulin regulates APP traffic and Aβ metabolism (Carro and Torres-Aleman 2004), but the physiological significance of these effects remains uncertain. Not coincidentally, IGF-I also modulates APP traffic and Aβ metabolism (Carro and Torres-Aleman 2004; Adlerz et al. 2007). A major point in the possible functional relationship between IGF-I and insulin that has been largely overlooked is that their respective dimeric receptors can form hybrid receptors able to recognize both ligands (Moxham et al. 1989). These hybrid receptors show greater affinity for IGF-I (Benyoucef et al. 2007), increase when insulin function is compromised (Slaaby et al. 2006) and show a pathogenic significance when dysregulated (Fernandez et al. 2001). Hence, under conditions of dysregulated insulin or IGF-I input to the brain, inevitably, both hormonal signals will eventually be affected.

Although the cause of AD remains uncertain, extracellular amyloidosis is generally considered the main pathogenic event. Whether amyloid peptide overproduction, decreased clearance, impaired degradation or a combination thereof are the culprits is not yet clear. A second putative pathogenic event in AD is tauopathy, the abnormal accumulation of intracellular deposits of tau and other proteins. Its etiological significance is more disputed. However, none of these approaches have yet led to any therapeutic solution. Therefore, new explanations are needed, causing a wealth of new proposals mostly to explain the processes leading to amyloidosis and to find new potential druggable targets.

As shown in Figure 1, we propose that, due to reduced insulin/IGF-I function, the primary cause of late-onset AD is brain metabolic impairment. This proposal must be understood in its broadest sense. Low insulin/IGF-I input will not only impair glucose handling by the brain; it will also disturb a wide array of essential tissue needs. These range from reduced energy supply (nutrients + oxygen) to altered APP trafficking. Blood flow will be reduced due to a compound problem of lower vessel renewal, impaired vessel functionality, brain-barrier malfunction, etc. Because an important set of brain actions of IGF-I involves astrocytes (Fernandez et al. 2007), astrocyte-neuron interactions will also be defective, leading to compromised nutrient and neuroprotective supply to the latter (Pellerin et al. 1998). Many other aspects of neuronal metabolism will also be affected, as the diversity of

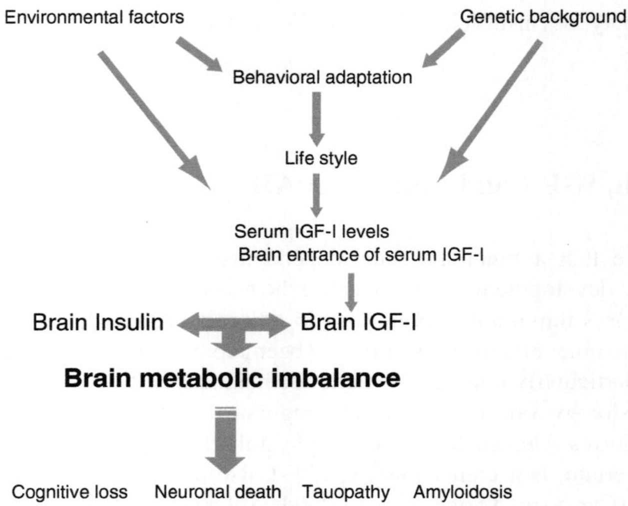


**Fig. 1** A functional interconnection between insulin and IGF-I underlies development of late-onset AD. The inter play between environment and genotype/phenotype modulates behavior that in turn conforms life style. Life style factors such as nutrition, physical and mental activity and stress levels, together with environmental and genetic factors influencing general health status and serum IGF-I levels, modulate serum IGF-I input to the brain. IGF-I input to the brain is determined by serum levels and by the mechanisms involved in transporting IGF-I across the BBBs. The latter can also be affected by life style factors, genetic background and environmental cues. Compromised IGF-I input to the brain will lead to reduced insulin sensitivity through mechanisms not yet known. The combined failure of IGF-I and insulin actions on the brain will lead to metabolic imbalance and eventually to the pathological changes found in AD

actions of insulin and IGF-I on brain cell function is indeed remarkable (see Davila et al. 2007 for a recent overview). Among these, deranged APP traffic and Aβ metabolism may be crucial. In addition, metabolic imbalance will eventually lead to AD neuropathology through various pathological cascades. The hierarchy of changes is still unknown, but from a clinical view the main one is cognitive loss, which, unfortunately, may appear late in the course of the disease.

In summary, impaired brain metabolism underlying AD would be the consequence of insufficient insulin and IGF-I input to the brain. As is true for any pathological condition, compromised insulin/IGF-I function stems from factors governed by inheritance and environment. Genetic information interacts with environmental influences to determine behavioral adaptation. A key role of insulin-like peptides in behavioral adaptation to environmental change is already found in primitive invertebrates (Braeckman et al. 2001). Similar to their putative conserved roles in longevity determination along phylogeny (Kenyon 2001), both insulin and IGF-I influence cognitive operations underlying behavioral adaptation in higher organisms such as mammals, including humans (Aleman et al. 1999; Strachan 2003). Because life style is determined by a specific set of behaviors, it follows that insulin/IGF-I contribute to life-style effects on brain function.

# 6 Conclusions

We propose that life style factors influence IGF-I availability to target cells in the brain. As insulin and IGF-I function are intertwined, defective IGF-I signaling will deteriorate insulin function and vice versa. Regardless of which is the primary pathogenic event, brain metabolic impairment will follow, leading to AD neuropathology. A better understanding of the mechanisms whereby life style factors determine insulin and IGF-I input to the brain will help clarify the role played by environmental factors in development of this devastating disease.

# References

Adlard PA, Perreau VM, Pop V, Cotman CW (2005) Voluntary dxercise Decreases amyloid load in a transgenic model of Alzheimer's disease. J Neurosci 25:4217–4221

Adlerz L, Holback S, Multhaup G, Iverfeldt K (2007) IGF-1-induced processing of the amyloid precursor protein family is mediated by different signaling pathways. J Biol Chem 282:10203–10209

Aleman A, Verhaar HJ, de Haan EH, de Vries WR, Samson MM, Drent ML, van der Veen EA, Koppeschaar HP (1999) Insulin-like growth factor-I and cognitive function in healthy older men. J Clin Endocrinol Metab 84:471–475

Baker J, Liu JP, Robertson EJ, Efstratiadis A (1993) Role of insulin-like growth factors in embryonic and postnatal growth. Cell 75:73–82

Baskin DG, Figlewicz DP, Woods SC, Porte D, Jr., Dorsa DM (1987) Insulin in the brain. Annu Rev Physiol 49:335–347

Benyoucef S, Surinya KH, Hadaschik D, Siddle K (2007) Characterisation of insulin/IGF hybrid receptors: contributions of the insulin receptor L2 and Fn1 domains and the alternatively spliced exon 11 sequence to ligand binding and receptor activation. Biochem J 403:603–13

Blum-Degen D, Frolich L, Hoyer S, Riederer P (1995) Altered regulation of brain glucose metabolism as a cause of neurodegenerative disorders? J Neural Transm Suppl 46:139–147

Bondy CA, Lee WH (1993) Patterns of insulin-like growth factor and IGF receptor gene expression in the brain. Functional implications. Ann NY Acad Sci 692:33–43

Braeckman BP, Houthoofd K, Vanfleteren JR (2001) Insulin-like signaling, metabolism, stress resistance and aging in Caenorhabditis elegans. Mech Ageing Dev 122:673–693

Busiguina S, Fernandez AM, Barrios V, Clark R, Tolbert DL, Berciano J, Torres-Aleman I (2000) Neurodegeneration Is associated to changes in serum insulin-like growth factors. Neurobiol Dis 7:657–665

Carro E, Nunez A, Busiguina S, Torres-Aleman I (2000) Circulating insulin-like growth factor I mediates effects of exercise on the brain. J Neurosci 20:2926–2933

Carro E, Trejo JL, Busiguina S, Torres-Aleman I (2001) Circulating insulin-like growth factor I mediates the protective effects of physical exercise against brain insults of different etiology and anatomy. J Neurosci 21:5678–5684

Carro E, Trejo JL, Gomez-Isla T, LeRoith D, Torres-Aleman I (2002) Serum insulin-like growth factor I regulates brain amyloid-β levels. Nature Med 8:1390–1397

Carro E, Torres-Aleman I (2004) The role of insulin and insulin-like growth factor I in the molecular and cellular mechanisms underlying the pathology of Alzheimer's disease. Eur J Pharmacol 490:127–133

Carro E, Spuch C, Trejo JL, Antequera D, Torres-Aleman I (2005) Choroid plexus megalin is involved in neuroprotection by serum insulin-like growth factor I. J Neurosci 25:10884–10893

Carro E, Trejo JL, Spuch C, Bohl D, Heard JM, Torres-Aleman I (2006) Blockade of the insulin-like growth factor I receptor in the choroid plexus originates Alzheimer's-like neuropathology in rodents: new cues into the human disease? Neurobiol Aging 27:1618–1631

Chan SJ, Cao QP, Steiner DF (1990) Evolution of the insulin superfamily: cloning of a hybrid insulin/insulin-like growth factor cDNA from amphioxus. Proc Natl Acad Sci USA 87:9319–9323

Chen HK, Fernandez-Funez P, Acevedo SF, Lam YC, Kaytor MD, Fernandez MH, Aitken A, Skoulakis EM, Orr HT, Botas J, Zoghbi HY (2003) Interaction of akt-phosphorylated ataxin-1 with 14-3-3 mediates neurodegeneration in spinocerebellar ataxia type 1. Cell 113:457–468

Clemmons DR (2004) Role of insulin-like growth factor lin maintaining normal glucose homeostasis. Horm Res 62 Suppl 1:77–82

Cracchiolo JR, Mori T, Nazian SJ, Tan J, Potter H, Arendash GW (2007) Enhanced cognitive activity–over and above social or physical activity–is required to protect Alzheimer's mice against cognitive impairment, reduce Aβ deposition, and increase synaptic immunoreactivity. Neurobiol Learning Memory 88:277–294

Daughaday WH, Rotwein P (1989) Insulin-like growth factors I and II. Peptide, messenger ribonucleic acid and gene structures, serum, and tissue concentrations. Endocr Rev 10:68–91

Davila D, Piriz J, Trejo JL, Nunez A, Torres-Aleman I (2007) Insulin and insulin-like growth factor I signalling in neurons. Front Biosci 12:3194–3202

De Felice FG, Vieira MN, Bomfim TR, Decker H, Velasco PT, Lambert MP, Viola KL, Zhao WQ, Ferreira ST, Klein WL (2009) Protection of synapses against Alzheimer's-linked toxins: Insulin signaling prevents the pathogenic binding of Aβ oligomers. Proc Natl Acad Sci USA 106:1971–1976

de Pablo F, de la Rosa EJ (1995) The developing CNS: a scenario for the action of proinsulin, insulin and insulin-like growth factors. Trends Neurosci 18:143–150

DeChiara TM, Efstratiadis A, Robertson EJ (1990) A growth-deficiency phenotype in heterozygous mice carrying an insulin-like growth factor II gene disrupted by targeting. Nature 345:78–80

Dietrich MO, Muller A, Bolos M, Carro E, Perry ML, Portela LV, Souza DO, Torres-Aleman I (2007) Western style diet impairs entrance of blood-borne insulin-like growth factor-1 into the brain. Neuromol Med 9:324–330

Dimopoulou I, Kouyialis AT, Orfanos S, Armaganidis A, Tzanela M, Thalassinos N, Tsagarakis S (2005) Endocrine alterations in critically ill patients with stroke during the early recovery period. Neurocrit Care 3:224–229

Dziegielewska KM, Saunders NR (2002) The ins and outs of brain-barrier mechanisms. Trends Neurosci 25:69–71

Eliakim A, Moromisato M, Moromisato D, Brasel JA, Roberts C, Cooper DM (1997) Increase in muscle IGF-I protein but not IGF-I mRNA after 5 days of endurance training in young rats. Am J Physiol 273: R1557-R1561

Eliakim A, Brasel JA, Mohan S, Wong WL, Cooper DM (1998) Increased physical activity and the growth hormone-IGF-I axis in adolescent males. Am J Physiol 275: R308-R314

Emamian ES, Hall D, Birnbaum MJ, Karayiorgou M, Gogos JA (2004) Convergent evidence for impaired AKT1-GSK3β signaling in schizophrenia. Nature Genet 36:131–137

Everson CA, Crowley WR (2004) Reductions in circulating anabolic hormones induced by sustained sleep deprivation in rats. Am J Physiol Endocrinol Metab 286: E1060-E1070

Fernandez AM, de la Vega AG, Torres-Aleman I (1998) Insulin-like growth factor I restores motor coordination in a rat model of cerebellar ataxia. Proc Natl Acad Sci USA 95:1253–1258

Fernandez AM, Kim JK, Yakar S, Dupont J, Hernandez-Sanchez C, Castle AL, Filmore J, Shulman GI, Le RD (2001) Functional inactivation of the IGF-I and insulin receptors in skeletal muscle causes type 2 diabetes. Genes Dev 15:1926–1934

Fernandez S, Fernandez AM, Lopez-Lopez C, Torres-Aleman I (2007) Emerging roles of insulin-like growth factor-I in the adult brain. Growth Horm IGF Res 17:89–95

Foster LA, Ames NK, Emery RS (1991) Food intake and serum insulin responses to intraventricular infusions of insulin and IGF-I. Physiol Behav 50:745–749

Fotenos AF, Mintun MA, Snyder AZ, Morris JC, Buckner RL (2008) Brain volume decline in aging: evidence for a relation between socioeconomic status, preclinical Alzheimer disease, and reserve. Arch Neurol 65:113–120

Gasparini L, Netzer WJ, Greengard P, Xu H (2002) Does insulin dysfunction play a role in Alzheimer's disease? Trends Pharmacol Sci 23:288–293

Gomez-Merino D, Drogou C, Chennaoui M, Tiollier E, Mathieu J, Guezennec CY (2005) Effects of combined stress during intense training on cellular immunity, hormones and respiratory infections. Neuroimmunomodulation 12:164–172

Grant WB, Campbell A, Itzhaki RF, Savory J (2002) The significance of environmental factors in the etiology of Alzheimer's disease. J Alzheimers Dis 4:179–189

Harrela M, Koistinen H, Kaprio J, Lehtovirta M, Tuomilehto J, Eriksson J, Toivanen L, Koskenvuo M, Leinonen P, Koistinen R, Seppala M (1996) Genetic and environmental components of interindividual variation in circulating levels of IGF-I, IGF-II, IGFBP-1, and IGFBP-3. J Clin Invest 98:2612–2615

Humbert S, Bryson EA, Cordelieres FP, Connors NC, Datta SR, Finkbeiner S, Greenberg ME, Saudou F (2002) The IGF-1/Akt pathway is neuroprotective in Huntington's disease and involves Huntingtin phosphorylation by Akt. Dev Cell 2:831–837

Itoh M, Ide S, Takashima S, Kudo S, Nomura Y, Segawa M, Kubota T, Mori H, Tanaka S, Horie H, Tanabe Y, Goto YI (2007) Methyl CpG-binding protein 2 (a mutation of which causes Rett syndrome) directly regulates insulin-like growth factor binding protein 3 in mouse and human brains. J Neuropathol Exp Neurol 66:117–123

Jain S, Golde DW, Bailey R, Geffner ME (1998) Insulin-like growth factor-I resistance. Endocr Rev 19:625–646

Jernstrom H, Deal C, Wilkin F, Chu W, Tao Y, Majeed N, Hudson T, Narod SA, Pollak M (2001) Genetic and nongenetic factors associated with variation of plasma levels of insulin-like growth factor-I and insulin-like growth factor-binding protein-3 in healthy premenopausal women. Cancer Epidemiol Biomarkers Prev 10:377–384

Kappeler L, De Magalhaes Filho CM, Dupont J, Leneuve P, Cervera P, Perin L, Loudes C, Blaise A, Klein R, Epelbaum J, Le BY, Holzenberger M (2008) Brain IGF-1 receptors control mammalian growth and lifespan through a neuroendocrine mechanism. PLoS Biol 6: e254

Katzman R (1993) Education and the prevalence of dementia and Alzheimer's disease. Neurology 43:13–20

Kenyon C (2001) A conserved regulatory system for aging. Cell 105:165–168

Kloting N, Koch L, Wunderlich T, Kern M, Ruschke K, Krone W, Bruning JC, Bluher M (2008) Autocrine IGF-1 action in adipocytes controls systemic IGF-1 concentrations and growth. Diabetes 57:2074–2082

Koopmans GC, Brans M, Gomez-Pinilla F, Duis S, Gispen WH, Torres-Aleman I, Joosten EA, Hamers FP (2006) Circulating insulin-like growth factor I and functional recovery from spinal cord injury under enriched housing conditions. Eur J Neurosci 23:1035–1046

Kumari M, Tabassum F, Clark C, Strachan D, Stansfeld S, Power C (2008) Social differences in insulin-like growth factor-1: findings from a British birth cohort. Ann Epidemiol 18:664–670

Lewitt MS, Hilding A, Ostenson CG, Efendic S, Brismar K, Hall K (2008) Insulin-like growth factor-binding protein-1 in the prediction and development of type 2 diabetes in middle-aged Swedish men. Diabetologia 51:1135–1145

Lopez-Lopez C, LeRoith D, Torres-Aleman I (2004) Insulin-like growth factor I is required for vessel remodeling in the adult brain. Proc Natl Acad Sci USA 101:9833–9838

Lorenzo-Zúñiga V, Bartoli R, Masnou H, Montoliu S, Morillas RM, Planas R (2007) Serum concentrations of insulin-like growth factor-I (IGF-I) as a marker of liver fibrosis in patients with chronic hepatitis C. Dig Dis Sci 52:3245–3250

Luo JM, Murphy LJ (1989) Dexamethasone inhibits growth hormone induction of insulin-like growth factor-I (IGF-I) messenger ribonucleic acid (mRNA) in hypophysectomized rats and reduces IGF-I mRNA abundance in the intact rat. Endocrinology 125:165–171

Malagelada C, Jin ZH, Greene LA (2008) RTP801 is induced in Parkinson's disease and mediates neuron death by inhibiting Akt phosphorylation/activation. J Neurosci 28:14363–14371

Miell JP, Taylor AM, Jones J, Holly JM, Gaillard RC, Pralong FP, Ross RJ, Blum WF (1993) The effects of dexamethasone treatment on immunoreactive and bioactive insulin-like growth factors (IGFs) and IGF-binding proteins in normal male volunteers. J Endocrinol 136:525–533

Milionis HJ, Florentin M, Giannopoulos S (2008) Metabolic syndrome and Alzheimer's disease: a link to a vascular hypothesis? CNS Spectr 13:606–613

Moxham CP, Duronio V, Jacobs S (1989) Insulin-like growth factor I receptor beta-subunit heterogeneity. Evidence for hybrid tetramers composed of insulin-like growth factor I and insulin receptor heterodimers. J Biol Chem 264:13238–13244

Muzumdar RH, Ma X, Fishman S, Yang X, Atzmon G, Vuguin P, Einstein FH, Hwang D, Cohen P, Barzilai N (2006) Central and opposing effects of IGF-I and IGF-binding protein-3 on systemic insulin action. Diabetes 55:2788–2796

Okamura-Oho Y, Miyashita T, Ohmi K, Yamada M (1999) Dentatorubral-pallidoluysian atrophy protein interacts through a proline-rich region near polyglutamine with the SH3 domain of an insulin receptor tyrosine kinase substrate. Human Mol Genet 8:947–957

Pardridge WM (1993) Transport of insulin-related peptides and glucose across the blood- brain barrier. Ann NY Acad Sci 692:126–137

Pellerin L, Pellegri G, Bittar PG, Charnay Y, Bouras C, Martin JL, Stella N, Magistretti PJ (1998) Evidence supporting the existence of an activity-dependent astrocyte-neuron lactate shuttle. Dev Neurosci 20:291–299

Peretz S, Jensen R, Baserga R, Glazer PM (2001) ATM-dependent expression of the insulin-like growth factor-I receptor in a pathway regulating radiation response. Proc Natl Acad Sci USA 98:1676–1681

Pierce SB, Costa M, Wisotzkey R, Devadhar S, Homburger SA, Buchman AR, Ferguson KC, Heller J, Platt DM, Pasquinelli AA, Liu LX, Doberstein SK, Ruvkun G (2001) Regulation of DAF-2 receptor signaling by human insulin and ins-1, a member of the unusually large and diverse C. elegans insulin gene family. Genes Dev 15:672–686

Refolo LM, Pappolla MA, Malester B, LaFrancois J, Bryant-Thomas T, Wang R, Tint GS, Sambamurti K, Duff K (2000) Hypercholesterolemia accelerates the Alzheimer's amyloid pathology in a transgenic mouse model. Neurobiol Dis 7:321–331

Richards M, Deary IJ (2005) A life course approach to cognitive reserve: a model for cognitive aging and development? Ann Neurol 58:617–622

Rönnemaa E, Zethelius B, Sundelöf J, Sundström J, Degerman-Gunnarsson M, Berne C, Lannfelt L, Kilander L (2008) Impaired insulin secretion increases the risk of Alzheimer disease. Neurology 71:1065–1071

Schwab S, Spranger M, Krempien S, Hacke W, Bettendorf M (1997) Plasma insulin-like growth factor I and IGF binding protein 3 levels in patients with acute cerebral ischemic injury. Stroke 28:1744–1748

Schwarz AJ, Brasel JA, Hintz RL, Mohan S, Cooper DM (1996) Acute effect of brief low- and high-intensity exercise on circulating insulin-like growth factor (IGF) I, II, and IGF-binding protein-3 and its proteolysis in young healthy men. J Clin Endocrinol Metab 81:3492–3497

Sharma S, Prasanthi RPJ, Schommer E, Feist G, Ghribi O (2008) Hypercholesterolemia-induced Abeta accumulation in rabbit brain is associated with alteration in IGF-1 signaling. Neurobiol Dis 32:426–432

Sierra-Johnson J, Romero-Corral A, Somers VK, Lopez-Jimenez F, Mälarstig A, Brismar K, Hamsten A, Fisher RM, Hellénius ML (2008) IGF-I/I GFBP-3 ratio: a mechanistic insight into the metabolic syndrome. Clin Sci 116:507–512

Slaaby R, Schaffer L, Lautrup-Larsen I, Andersen AS, Shaw AC, Mathiasen IS, Brandt J (2006) Hybrid receptors formed by insulin receptor (IR) and insulin-like growth factor I receptor (IGF-IR) have low insulin and high IGF-1 affinity irrespective of the IR splice variant. J Biol Chem 281:25869–25874

Strachan MW (2003) Insulin and cognitive function. The Lancet 362:1253

Thissen JP, Ketelslegers JM, Underwood LE (1994) Nutritional regulation of the insulin-like growth factors. Endocr Rev 15:80–101

Torres Aleman I (2005) Role of insulin-like growth factors in neuronal plasticity and neuroprotection. Adv Exp Med Biol 567:243–258

Torres Aleman I, Pons S, Arevalo MA (1994) The insulin-like growth factor I system in the rat cerebellum: developmental regulation and role in neuronal survival and differentiation. J Neurosci Res 39:117–126

Trejo JL, Carro E, Torres-Aleman I (2001) Circulating insulin-like growth factor I mediates exercise-induced increases in the number of new neurons in the adult hippocampus. J Neurosci 21:1628–1634

Trejo JL, Carro E, Nunez A, Torres-Aleman I (2002) Sedentary life impairs self-reparative processes in the brain: the role of serum insulin-like growth factor-I. Rev Neurosci 13:365–374

Trejo J, Piriz J, Llorens-Martin MV, Fernandez AM, Bolos M, LeRoith D, Nunez A, Torres-Aleman I (2007) Central actions of liver-derived insulin-like growth factor I underlying its pro-cognitive effects. Mol Psychiat 12:1118–1128

van Praag H, Kempermann G, Gage FH (2000) Neural consequences of environmental enrichment. Nature Rev Neurosci 1:191–198

Vardy ER, Rice PJ, Bowie PC, Holmes JD, Grant PJ, Hooper NM (2007) Increased circulating insulin-like growth factor-1 in late-onset Alzheimer's disease. J Alzheimers Dis 12:285–290

Walter HJ, Berry M, Hill DJ, Logan A (1997) Spatial and temporal changes in the insulin-like growth factor (IGF) axis indicate autocrine/paracrine actions of IGF-I within wounds of the rat brain. Endocrinology 138:3024–3034

Wilczak N, de Vos RA, De Keyser J (2003) Free insulin-like growth factor (IGF)-I and IGF binding proteins 2, 5, and 6 in spinal motor neurons in amyotrophic lateral sclerosis. Lancet 361:1007–1011

Woods KA, Camacho-Hubner C, Barter D, Clark AJ, Savage MO (1997) Insulin-like growth factor I gene deletion causing intrauterine growth retardation and severe short stature. Acta Paediatr Suppl 423:39–45

Yakar S, Liu JL, Stannard B, Butler A, Accili D, Sauer B, LeRoith D (1999) Normal growth and development in the absence of hepatic insulin-like growth factor I. Proc Natl Acad Sci USA 96:7324–7329

# Index

MIX
Papier aus verantwortungsvollen Quellen
Paper from responsible sources
FSC® C105338

If you have any concerns about our products,
you can contact us on
**ProductSafety@springernature.com**

In case Publisher is established outside the EU,
the EU authorized representative is:
**Springer Nature Customer Service Center GmbH**
**Europaplatz 3, 69115 Heidelberg, Germany**

Printed by Libri Plureos GmbH
in Hamburg, Germany